VECTOR DERIVATIVE FORMULAS

$$\text{curl grad } f = 0$$

$$\text{div curl } \mathbf{F} = 0$$

$$\text{curl curl } \mathbf{F} = \text{grad div } \mathbf{F} - \nabla^2 \mathbf{F}$$

$$\text{div} (f\mathbf{F}) = f \text{ div } \mathbf{F} + \mathbf{F} \cdot \text{grad } f$$

$$\text{curl} (f\mathbf{F}) = f \text{ curl } \mathbf{F} + (\text{grad } f) \times \mathbf{F}$$

$$\text{grad} (\mathbf{F} \cdot \mathbf{G}) = (\mathbf{F} \cdot \nabla)\mathbf{G} + (\mathbf{G} \cdot \nabla)\mathbf{F} + \mathbf{F} \times \text{curl } \mathbf{G} + \mathbf{G} \times \text{curl } \mathbf{F}$$

$$\text{div} (\mathbf{F} \times \mathbf{G}) = \mathbf{G} \cdot \text{curl } \mathbf{F} - \mathbf{F} \cdot \text{curl } \mathbf{G}$$

$$\text{curl} (\mathbf{F} \times \mathbf{G}) = \mathbf{F} \text{ div } \mathbf{G} - \mathbf{G} \text{ div } \mathbf{F} + (\mathbf{G} \cdot \nabla)\mathbf{F} - (\mathbf{F} \cdot \nabla)\mathbf{G}$$

$$\text{grad } f(R) = -\text{grad}' f(R), \text{ where } R = |\mathbf{r} - \mathbf{r}'|$$

$$\text{div } \mathbf{r} = 3$$

$$\text{curl } \mathbf{r} = 0$$

$$\text{grad } f(r) = \hat{\mathbf{a}}_r \, df/dr$$

$$(\mathbf{m} \cdot \nabla)\mathbf{r} = \mathbf{m}$$

ELECTROMAGNETIC THEORY

ELECTROMAGNETIC THEORY

Daniel R. Frankl

The Pennsylvania State University

Prentice-Hall, Inc., Englewood Cliffs, New Jersey 07632

Library of Congress Cataloging in Publication Data

FRANKL, DANIEL R.
  Electromagnetic theory.

  Bibliography: p.
  Includes index.
  1. Electromagnetic theory. I. Title.
QC670.F68 1986   530.1′41   85–6425
ISBN 0-13-249095-1

Editorial/production supervision
    and interior design: *Kathleen M. Lafferty*
Cover design: *20/20 Services, Inc.*
Manufacturing buyer: *John B. Hall*

Printed in the United States of America

10 9 8 7 6 5 4 3 2 1

ISBN 0-13-249095-1   01

PRENTICE-HALL INTERNATIONAL (UK) LIMITED, *London*
PRENTICE-HALL OF AUSTRALIA PTY. LIMITED, *Sydney*
PRENTICE-HALL CANADA INC., *Toronto*
PRENTICE-HALL HISPANOAMERICANA, S.A., *Mexico*
PRENTICE-HALL OF INDIA PRIVATE LIMITED, *New Delhi*
PRENTICE-HALL OF JAPAN, INC., *Tokyo*
PRENTICE-HALL OF SOUTHEAST ASIA PTE. LTD., *Singapore*
EDITORA PRENTICE-HALL DO BRASIL, LTDA., *Rio de Janeiro*
WHITEHALL BOOKS LIMITED, *Wellington, New Zealand*

Contents

Part II
ELECTROMAGNETIC FIELDS
IN MATTER   69

Preface

Electromagnetic Theory is intended as a basic text for junior and senior-level courses in physics and engineering. Graduate students may also find it useful for review or brush-up. Readers are expected to have had some introduction to the subject in the usual elementary physics courses, but this is not absolutely necessary as it is developed fully from the ground up. The approach taken is to base the development on the experimental laws of Coulomb, Ampère, and Faraday (and, of course, many other contributors) and the theoretical insights of Maxwell, culminating in the special theory of relativity of Einstein.

The arrangement of the material is somewhat unique. In Part I, comprising the first three chapters, the basic laws for fields in vacuum are developed in an uninterrupted sweep without the usual digressions into material responses or boundary value problems. This unification should help the student to grasp the essential structure of the subject and gain a better understanding of how the various sub-areas fit in. Part II, the next four chapters, takes up the question of fields in matter. Dielectrics, conductors, and magnetic materials including superconductors are treated in considerably more detail than is usual at this level. This reflects the current scientific interest and technological importance of many different types of materials, and may serve as an introduction into various aspects of materials science, device engineering, and so on. Chapter 7 is a resumé of the field equations in the presence of matter and includes a thorough discussion of units and dimensionalities of electromagnetic quantities.

Part III, Chapters 8, 9, and 10, is devoted to energy relations and forces in fields, thus elucidating the essential connections of electromagnetism to mechanics and thermodynamics. Part IV, two chapters, returns to the by-passed subject of static boundary value problems. In electrostatics, solutions of the Laplace equation for rectangular, infinitely long cylinder, and axially symmetrical spherical boundaries are discussed. The method of images is also included but, in keeping with what appears to be the current trend, conformal mapping is omitted. In magnetostatics, scalar as well as vector potential methods are used, and the concepts of magnetic poles and magnetic circuits are introduced.

Part V, dealing with time-varying electromagnetic fields, comprises nearly half the book. Chapter 13 develops the wave equation and discusses plane wave propagation in unbounded media including conductors and plasmas. Frequency-dependent response functions are illustrated by the classical harmonic oscillator model, and the different types of behavior in the various significant frequency ranges are systematically worked out. Boundaries are introduced in Chapter 14 via the usual Fresnel relations and rectangular ideal waveguides and cavities. Attenuation in real guides and cavities and the significance of the Q-factor are also discussed, as is the increasingly important subject of dielectric waveguides. Chapter 15 deals with the production of electromagnetic waves by the processes of radiation and scattering. In addition to the usual classical formulas for simple macroscopic sources, there is a brief qualitative description of the major differences caused by quantum effects in microscopic sources. Finally, Chapter 16 is devoted to special relativity, starting with the experimental dilemmas, Einstein's revolutionary resolution of them, the covariance of the Maxwell theory, and a tiny taste of the reformulation of classical mechanics leading to the famous mass-energy equivalence.

One of the major stumbling blocks in the teaching and learning of this material is frequently the lack of sufficient mathematical expertise on the part of the students. Generally, they are supposed to have had courses including such matters as partial derivatives, multiple integrals, and vector calculus. Realistically, however, every instructor knows that a very generous amount of review is always needed. Accordingly, the appendix summarizes most of the necessary material, including a detailed description of the three coordinate systems employed and of the relations among them. The main mathematical formulas are collected on the inside covers for handy reference.

Care has been taken to keep the mathematical symbolism clear and consistent throughout. The cylindrical coordinates are designated s, ϕ, and z to avoid the common ambiguity in the meanings of r and ρ. Also, the dimensionality of multiple integrals is always designated by the number of integral signs as well as the form of the differential element, and closed manifolds are distinguished by circles through the integral signs. The use of subscripts, superscripts, and unfamiliar letters is held to a minimum to promote readability, but not, it is hoped, at the expense of clarity.

All chapters are generously supplied with worked examples. Many of these also advance the development and provide important results for later use. Thus they constitute an integral part of the text and should not be skipped. However, by being set off in the problem-and-solution format, they give the serious student an opportunity to test understanding and insight while going along. There are, of course, additional problems at the ends of the chapters and appendix. The perennial question of whether or not to give answers is ducked in the usual way.

The subject of dc and ac circuit theory has been deliberately omitted, in line with the modern trend toward emphasis on fields and waves. Circuit theory, especially as regards pulse propagation in digital devices, has grown so complex that at least a full-semester course is devoted to it in most physics and engineering curricula, so a single chapter in a book such as this can scarcely do it justice. However, by way of compensation, various types of non-ohmic current flow are treated rather more thoroughly than usual (Chap. 5).

Thanks are owed to several of my colleagues, Peter B. Shaw, Roland H. Good, Jr., and Emil Kazes, for numerous helpful and enlightening discussions over many years on various aspects of the subject. Three reviewers, T. R. Palfrey, Jr., of Purdue University, Bernard J. Feldman of the University of Missouri, St. Louis, and Lowell Wood of the University of Houston offered many perceptive suggestions, most of which were adopted. Santiago R. Polo suggested the name "chargeamentum" for the product of charge and velocity. Sidney A. Davis stimulated the writing of an earlier version that slowly evolved into the present book, and my daughter Phyllis G. Frankl helped with some of the initial revisions. Doug Humphrey and Kathleen Lafferty of Prentice-Hall contributed editorial and production support. The typing was handled with skill and grace by Sabrina Glasgow. To all these people, my heartiest thanks.

Daniel R. Frankl

Part I

BASIC LAWS OF ELECTROMAGNETIC FIELDS IN VACUUM

chapter 1

Electric Charge and the Electrostatic Field

1-1 CHARGE

All matter is made up (as far as is now known) of a limited number of types of particles. The electron, proton, and neutron are the most familiar of these and may be regarded (in a crude sense) as the basic constituents of ordinary matter. This is not precisely correct since the proton and neutron are, themselves, composed of other species of particles called *quarks*. Nevertheless, it is a sufficiently accurate picture for phenomena at ordinary energies, which are all that shall concern us in this book.

Particles, whether *elementary* (i.e., indivisible) like the electron, or *composite* like the proton and neutron, interact in various ways. There are four so-called *fundamental interactions* known. Two of them, the *gravitational* and *electromagnetic*, lead to long-range forces and are thus directly observable. The other two, called the *strong* and *weak nuclear* interactions, are confined to distances of some 10^{-15} meter (m) or less.

Each species of particle has a precise set of properties that characterize the strengths of its interactions. For the gravitational interaction, the property is called the *mass*. Many different mass values ranging from zero up (i.e., all *positive*) are known for various particles. The corresponding force is of infinite range, falling off only as the inverse square of the distance between particles.

For the electromagnetic interaction, which is of primary concern in this book, the relevant particle property is called the *electric charge*. The only values observed for this property are (1) zero; (2) the proton charge (called e); (3) its exact opposite, the electron charge (called $-e$); or (4) integer multiples of these. Various kinds of quarks may have charges of $\pm\frac{1}{3}e$ or $\pm\frac{2}{3}e$, but these have so far been only indirectly inferred. The electric force between stationary charged particles is, like the gravitational force, of long range. For a proton and electron at a given distance, the electric force overwhelms the gravitational by a factor of 10^{37}, an unimaginably huge number! However, since both positive and negative charges exist, large aggregates of matter may be nearly or exactly neutral, so that the gravitational force may then be dominant, as is observed in everyday life. Only when the charge balance is disturbed, as, for example, by rubbing or otherwise bringing dissimilar materials into intimate contact, do macroscopic objects acquire net charges and experience electric forces of attraction or repulsion. The gravitational force is, of course, always attractive since only one sign of mass exists.

In SI units, which we use throughout, charge is measured in an arbitrary unit called the *coulomb*. This will be operationally defined in Section 7-4. For the present, we note that it is $6.2418 \times 10^{18}e$; in other words, $e = 1.6021 \times 10^{-19}$ coulomb (C).

1-2 MACROSCOPIC CHARGE DENSITIES

It is inconvenient to deal mathematically with the highly localized regions of charge (the particles) separated by relatively vast stretches of empty space within atoms. The techniques of mathematical analysis are much better suited to quantities that vary in a continuous manner in space and time. Accordingly, we define a continuous function, the *charge density*, as

$$\rho(\mathbf{r}) = \text{net charge per unit volume, averaged over a}$$
$$\text{suitable region around point } \mathbf{r}$$

The size of the "suitable region" for the averaging requires some discussion. Clearly, if it be chosen too small, ρ becomes very erratic, while if too large, all the interesting structure is averaged away. In condensed matter, atomic spacings are of the order of a few angstrom units (Å). Thus regions of the order of 100 to 1000 Å in size contain many thousands of atoms, yet are so tiny on an everyday scale as to be virtually infinitesimal. For most purposes, therefore, such a size scale is suitable, and the macroscopic effects will not be sensitive to the exact choice.

Since the net charge in any region is the algebraic sum of the charges of the particles within it,

$$\rho(\mathbf{r}) = \sum_i q_i n_i(\mathbf{r}) \tag{1-1}$$

where q_i is the charge of a particle of the ith species and n_i is the number of such particles per unit volume, again averaged over a suitable region around point $\mathbf{r}$. The situation is illustrated in Figure 1-1.

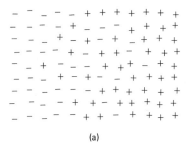

(a)

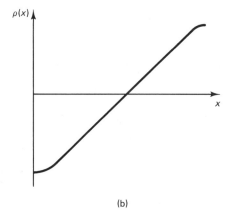

(b)

Figure 1-1 (a) Hypothetical distribution of charged particles; (b) corresponding macroscopic charge density (schematic).

It is often convenient to endow ρ with various types of discontinuity or singularity, just as if it were a true mathematical point function. The most extreme of these is the familiar *point charge*, a charge that occupies no volume. The electron is, in fact, regarded as a point charge in modern quantum theory. We do not, however, limit ourselves to elementary charges, but imagine arbitrarily strong point charges. The concept is very useful as an idealization.

One can also imagine a slab of charge with infinitesimal thickness. This gives rise to the concept of the *surface charge*. Similarly, a surface of infinitesimal width gives a *line charge*. The Greek letters σ and λ are used to denote surface charge density and line charge density, respectively. Figure 1-2 illustrates the transitions from continuous to singular distributions.

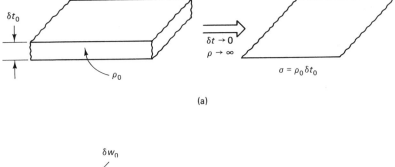

(a)

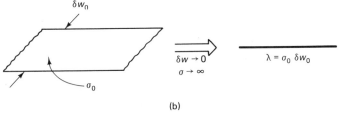

(b)

Figure 1-2   (a) Surface charge density σ as a limit of volume charge density ρ; (b) line charge density λ as a limit of surface charge density σ.

These idealized continuous functions, $\rho(\mathbf{r})$, $\sigma(\mathbf{r})$, and $\lambda(\mathbf{r})$, can be used to approximate the charge in objects actually made up of discrete charged particles.

Example 1-1

Estimate the charge density that would be present in an NaCl crystal if 1% of the Cl^- ions were neutralized. The spacing between like ions is about 5.5 Å. If the neutralized atoms were present only in a surface layer 10 Å thick, what would the surface charge density be?

Solution   There is one Cl^- ion per $(5.5 \times 10^{-10} \text{ m})^3$ or 6.0×10^{27} Cl^- ions/m³. Neutralization of 1% leaves 6.0×10^{25} excess Na^+ ions per cubic meter, each with a charge of e. So

$$\rho = (6.0 \times 10^{25} \text{ ions/m}^3)\,(1.6 \times 10^{-19} \text{ C/ion})$$
$$= 9.6 \times 10^6 \text{ C/m}^3$$
$$\sigma = (9.6 \times 10^6 \text{ C/m}^3)\,(10 \times 10^{-10} \text{ m})$$
$$= 9.6 \times 10^{-3} \text{ C/m}^2$$

1-3 COULOMB'S LAW

The force law between charges was ascertained originally by direct measurement (Charles Augustin de Coulomb, 1785). This method is inherently not very accurate, but the law has since been verified to a very high order of precision by indirect means (see Section 1-10). The statement of the law comprises three distinct parts as follows:

For point charges (in practice, for charged bodies that are small relative to their separation), the force on one due to the other is

1. Proportional in magnitude to the product of the magnitudes of the charges divided by the square of the distance of separation
2. Directed along the line of the centers, as is evident from the symmetry of the situation
3. Away from the other body if the charges are of like sign, and toward the other body if they are of opposite signs.

In mathematical terms, statement 1 is

$$|\mathbf{F}_{A,B}| = K_f \frac{|q_A q_B|}{|\mathbf{r}_A - \mathbf{r}_B|^2} \tag{1-2}$$

where $\mathbf{F}_{A,B}$ is the force on A due to B and K_f is a constant whose value depends on the units of measurement. In the MKS system, $K_f = 8.9874 \times 10^9$ newton-meter²/coulomb² (N-m²/C²).* The direction from B toward A is specified by the unit vector $(\mathbf{r}_A - \mathbf{r}_B)/|\mathbf{r}_A - \mathbf{r}_B|$. Hence the complete statement of the force law, known as *Coulomb's law*, is

$$\mathbf{F}_{A,B} = K_f q_A q_B \frac{\mathbf{r}_A - \mathbf{r}_B}{|\mathbf{r}_A - \mathbf{r}_B|^3} \tag{1-3}$$

Note that $\mathbf{F}_{A,B} = -\mathbf{F}_{B,A}$, so that Coulomb's law is clearly in accord with Newton's third law.

If a third charge, q_C, is added to the system while q_A and q_B are held in place, the force on q_A is equal to the *vector sum* of the forces that would be exerted by q_B and q_C independently.

The total force on q_A can be expressed as

$$\mathbf{F}_A = \mathbf{F}_{A,B} + \mathbf{F}_{A,C}$$
$$= K_f q_A \left[\frac{q_B(\mathbf{r}_A - \mathbf{r}_B)}{|\mathbf{r}_A - \mathbf{r}_B|^3} + \frac{q_C(\mathbf{r}_A - \mathbf{r}_C)}{|\mathbf{r}_A - \mathbf{r}_C|^3} \right]$$

In general, the force on q_A due to n other point charges is

$$\mathbf{F}_A = K_f q_A \sum_{i=1}^{n} \frac{q_i(\mathbf{r}_A - \mathbf{r}_i)}{|\mathbf{r}_A - \mathbf{r}_i|^3} \tag{1-4}$$

1-4 ELECTRIC FIELD

The underlying origin of the force described in Eq. (1-3) can be viewed in two possible ways: (1) the *action at a distance* point of view holds that the force on A emanates directly and instantaneously from B, while (2) the *field* point of view is that B causes

---

*Later (Section 3-5) we shall see that it is convenient to write K_f as $(4\pi\epsilon_0)^{-1}$, where ϵ_0 is another constant.

some *condition* in the space surrounding it, such that a force will be exerted at $\mathbf{r}_A$ if and when a *test charge* q_A is placed there. If only static phenomena are considered, the two viewpoints are indistinguishable. However, as will be seen later, time-dependent effects clearly support the field viewpoint.

Accordingly, the *electric field strength* at any point is defined as the force per unit charge on a small (in both charge and extension) test charge held stationary at that point. The stipulation that the test body have only a small charge is made to ensure that it does not cause any motion of the source charges responsible for the field; hence it is usual to take the limit as the charge of the test body goes to zero.

The electric field strength is denoted $\mathbf{E}$ and its MKS unit is newtons per coulomb. The latter is equivalent to volts per meter, where 1 volt = 1 joule per coulomb. For the present, we shall discuss only static (i.e., time-independent) electric fields, and shall use a subscript S to emphasize this.

Thus, for the case in which the field is due to a single charge, q_B, located at $\mathbf{r}_B$, the field at $\mathbf{r}_A$ is

$$\mathbf{E}_S(\mathbf{r}_A; q_B, \mathbf{r}_B) \equiv \frac{\mathbf{F}_{A,B}}{q_A} = K_f q_B \frac{\mathbf{r}_A - \mathbf{r}_B}{|\mathbf{r}_A - \mathbf{r}_B|^3} \tag{1-5}$$

The notation used here emphasizes that the field at *field point* $\mathbf{r}_A$ is a function of the charge q_B at the *source point* $\mathbf{r}_B$. If numerous sources are present, it becomes too cumbersome to list them all as variables, but it is *essential* to keep in mind the distinction between field points and source points. Accordingly, when both are present in an expression, we designate the *source points by primes*. Thus, for a general field point, $\mathbf{r}$, Eq. (1-5) is written

$$\mathbf{E}_S(\mathbf{r}) = K_f q \frac{\mathbf{r} - \mathbf{r}'}{|\mathbf{r} - \mathbf{r}'|^3} \tag{1-6}$$

If several sources are present, Eq. (1-5) gives

$$\mathbf{E}_S(\mathbf{r}) = K_f \sum_i q_i \frac{\mathbf{r} - \mathbf{r}'_i}{|\mathbf{r} - \mathbf{r}'_i|^3} \tag{1-7}$$

When the charge distribution is continuous, the summation is replaced by an integral, with $d\mathbf{r}'$ (*not* $d\mathbf{r}$) as the variable of integration. Since the sources can be discrete, or can be distributed continuously in either one, two, or three dimensions, the most general expression for the electrostatic field is

$$\mathbf{E}_S(\mathbf{r}) = K_f \left[\sum_i q_i \frac{\mathbf{r} - \mathbf{r}'_i}{|\mathbf{r} - \mathbf{r}'_i|^3} + \int \lambda(\mathbf{r}') \frac{\mathbf{r} - \mathbf{r}'}{|\mathbf{r} - \mathbf{r}'|^3} \, d\mathbf{r}' \right.$$
$$\left. + \iint \sigma(\mathbf{r}') \frac{\mathbf{r} - \mathbf{r}'}{|\mathbf{r} - \mathbf{r}'|^3} \, d^2\mathbf{r}' + \iiint \rho(\mathbf{r}') \frac{\mathbf{r} - \mathbf{r}'}{|\mathbf{r} - \mathbf{r}'|^3} \, d^3\mathbf{r}' \right] \tag{1-8}$$

Since this equation is clearly too cumbersome to write down every time we wish to describe an electrostatic field, we shall henceforth adopt the shorthand notation of writing the volume integral term only. This is to be understood to include the other forms where appropriate. The practice is consistent with our understanding that the singular distributions are idealizations anyway.

Given any charge distribution, the electrostatic field at any point in space can, at least in theory, always be found by evaluating the appropriate terms of Eq. (1-8).

Example 1-2

Point charges of 1 and 2 C are located at the origin and at $z = 3$, respectively. (a) Find the electrostatic field strength everywhere, (b) find any points where the field strength vanishes, and (c) plot the field strength at points along the z axis.

Solution   (a) $\mathbf{r} = x\hat{\mathbf{i}} + y\hat{\mathbf{j}} + z\hat{\mathbf{k}}$ (a general field point); $\mathbf{r}'_1 = 0$; $\mathbf{r}'_2 = 3\hat{\mathbf{k}}$.

$$\mathbf{E}_S(x, y, z) = K_f \left[\frac{x\hat{\mathbf{i}} + y\hat{\mathbf{j}} + z\hat{\mathbf{k}}}{[x^2 + y^2 + z^2]^{3/2}} + 2 \frac{x\hat{\mathbf{i}} + y\hat{\mathbf{j}} + (z - 3)\hat{\mathbf{k}}}{[x^2 + y^2 + (z - 3)^2]^{3/2}} \right]$$

(b) From Figure 1-3a, it is clear that at any field point off the z axis, the field will have at least a transverse component that is nonzero.

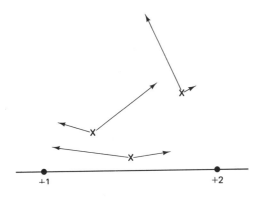

Figure 1-3   (a) Field contributions of the charges in Example 1-2 at three field points.

Hence the field strength can vanish only at points for which $x = y = 0$. At such points,

$$\mathbf{E}_S(0, 0, z) = K_f \left[\frac{z}{|z|^3} + 2 \frac{z - 3}{|z - 3|^3} \right] \hat{\mathbf{k}}$$

This can vanish only if $0 < z < 3$, so that $|z| = z$ and $|z - 3| = 3 - z$. Thus the condition for $E_S = 0$ is

$$z^{-2} - 2(3 - z)^{-2} = 0 \qquad z = \frac{3}{\sqrt{2} + 1}$$

The only point at which the field strength vanishes is $\left(0, 0, \dfrac{3}{\sqrt{2} + 1} \right)$.

(c) Since each term becomes infinite *and changes sign* where its denominator vanishes, the sum is as shown in Figure 1-3b.

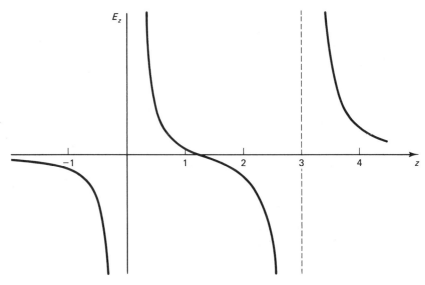

Figure 1-3 (b) Plot of E_z at points on the z axis.

Example 1-3

Find the field due to an infinitely long uniform line charge (Figure 1-4).

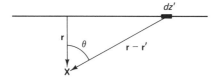

Figure 1-4 Geometry for Example 1-3.

Solution It is evident from the symmetry of the charge distribution that the field is radial, so we need only consider the radial component of the contribution of each source element to the field. Letting θ represent the angle between $\mathbf{r}$ and $\mathbf{r} - \mathbf{r}'$, and using the cylindrical coordinates $\mathbf{r} = s\hat{\mathbf{a}}_s$ and $\mathbf{r}' = z'\hat{\mathbf{k}}$, these radial contributions are of the form

$$K_f \frac{\lambda dz'}{s^2 + z'^2} \cos \theta$$

Substituting $s/\sqrt{s^2 + z'^2}$ for $\cos \theta$ and integrating over all the source elements gives the expression for the field.

$$E_s(s) = K_f \int_{-\infty}^{\infty} \frac{\lambda s \, dz'}{(s^2 + z'^2)^{3/2}}$$

This integral can be evaluated by making the substitution $z' = x \tan \alpha$, the result being

$$\mathbf{E}_S = \frac{2K_f\lambda}{s}\hat{\mathbf{a}}_s$$

Example 1-4

Find the field due to an infinite uniform plane of charge (Figure 1-5).

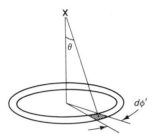

Figure 1-5 Geometry for Example 1-4.

Solution If we choose the plane of charge to be the xy plane, symmetry dictates that $\mathbf{E}$ must be in the $\pm\hat{\mathbf{k}}$ direction. Consider a ring of inner radius s' and width ds' lying in the xy plane with its center at the origin. With θ defined as in Example 1-3, each charge element, $\sigma s'ds'\,d\phi'$, contributes

$$K_f\frac{\sigma s'ds'\,d\phi'\,\hat{\mathbf{k}}}{s'^2 + z^2}\cos\theta = K_f\frac{\sigma s'ds'\,d\phi'\,z\hat{\mathbf{k}}}{(s'^2 + z^2)^{3/2}}$$

The field is then

$$\mathbf{E}_S(z) = K_f\hat{\mathbf{k}}\int_0^\infty\int_0^{2\pi}\frac{\sigma s'ds'\,d\phi'\,z}{(s'^2 + z^2)^{3/2}}$$

$$= 2\pi K_f\sigma z\hat{\mathbf{k}}\left[\frac{-1}{(z^2 + s'^2)^{1/2}}\right]_0^\infty$$

$$= 2\pi K_f\sigma\frac{z}{|z|}\hat{\mathbf{k}}$$

Note that this field is of constant magnitude throughout space, but changes direction at the source plane. If two equally but oppositely charged infinite planes are placed parallel to one another, the field contributions will be parallel at points between, so that the total field strength will be $4\pi K_f\sigma$ between the planes and will vanish everywhere else. This result is of great practical importance; to form a uniform field in a region, one need only use a pair of uniformly charged plates of much greater area than the region.

Example 1-5

A uniform line charge is bent into a ring of radius a. Find the field at points on the axis.

Solution $\mathbf{r} = z\hat{\mathbf{k}},\ \mathbf{r}' = a\hat{\mathbf{a}}_{s'}$

$$\mathbf{E}_S(0, 0, z) = K_f\int_0^{2\pi}\frac{z\hat{\mathbf{k}} - a\hat{\mathbf{a}}_{s'}}{(z^2 + a^2)^{3/2}}\,\lambda a\,d\phi'$$

Now

$$\int_0^{2\pi} \hat{\mathbf{a}}_{s'} \, d\phi' = 0$$

as may be seen by noting that $\hat{\mathbf{a}}_{s'} = \hat{\mathbf{i}} \cos \phi' + \hat{\mathbf{j}} \sin \phi'$. Thus only the first term of the integral for $\mathbf{E}_S$ survives.

$$\mathbf{E}_S(0, 0, z) = 2\pi K_f \lambda \frac{az}{(z^2 + a^2)^{3/2}} \hat{\mathbf{k}}$$

Note that the field is directed away from the plane of the ring for all values of z. Note also that as $z \longrightarrow \pm\infty$, $E_S \longrightarrow K_f q / |z|^2$. This illustrates that at sufficiently great distance the details of the charge distribution become irrelevant, and only the total amount of charge matters.

The final example of this section is rather more complicated, but gives an important result that we shall need in Chapter 4.

Example 1-6

A spherical surface (Figure 1-6) has a charge density that varies as $\sigma_0 \cos \theta$. Find the field at the center.

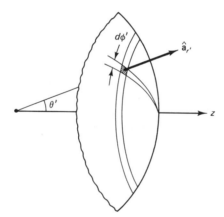

Figure 1-6   Geometry for Example 1-6.

Solution   The charge of an area element at (θ', ϕ') is $(\sigma_0 \cos \theta')(a \, d\theta')(a \sin \theta' \, d\phi')$. Since $\mathbf{r}' = a\hat{\mathbf{a}}_{r'}$, the field at $r = 0$ is

$$\mathbf{E}_S(0) = K_f \oiint \sigma_0 a^2 \cos \theta' \sin \theta' \frac{-a\hat{\mathbf{a}}_{r'}}{a^3} \, d\theta' \, d\phi'$$

$$= -\sigma_0 K_f \int_0^{\pi} \cos \theta' \sin \theta' \, d\theta' \int_0^{2\pi} \hat{\mathbf{a}}_{r'} \, d\phi'$$

Now

$$\hat{\mathbf{a}}_{r'} = \cos \theta' \hat{\mathbf{k}} + \sin \theta' \hat{\mathbf{a}}_{s'}$$

so since the second term integrates to zero over $d\phi'$ (see Example 1-5), and

$$\int_0^{2\pi} \hat{\mathbf{a}}_{r'} \, d\phi' = 2\pi \cos \theta' \hat{\mathbf{k}}$$

Then

$$\mathbf{E}_S(0) = -2\pi K_f \sigma_0 \hat{\mathbf{k}} \int_0^\pi \cos^2 \theta' \sin \theta' \, d\theta'$$

$$= -2\pi K_f \sigma_0 \hat{\mathbf{k}} \left(\tfrac{2}{3}\right)$$

$$= \tfrac{1}{3}(4\pi K_f)\sigma_0(-\hat{\mathbf{k}})$$

1-5 GAUSS'S LAW

We saw in Section 1-4 that the electrostatic field at any field point is an integral (or sum) over contributions from *all source points*. We now prove an important theorem concerning the integral, over a set of field points, of a certain function of the field.

The *outflow*, $O(S)$, through a *closed* surface S is defined as the integral, over the surface, of the component of the field in the direction of the normal. In mathematical notation

$$O(S) = \oiint_S \mathbf{E}_S \cdot \hat{\mathbf{n}} \, dS \tag{1-9}$$

S can be *any* closed surface on which the integrand is defined—that is, any closed surface which does not pass through a point charge or lie tangent to a line or surface charge. *S need not correspond to any physical surface.*

Gauss's law states that

$$O(S) = 4\pi K_f \times \text{(net charge enclosed by } S) \tag{1-10}$$

This result follows readily from Coulomb's law, with the aid of a few lemmas.

Lemma 1. Any charge outside of S makes no contribution to $O(S)$.

Proof. From the charge point, draw an infinitesimal cone of solid angle $d\Omega$ as shown in Figure 1-7. Let the area elements cut out be dS_1 and dS_2 at distances r_1 and r_2. Then the contributions to the integral are $-(K_f q/r_1^2) \cos \theta_1 \, dS_1$ and $(K_f q/r_2^2) \cos \theta_2 \, dS_2$, respectively. But $dS_1 \cos \theta_1 = r_1^2 \, d\Omega$ and $dS_2 \cos \theta_2 = r_2^2 \, d\Omega$. Hence the two contributions cancel. If the surface is partly concave, there may be additional pairs of intersections, but these will cancel in the same way. Since the

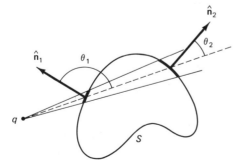

Figure 1-7 Areas cut out of a closed surface S by a small cone with apex at an external point charge q.

entire surface S may be divided into pairs of area elements in the same manner, the sum of all the contributions must vanish.

Lemma 2. If a surface S_2 completely encloses surface S_1, but encloses *no additional charges*, then $O(S_2) = O(S_1)$.

Proof. Apply Lemma 1 to the shell-like region between the two surfaces. Since the outward normal to the shell points outward from S_2 but inward from S_1,

$$O(\text{shell}) = O(S_2) - O(S_1) = 0 \tag{1-11}$$

by Lemma 1, since there is no charge in the shell.

Lemma 3. The outflow through a spherical surface with a point charge q, at its center is $4\pi K_f q$.

Proof (a simple direct calculation). The field is $K_f q/r^2$, directed along the normal and constant in magnitude over the surface of the sphere. So

$$O(S) = \oiint_S \mathbf{E}_s \cdot \hat{\mathbf{n}} \, dS = \oiint_S E_s \, dS = \frac{K_f q}{r^2} 4\pi r^2$$

$$= 4\pi K_f q \tag{1-12}$$

We are now in a position to prove Gauss's law. Consider surface S, which surrounds some charges q_i (Figure 1-8). By Lemma 1, these charges are the only sources of $O(S)$, although *all* charges, both outside and inside S, contribute to $\mathbf{E}_S$ at every point. Now let us surround each of the q_i by a spherical surface S_i, lying within S. Since the field contributions at any point are additive, so are the integrals.

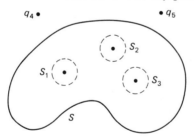

Figure 1-8   Geometry for proof of Gauss's law.

Hence

$$O(S) = \sum_i O(S_i) \quad \text{by Lemma 2}$$

Application of Lemma 3 then gives

$$O(S) = 4\pi K_f \sum_i q_i = 4\pi K_f \times (\text{net charge enclosed})$$

which completes the proof of Eq. (1-10).

Gauss's law can be written in another form, which gives a point-by-point relationship between the field and the charge density. By the divergence theorem of vector calculus (see the Appendix), the outflow is

$$O(S) = \oint_S \mathbf{E}_s \cdot \hat{\mathbf{n}} \, dS = \iiint_{V(S)} \text{div } \mathbf{E}_s \, d^3\mathbf{r} \qquad (1\text{-}13)$$

The charge enclosed is also a volume integral.

$$4\pi K_f \sum_i q_i = \iiint_{V(S)} 4\pi K_f \rho(\mathbf{r}) \, d^3\mathbf{r} \qquad (1\text{-}14)$$

Gauss's law states that the volume integrals in Eqs. (1-13) and (1-14) are equal for *any* surface. The only way this can come about is for the two integrands to be equal. Hence an equivalent statement of the law is

$$\text{div } \mathbf{E}_s(\mathbf{r}) = 4\pi K_f \rho(\mathbf{r}) \qquad (1\text{-}15)$$

This relation, known as the *differential form of Gauss's law*, holds at *every point*. If the point happens to lie at a singularity of the charge distribution, div $\mathbf{E}_s$ is infinite there. If the point lies in empty space, div $\mathbf{E}_s$ is zero there.

Applications of Gauss's Law

Aside from its fundamental importance as a point relation between fields and sources, there are two classes of direct applications of Gauss's law: (1) calculation of the source distribution responsible for a given field, and (2) simplified calculation of fields in cases of high symmetry. The following examples illustrate these applications.

Example 1-7

Find the charge distribution that gives a field of the form $\mathbf{E}_s(x, y, z) = Ax^2\hat{\mathbf{i}}$.

Solution $\rho(x, y, z) = (4\pi K_f)^{-1} \text{div } \mathbf{E}_s(x, y, z)$

$$= \frac{2Ax}{4\pi K_f}$$

Example 1-8

Find the field due to a uniform infinite plane sheet of positive charge.

Solution   Symmetry dictates that the field must be everywhere normal to the sheet, directed away from it, and of uniform strength (since the source configuration "looks" the same from all field points). Hence for surface S in the shape of the cylinder shown in Figure 1-9, only the ends contribute to the outflow

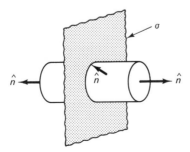

Figure 1-9 Cylindrical Gaussian surface for Example 1-8.

integral, and

$$2E_s A = 4\pi K_f \sigma A, \qquad \text{the same result as in Example 1-4.}$$

Example 1-9

Find the field due to a spherical shell of radius a with uniform surface charge σ.

Solution Symmetry dictates that the field must be radial everywhere and of uniform magnitude on any sphere concentric with the charge distribution. We evaluate the integral over such a sphere, of radius r.
For $r > a$,

$$4\pi K_f(4\pi a^2 \sigma) = \oiint_S \mathbf{E}_s \cdot \hat{\mathbf{n}} \, dS = E_s \oiint_S dS = 4\pi r^2 E_s$$

so

$$\mathbf{E}_s = \frac{4\pi K_f \sigma a^2}{r^2} \hat{\mathbf{a}}_r$$

For $r < a$,

$$\oiint_S E_s \cdot \hat{\mathbf{n}} \, dS = 0 \qquad \text{since no charge is enclosed}$$

so $E_s = 0$.

Note that Gauss's law is useful for field calculations *only in highly symmetrical problems*. We must be able to infer the field direction in advance and to find a surface such that the field is either parallel or perpendicular to it at every point. Furthermore, the field magnitude must be uniform over the entire perpendicular part, so that this magnitude is the only unknown quantity remaining in the problem.

The direct method, on the other hand, can always be used, although in complicated cases the integrals may have to be evaluated numerically. Even in the direct method, the use of any symmetry that may be present can help to simplify the calculation or to check the result, as was illustrated in several examples in Section 1-4.

1-6 ELECTROSTATIC SCALAR POTENTIAL

We now show that the electrostatic field can be expressed as $(-)$ the gradient of a certain scalar point function, $U(\mathbf{r})$, called the *electrostatic potential*.
For a point charge at the origin, we have

$$\mathbf{E}_s(\mathbf{r}; q, 0) = \frac{K_f q \mathbf{r}}{r^3} = -\text{grad}\, \frac{K_f q}{r}$$

and for a point charge at point $\mathbf{r}'$,

$$\mathbf{E}_s(\mathbf{r}; q, \mathbf{r}') = -\text{grad}\, \frac{K_f q}{|\mathbf{r} - \mathbf{r}'|}$$

Thus, on summing over all charge elements, we get

$$\mathbf{E}_s(\mathbf{r}) = -\text{grad } U(r) \tag{1-16}$$

where

$$U(\mathbf{r}) = K_f \iiint \frac{\rho(\mathbf{r}')}{|\mathbf{r} - \mathbf{r}'|} d^3\mathbf{r}' \tag{1-17}$$

Here, as in Eq. (1-8), we write only a volume integral term, but keep in mind that surface integral, line integral, and discrete sum terms will also exist if the corresponding charge distributions are present. Note that each source element contributes in inverse proportion to its distance from the field point.

The correctness of Eq. (1-16) may be verified by calculating the gradient of the right side of Eq. (1-17). This is not nearly as formidable as it looks, because the gradient involves derivatives with respect to the coordinates of the *field point* **r**, while the variables of integration are the coordinates of the *source point* **r**'. Thus (1) the operations may freely be interchanged and (2) the derivatives do not operate on $\rho(\mathbf{r}')$, so

$$\text{grad } U(\mathbf{r}) = K_f \iiint \rho(\mathbf{r}') \text{ grad } \frac{1}{|\mathbf{r} - \mathbf{r}'|} d^3\mathbf{r}'$$

$$= K_f \iiint \rho(\mathbf{r}') \frac{-(\mathbf{r} - \mathbf{r}')}{|\mathbf{r} - \mathbf{r}'|^3} d^3\mathbf{r}'$$

which agrees with Eq. (1-8). This calculation illustrates why it is *essential* to keep the two sets of coordinates separated.

Equations (1-16) and (1-17) provide an alternative way to calculate the field of a given source distribution. One calculates the potential and takes its gradient. The scalar integration for the potential is usually simpler than the vector integration for the field. It is important to keep in mind, though, that the potential must be obtained as a *function of the coordinates of the field point*, so that the derivatives can be taken. The following examples illustrate the method.

Example 1-10

Use the potential to calculate the field due to the charge distribution of Example 1-2. Evaluate the potential and the field at point (1, 1, 1).

Solution

$$U(x, y, z) = K_f \left[\frac{1}{(x^2 + y^2 + z^2)^{1/2}} + \frac{2}{[x^2 + y^2 + (z - 3)^2]^{1/2}} \right]$$

$$\mathbf{E}_s(x, y, z) = -\text{grad } U(x, y, z)$$

$$= K_f \left[\frac{x\hat{\mathbf{i}} + y\hat{\mathbf{j}} + z\hat{\mathbf{k}}}{(x^2 + y^2 + z^2)^{3/2}} + 2\frac{x\hat{\mathbf{i}} + y\hat{\mathbf{j}} + (z - 3)\hat{\mathbf{k}}}{[x^2 + y^2 + (z - 3)^2]^{3/2}} \right]$$

(the same result as was found directly in Example 1-2).

$$U(1, 1, 1) = K_f \left(\frac{1}{\sqrt{3}} + \frac{2}{\sqrt{6}} \right)$$

$$\mathbf{E}_s(1, 1, 1) = K_f \left(\frac{\hat{\mathbf{i}} + \hat{\mathbf{j}} + \hat{\mathbf{k}}}{3\sqrt{3}} + 2\frac{\hat{\mathbf{i}} + \hat{\mathbf{j}} - 2\hat{\mathbf{k}}}{6\sqrt{6}} \right)$$

Note that *the field at a given point cannot be obtained from the potential at that point alone.*

Example 1-11

Use the potential to calculate the field due to an infinite uniformly charged plane (see Example 1-4).

Solution   We consider first the potential on the axis of a uniformly charged disk, an easier problem to handle because of its finite extent.

$$U(0, 0, z) = K_f \int_0^{2\pi} \int_0^a \frac{\sigma s' \, ds' \, d\phi'}{(s'^2 + z^2)^{1/2}}$$

$$= 2\pi K_f \sigma(\sqrt{a^2 + z^2} - \sqrt{z^2})$$

To find the potential for the infinite sheet, simply let $a \to \infty$.

$$U(z) = \lim_{a \to \infty} 2\pi K_f \sigma(\sqrt{a^2 + z^2} - \sqrt{z^2})$$

Then

$$\mathbf{E}_s(x, y, z) = -\text{grad } U(x, y, z) = 2\pi K_f \sigma \lim_{a \to \infty} \left(\frac{-z}{\sqrt{a^2 + z^2}} + \frac{z}{\sqrt{z^2}} \right) \hat{\mathbf{k}}$$

$$= 2\pi K_f \frac{z}{|z|} \hat{\mathbf{k}}$$

This is the same result that was obtained in Examples 1-4 and 1-8.

Irrotational Nature of E_s

Since $\mathbf{E}_s$ is a gradient, it follows that its curl vanishes, that is,

$$\text{curl } \mathbf{E}_s(\mathbf{r}) = 0 \tag{1-18}$$

This important result could also have been obtained by direct calculation of the curl of the expression in Eq. (1-8). Here again the separateness of the variables of differentiation ($\mathbf{r}$) and of integration ($\mathbf{r}'$) comes into play. The result is Eq. (1-18). The fact that $\mathbf{E}_s$ is a gradient could have been deduced from this.

From Eq. (1-18) we deduce another important property of $\mathbf{E}_s$. By integrating both sides of the equation over any open (i.e., bounded) area and applying Stokes' theorem, we get

$$\iint_S \text{curl } \mathbf{E}_s \cdot \hat{\mathbf{n}} \, dS = \oint_{C(S)} \mathbf{E}_s \cdot d\mathbf{l} \tag{1-19}$$

The last integral may be taken around *any* closed contour. The integrand $\mathbf{E}_s \cdot d\mathbf{l}$ is obviously the work per unit charge done by the field when a charged particle is displaced by $d\mathbf{l}$. Equation (1-19) states that when the particle gets back to its starting point *by any path whatever*, the net amount of work done will be zero. For this reason, $\mathbf{E}_s$ is said to be *conservative*.

Another way of stating Eq. (1-19) is that the line integral of $\mathbf{E}_s \cdot d\mathbf{l}$ between any two points is *independent of the path*. This follows because any path from $\mathbf{r}_1$ to $\mathbf{r}_2$ plus any other path from $\mathbf{r}_2$ to $\mathbf{r}_1$ constitute a closed loop.

Potential in Terms of Field

From Eq. (1-17) we find

$$-\mathbf{E}_S \cdot dl = \text{grad } U \cdot \mathbf{dl} = dU \qquad (1\text{-}20)$$

Hence

$$U(r_2) - U(r_1) = \int_{r_1}^{r_2} -\mathbf{E}_S \cdot \mathbf{dl} \qquad (1\text{-}21)$$

According to the preceding discussion, the integral may be evaluated along any path whatever. This is often useful for calculation when $\mathbf{E}_S$ can be found in some independent way, such as the integrated form of Gauss's law.

Example 1-12

Find the potential due to a uniform spherical shell of charge.

Solution 1   Direct integration: Take the direction from the center to the field point as the polar axis. Then

$$U(r, 0, 0) = K_f \int_0^{\pi} \frac{\sigma(2\pi a \sin \theta')a\, d\theta'}{(r^2 + a^2 - 2ar \cos \theta')^{1/2}}$$

$$= 2\pi K_f \sigma a^2 \int_{-1}^{1} \frac{dw}{(A + Bw)^{1/2}}$$

$$(w = \cos \theta'; A = r^2 + a^2; B = -2ar)$$

$$= 2\pi K_f \sigma a^2 \frac{2}{-2ar}[\sqrt{(r-a)^2} - \sqrt{(r+a)^2}]$$

For $r > a$,

$$U = K_f q \frac{(r+a) - (r-a)}{2ar} = \frac{K_f q}{r}$$

For $r < a$,

$$U = K_f q \frac{(r+a) - (a-r)}{2ar} = \frac{K_f q}{a}$$

Solution 2   Indirect:

$$U(r_1) = -\int_{\infty}^{r_1} \mathbf{E}_S \cdot \mathbf{dr}$$

The field strength was found using Gauss's law in Example 1-9.

For $r_1 > a$,

$$\mathbf{E}_S = \frac{K_f q}{r^2} \, \hat{\mathbf{a}}_r$$

$$U(r_1) = -K_f q \int_{\infty}^{r_1} \frac{dr}{r^2} = \frac{K_f q}{r_1}$$

For $r_1 < a$,

$$\mathbf{E}_S = \begin{cases} \dfrac{K_f q}{r^2}\,\hat{\mathbf{a}}_r & \text{for } r > a \\[2mm] 0 & \text{for } r < a \end{cases}$$

$$U(r_1) = -K_f q \int_{\infty}^{a} \frac{dr}{r^2} = \frac{K_f q}{a}$$

The reader should study Figure 1-10 carefully, noting the discontinuity in the field but not in the potential.

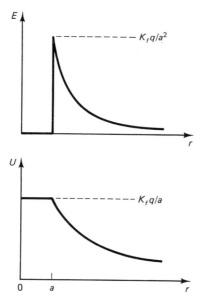

Figure 1-10   Field and potential due to uniform spherical shell of charge.

Physical Significance of the Potential

Equation (1-21) may be interpreted as stating that the difference of potential between two points is the work per unit charge required of an *outside agency* to move the particle from $\mathbf{r}_1$ to $\mathbf{r}_2$ against the force of the field. This work goes into *electrostatic potential energy.* Thus $U(\mathbf{r})$ may be interpreted as *potential energy per unit charge.* The units in the MKS system are therefore joules per coulomb. This unit is also called the *volt*:

$$1 \text{ volt} \equiv 1 \text{ joule per coulomb}$$

Like any other form of potential energy, (e.g., gravitational), only differences are physically meaningful. Any arbitrary constant may be added. This is usually chosen so that the potential is zero at some suitable "reference point." In cases where the charge distribution is finite in extent, the reference point is usually taken at infinite distance away, which is consistent with Eq. (1-17). For those problems in which an infinitely extended distribution is given (e.g., Example 1-11), U does not die off to zero anywhere, and an arbitrary reference point may be chosen. In Example 1-11 we could take $z = 0$ as a reference, so that

$$U(z) - U(0) = 2\pi K_f \sigma \lim_{a \to \infty} (\sqrt{a^2 + z^2} - \sqrt{z^2} - a)$$

$$= -2\pi K_f \sigma \sqrt{z^2}$$

1-7 POISSON AND LAPLACE EQUATIONS

By inserting Eq. (1-16) into the differential form of Gauss's law [Eq. (1-15)] we obtain two useful equations.

$$4\pi K_f \rho(\mathbf{r}) = \text{div}\,(-\text{grad}\,U(\mathbf{r}))$$

$$= -\left(\frac{\partial^2 U}{\partial x^2} + \frac{\partial^2 U}{\partial y^2} + \frac{\partial^2 U}{\partial z^2}\right)$$

Using the symbol ∇^2 to denote the differential operator, the expression above can be written

$$\nabla^2 U(\mathbf{r}) = -4\pi K_f \rho(\mathbf{r}) \qquad \text{(Poisson equation)} \qquad (1\text{-}22)$$

At any point where there is no charge, this becomes

$$\nabla^2 U(\mathbf{r}) = 0 \qquad \text{(Laplace equation)} \qquad (1\text{-}23)$$

These two equations are seen to be nothing more than Gauss's law expressed in terms of the potential. Their importance, however, far transcends this simple origin, since they provide an avenue for the solution of problems of an entirely different sort than we have considered thus far, namely problems in which the charge distribution is not completely specified but depends in part on the properties of pieces of matter in the field. These are called *boundary-value problems*. We shall defer their discussion until Chapter 11 so as not to interrupt the development of the main ideas of the theory.

Since the Poisson and Laplace equations are partial differential equations, their solutions tend to be rather complicated. However, in cases where the charge distribution can be expressed as a function of a *single variable*, they reduce to ordinary differential equations and can then be solved straightforwardly.

Example 1-13

Find the field and potential within a uniform infinite slab of charge.

Solution   Since the potential can depend only on the thickness coordinate (call it x),

$$\frac{d^2 U}{dx^2} = -4\pi K_f \rho_0$$

Thus

$$U(x) = -2\pi K_f \rho_0 x^2 + C_1 + C_2 x$$

and

$$\mathbf{E}_S(x) = (4\pi K_f \rho_0 x - C_2)\hat{\mathbf{i}}$$

Thus we see that the field varies linearly and the potential quadratically with distance through the slab.

Summary on Field Calculations

The following methods are available:

1. Calculate $\mathbf{E}_S$ directly at desired points by Eq. (1-8). This is always possible in principle, although the integral may not be workable analytically.

2. Calculate U as function of coordinates of $\mathbf{r}$ by Eq. (1-17). Then take the gradient. Finally, evaluate at desired points. This is always possible in principle; the integral may be easier than that for $\mathbf{E}_S$.

3. Use the integrated form of Gauss's law [Eq. (1-10)] to get $\mathbf{E}_S$. This is possible only with highly symmetrical charge distributions.

4. If $\mathbf{E}_S(\mathbf{r})$ is known, $U(\mathbf{r})$ may be obtained as a line integral [Eq. (1-21)] and the charge density may be calculated from the differential form of Gauss's law [Eq. (1-15)].

5. U may also be obtained by solving the Poisson equation (1-22), which reduces to the Laplace equation (1-23) in charge-free regions. This method will be taken up in Chapter 11.

1-8 MULTIPOLE EXPANSION

We have seen that whenever the charge distribution is given, the field and potential can be straightforwardly calculated by evaluating the appropriate integrals. However, the resulting expressions are often quite cumbersome, and it is desirable to have simpler *approximate* expressions available. In this section, such an approximation is developed.

First, we find a power series expansion for $1/|\mathbf{r} - \mathbf{r}'|$. To simplify the notation, let $R = |\mathbf{r} - \mathbf{r}'|$. By the law of cosines, $R^2 = r^2 + r'^2 - 2\mathbf{r}\cdot\mathbf{r}'$, so

$$R^{-1} = (r^2 + r'^2 - 2\mathbf{r}\cdot\mathbf{r}')^{-1/2}$$

$$= r^{-1}\left[1 + \left(\frac{r'}{r}\right)^2 - 2\hat{\mathbf{a}}_r \cdot \frac{\mathbf{r}'}{r}\right]^{-1/2}$$

If $r > r'$, the second plus third terms are less than 1, so a binomial expansion will converge. Thus

$$R^{-1} = r^{-1}\left\{1 - \frac{1}{2}\left[\left(\frac{r'}{r}\right)^2 - 2\frac{\hat{\mathbf{a}}_r \cdot \mathbf{r}'}{r}\right] + \frac{3}{8}\left[\left(\frac{r'}{r}\right)^2 - 2\frac{\hat{\mathbf{a}}_r \cdot \mathbf{r}'}{r}\right]^2 + \cdots\right\}$$

Gathering up like powers of r'/r, we find

$$R^{-1} = r^{-1}\left\{1 + \frac{\hat{\mathbf{a}}_r \cdot \mathbf{r}'}{r} + \left[-\frac{1}{2}\left(\frac{r'}{r}\right)^2 + \frac{3}{2}\left(\frac{\hat{\mathbf{a}}_r \cdot \mathbf{r}'}{r}\right)^2\right] + \cdots\right\} \tag{1-24}$$

(Remember that this expansion is valid only for $r' < r$. A similar one can be derived for $r' > r$, but it is less generally useful.)

Equation (1-24) can now be used to develop a power series expansion for the potential at points sufficiently distant from the source charge distribution. By choosing the origin to lie within the source distribution, and restricting **r** to the region well outside, we ensure that $r > r'$ for all **r**' in the distribution. We need only substitute Eq. (1-24) into Eq. (1-16), and integrate term by term. All functions of the field point coordinates can, of course, be taken out of the integrals. Thus

$$U(\mathbf{r}) = K_f \iiint \rho \frac{r'}{r}\left(1 + \frac{\hat{\mathbf{a}}_r}{r} \cdot \mathbf{r}' + \cdots\right) d^3r'$$

$$= K_f\left[\frac{1}{r} \iiint \rho(\mathbf{r}') \, d^3r' + \frac{1}{r^2}\hat{\mathbf{a}}_r \cdot \iiint \mathbf{r}'\rho(\mathbf{r}') \, d^3r' + \cdots\right]$$

The first integral is simply the total charge q of the distribution. The second has each charge element multiplied by its vector displacement from the origin. This defines the *first moment*, usually called the *dipole moment* **p** of the distribution. That is,

$$\mathbf{p} \equiv \iiint \mathbf{r}'\rho(\mathbf{r}') \, d^3r' \qquad (1\text{-}25)$$

The subsequent terms have higher powers of the displacements multiplying the charge elements, giving higher moments of the distribution. These get quite complicated and we shall not deal with them explicitly. To two terms, the series then is

$$U(\mathbf{r}) = K_f\left(\frac{q}{r} + \frac{\hat{\mathbf{a}}_r \cdot \mathbf{p}}{r^2} + \cdots\right) \qquad (1\text{-}26)$$

Note that the successive terms fall off more and more rapidly with distance from the source. Often the first term with a nonzero numerator provides an adequate approximation of the potential.

Example 1-14

Find the lowest nonvanishing moment of the following distribution of point charges (charges given in microcoulombs, distances in centimeters): -3 at $(0, 0, 0)$, 2 at $(1, 0, 0)$, $\frac{1}{2}$ at $(0, 1, 0)$, $\frac{1}{2}$ at $(0, -1, 0)$.

Solution

$$q = -3 + 2 + \tfrac{1}{2} + \tfrac{1}{2} = 0$$
$$\mathbf{p} = -3(0) + 2(\hat{\mathbf{i}}) + \tfrac{1}{2}(\hat{\mathbf{j}}) + \tfrac{1}{2}(-\hat{\mathbf{j}})$$
$$= 2\hat{\mathbf{i}} \text{ in units of } 10^{-8} \text{ C-m}$$

Example 1-15

Verify the correctness of Eq. (1-26) for the charge distribution just considered for field points far out on the positive x axis.

Solution   The exact potential function is

$$U(x, 0, 0) = K_f\left(\frac{-3}{x} + \frac{2}{x - 1} + 2\frac{\frac{1}{2}}{\sqrt{x^2 + 1}}\right)$$

To obtain the approximate form for large x, we factor x out of each denominator and make binomial expansions of what remains.

$$U(x, 0, 0) = \frac{K_f}{x}\left[-3 + \frac{2}{1 - 1/x} + \frac{1}{(1 + 1/x^2)^{1/2}}\right]$$

$$\simeq \frac{K_f}{x}\left[-3 + 2\left(1 + \frac{1}{x} + \cdots\right) + 1\left(1 - \frac{1}{2x^2} + \cdots\right)\right]$$

$$\simeq \frac{K_f}{x}\left(-3 + 2 + 1 + \frac{2}{x} + \cdots\right)$$

$$= 2K_f/x^2 \qquad \text{which agrees with Eq. (1-26)}$$

As long as the total charge in the distribution is zero, the dipole moment is independent of the choice of origin. This is easily verified by calculating $\mathbf{p}'$, the dipole moment with respect to an origin O' displaced by $\mathbf{a}$ from O. The position vectors are $\mathbf{r}' - \mathbf{a}$, so

$$\mathbf{p}' = \iiint (\mathbf{r}' - \mathbf{a})\rho(\mathbf{r}')\, d^3\mathbf{r}'$$

$$= \mathbf{p} - \mathbf{a} \iiint \rho(\mathbf{r}')\, d^3\mathbf{r}'$$

$$= \mathbf{p} - \mathbf{a}q = \mathbf{p} \qquad \text{if } q = 0$$

Ideal Dipoles

It is useful to define an entity called an *ideal dipole* or *point dipole* as an analog of the point charge. We start with a pair of point charges, $\pm q$, with $+q$ displaced by some vector $\mathbf{b}$ from $-q$. The dipole moment of this charge distribution is $q\mathbf{b}$. There will also be higher moments, whose values will depend on where the origin is chosen but will be minimized by choosing the origin at the midpoint between the charges. Now we repeatedly double q while halving $\mathbf{b}$. At each step, $\mathbf{p}$ remains unchanged but all higher moments diminish. In the limit, we reach a point distribution which has *only* a dipole moment. Its potential is given *exactly* by the second term of Eq. (1-26).

Example 1-16

Calculate the field due to an ideal dipole.

Solution 1

$$K_f^{-1}\mathbf{E}_S(\mathbf{r}) = -\text{grad}\,\frac{\hat{\mathbf{a}}_r \cdot \mathbf{p}}{r^2} = -\text{grad}\,\frac{\mathbf{r} \cdot \mathbf{p}}{r^3}$$

$$= -r^{-3}\,\text{grad}\,(\mathbf{r} \cdot \mathbf{p}) - (\mathbf{r} \cdot \mathbf{p})\,\text{grad}\,r^{-3}$$

$$= -\mathbf{p}r^{-3} - (\mathbf{r} \cdot \mathbf{p})\frac{-3\mathbf{r}}{r^5}$$

$$= \frac{3(\mathbf{p} \cdot \hat{\mathbf{a}}_r)\hat{\mathbf{a}}_r - \mathbf{p}}{r^3} \tag{1-27a}$$

Solution 2   In spherical coordinates,

$$K_f^{-1} \mathbf{E}_S(r, \theta) = -\operatorname{grad} \frac{p \cos \theta}{r^2}$$

$$= \frac{p}{r^3} (2 \cos \theta \, \hat{\mathbf{a}}_r + \sin \theta \, \hat{\mathbf{a}}_\theta) \qquad (1\text{-}27\text{b})$$

The third term in Eq. (1-26) is generally of importance only if both the net charge and the dipole moment of the distribution are zero. The factor that depends on the charge distribution is called the *quadrupole moment*. It is neither a scalar nor a vector, but belongs to the class of mathematical quantities known as *second-rank tensors*. As such, it lies beyond the scope of this book.

1-9 FIELD LINES AND EQUIPOTENTIAL SURFACES

Since fields and potentials are often described by fairly complicated functions, it is helpful to have some means of depicting them graphically. One method, which we have already used, is simply to plot a particular component along some line in space. This, however, gives only a small part of the overall picture. In this section we discuss other methods of visualizing fields and potentials.

An *equipotential* is a surface containing all points at which the potential has the same value. Its equation is

$$U(\mathbf{r}) = \text{constant} \qquad (1\text{-}28)$$

By choosing different values of the constant, one obtains a family of surfaces filling the space.

A *field line* is a line which at every point is tangent to the field vector at that point. Its equations are

$$dx : dy : dz = E_x : E_y : E_z \qquad (1\text{-}29)$$

or the equivalent in some other coordinate system. Since the field is (−) the gradient of the potential, the field line is perpendicular to the equipotential at every point.

A good picture of the field behavior can often be obtained by drawing a sampling of the field lines and the traces of the equipotential surfaces in some suitably choosen plane.

Example 1-17

Sketch the field pattern due to an ideal dipole (Figure 1-11).

Solution   Obviously, we choose a plane containing the dipole; call this the xz plane, with the dipole along the z axis. The family of equipotential lines in this plane is

$$p \cos \theta = Cr^2$$

The field lines obey the equation

$$\frac{dr}{r\,d\theta} = \frac{E_r}{E_\theta} = \frac{2\cos\theta}{\sin\theta} = 2\cot\theta$$

$$\ln r = 2\ln\sin\theta + \text{const.}$$

$$r = C'\sin^2\theta$$

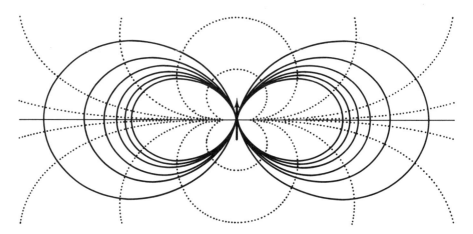

Figure 1-11   Field lines (solid) and equipotentials (dotted) for a dipole. The field lines are spaced at equal increments of E at $\theta = 90$ degrees; increasing inward. The equipotentials are spaced by factors of 2, increasing inward.

It is seen that the information contained in this form of field plot relates primarily to the *direction* of the field. However, some information relating to the *magnitude* can also be incorporated by choosing the values of C' such that the field lines drawn cut some particular equipotential (e.g., the x axis in the foregoing example) at equal increments of strength. Then the magnitude of the field at any point will be represented by the density of lines in the vicinity. (We leave it as an exercise for the reader to prove this; he or she should first show that field lines can never cross in a charge-free region.)

The quantitative aspects of the field plot are sometimes overemphasized. In fact, the field lines were formerly often called *lines of force* and endowed with an aura of physical reality. It is important to recognize that they are purely a pictorial device and have *no physical content* whatever. However, they are a very useful aid in visualization.

1-10 IDEAL CONDUCTORS

An *ideal conductor* is a material whose properties make it impossible for an electric field to exist at any point within it. No such material exists in nature, but real conductors (e.g., metals) approximate the behavior to some extent. As we shall see in

Chapter 5, metals are good conductors because they contain a large number of *free* electrons. Any attempt to impose a field causes these electrons to move and reposition themselves so as to tend to cancel it. In a real metal, this cancellation is complete except for a fairly thin layer at the surface. In an ideal conductor, we postulate that the layer is *infinitesimally* thin, so that the cancellation is 100% complete at every interior point.

An ideal conductive body is an equipotential region, and its surface is an equipotential surface. This is obvious, since grad $U = 0$. Since field lines must be perpendicular to equipotential surfaces, any field at a point just outside the body must be perpendicular to the surface.

A number of important consequences can be derived from Gauss's law. If we draw a Gaussian surface S lying entirely within the body, then $O(S) = 0$, since $E_s = 0$ at every point. Thus no such surface can enclose any net charge. In other words, $\rho = 0$ at every interior point of an ideal conductor. There can, of course, be charge on the *surface*. Indeed, it is the surface charge distribution that cancels out any applied field at the interior points. These surface charges will make their full contributions to fields at exterior points. The relation of surface charge density $\sigma(\mathbf{r}_s)$ at any surface point to the field just adjacent to this point can also be found from Gauss's law. We use a cylindrical Gaussian surface, as in Example 1-8, but now the field is zero on one face (inside the material), so we get twice as much at the other, that is,

$$\mathbf{E}_S(\mathbf{r}_s) = 4\pi K_f \sigma(\mathbf{r}_s)\hat{\mathbf{n}}(\mathbf{r}_s) \tag{1-30}$$

where $\hat{\mathbf{n}}$ is the outward-directed normal to the surface at the point in question.

Example 1-18

Charge q is deposited on an ideal conductive sphere, and no other charges are present anywhere. Find the field and potential at all points.

Solution   The charge will reside entirely on the surface, and the symmetry assures us that it will be distributed uniformly. Thus we have a spherical shell of charge, $\sigma = q/4\pi a^2$. This problem was solved in Example 1-12, which should be carefully reviewed. It will be seen that the field strength just outside the sphere agrees with Eq. (1-30). The potential of the sphere is $K_f q/a$.

Example 1-19

Suppose that a conductive sphere is surrounded by a medium (such as air) that is nonconductive at low field strengths but which "breaks down" and becomes conductive if the field strength exceeds some critical value E_{max}. What is the highest potential to which the sphere can be charged?

Solution

$$q_{max} = \frac{E_{max}a^2}{K_f}$$

$$U_{max} = \frac{K_f q_{max}}{a} = E_{max}\, a$$

This example shows that for very high potentials, one must use large smoothly curved conductors, as is commonly done in electrostatic generators such as the van de Graaf generator. These are often surrounded by tanks of dry pressurized gases to increase the breakdown strength, but still require spheres of several meters diameter to reach potentials of some millions of volts.

Conversely, if one wants breakdown to take place readily, a very small radius of curvature is used (Figure 1-12). This is applied in the *lightning rod*, a pointed metal rod sticking up in the air and running down to a good contact with the moist earth. The sharp point causes breakdown in the field arising from charged clouds in a thunderstorm. The charge is thus harmlessly conducted to earth rather than striking some other part of the building and possibly doing damage.

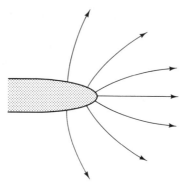

Figure 1-12 Field pattern near a sharply curved charged conductor (schematic).

Hollow Conductors

A very important configuration is a region completely surrounded by a conductor. We treat here only ideal conductors, but in practice a moderate thickness (a fraction of a millimeter) of a good metal (e.g., copper) will be virtually the same. The importance lies in the fact that the interior region is completely shielded from any fields that may be present outside. This follows because the surface charge distributed over the outer surface cancels the field at all points *within this surface* independently of whether the material is present at those points or not (Figure 1-13).

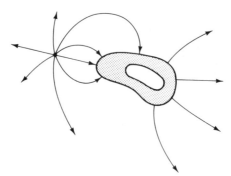

Figure 1-13 Field pattern near a hollow conductor with a positive point charge outside (schematic).

The converse, however, is not true. There may well be charges in the inner space and these will produce fields both in the cavity and outside the outer surface (Figure 1-14). The reason is that, by Gauss's law, a charge q in the cavity will cause a surface charge totaling $-q$ to be drawn to the inner surface, in order that $O(S)$ will be zero on any surface S lying in the body of the conductor. Thus a surface charge totaling q is left on the outer surface and produces a field outside.

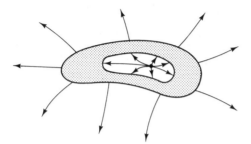

Figure 1-14 Field pattern near a hollow conductor with a positive point charge inside (schematic).

Example 1-20

A hollow sphere of ideal conductor has inner and outer radii a and b. A point charge q is placed at the center. What is the potential of the conductor?

Solution The surface charge density on the outer surface is $q/4\pi b^2$ and this results in potential $K_f q/b$, as in Example 1-18.

Example 1-21

The point charge in the preceding example is slowly moved off center and finally brought into contact with the inner surface of the conductor. What happens to the potential of the conductor?

Solution Nothing! The external field is due to the surface charge distribution on the outer surface. As the point charge is moved, the charge on the inner surface tends to "clump up" near it, and where the two come into contact they exactly neutralize each other. We will analyze the situation quantitatively in Chapter 11.

Notice that the net effect of inserting charge q into a cavity in a conductor (e.g., via a trapdoor) and touching it to the inner wall is to transfer charge q *completely* to the outer surface. This may be repeated time after time, regardless of how high a potential is built up on the conductor, because the inside is completely screened off from the outside. This is the basic way that very high potentials are built up in van de Graaf generators. In these machines, the charge is transported into the hollow conductor on a motor-driven belt, and transferred to the inner wall by means of sharp-pointed wire brushes (like mini-lightning rods).

Charging by Induction

The next example deals with another method of charging a conductor. Again we defer the quantitative treatment to Chapter 11, but a qualitative discussion is appropriate now.

Example 1-22

Consider a conductive sphere, initially isolated and uncharged. Qualitatively discuss its potential, total charge, and charge distribution after each of the following successive steps: (a) a point charge $+q_0$ is brought near, (b) the sphere is connected by a wire to the earth (which acts as an infinite charge reservoir), (c) the wire is removed, and (d) q_0 is removed.

Solution   After (a), the sphere is at a positive potential due to the proximity of q_0 but is still uncharged, since there was no way for charge to get on or off. The surface charge density is negative on the side toward q_0 and positive on the opposite side.

After (b), the potential is brought to zero by connection to the earth. This requires negative charge to flow to the sphere through the wire. The surface charge density is still nonuniform, but the *added* part is uniformly distributed over the surface.

After (c) nothing changes.

After (d), the potential becomes negative due to the negative charge, which is now distributed uniformly over the surface.

It is noteworthy that the charge q_0 is not tranferred or "used up" in any way. If desired, the sphere could be discharged into another body, by the method of Example 1-21, for example, and the process repeated many times over to build up a large potential in the receiver. This is the basis of many older types of electrostatic generators, now largely supplanted by the van de Graaf.

Verification of Coulomb's Law

As noted in Section 1-3, direct force measurement is not a very good way to substantiate Coulomb's law. Obviously, for a law as basic as this we want the utmost precision in experimental tests. This has been achieved by testing the consequences of Gauss's law, which, as we saw in Section 1-5, is itself a consequence of Colomb's law. The methods fall into two classes:

1. Tests of the exactness of neutralization of charged conductors touched to the inner surface of a hollow conductor, as in Example 1-21.
2. Tests of the exactness of screening of an external field from a cavity in a conductor, as discussed before Example 1-20.

For these purposes, a modest thickness of a real metal is a perfectly adequate approximation to an ideal conductor, as we shall see in Chapter 5.

Method 1 was used as early as 1755 by Benjamin Franklin, who fully understood its implications for an inverse-square force law. The best application, by Maxwell and McAlister in 1878, led to a verification of the distance exponent as 2.0 to better than 1 part in 20,000. Method 2 is inherently more accurate. The most recent determination* sets the limit at 5 parts in 10^{16}, a fantastic accuracy indeed!

The final example of this chapter illustrates the principle of these methods.

Example 1-23

Suppose the exponent in Coulomb's law to be $2 + x$, where x is some very small number. Calculate the charge q_b induced on the inner surface, radius b, of a hollow conductive sphere by a spherical charge q_a, radius a, at its center.

Solution Equation (1-30) will still hold true at points very close to each surface, since the distant charges on that surface (or elsewhere) do not contribute much anyway. Thus

$$\frac{\sigma_b}{\sigma_a} = \frac{E_b}{E_a} = \left(\frac{a}{b}\right)^{2+x}$$

$$\frac{q_b}{q_a} = \frac{\pi b^2 \sigma_b}{\pi a^2 \sigma_a} = \left(\frac{a}{b}\right)^x = [e^{\ln (b/a)}]^{-x}$$

$$\cong 1 - \left(\ln \frac{b}{a}\right)x$$

The failure of exact neutralization then amounts to

$$\frac{q_a - q_b}{q_a} = x \ln \frac{b}{a}$$

PROBLEMS

1-1. Two point charges of 1 μC are 1 cm apart. Find the magnitude of the force on each.

1-2. Two point particles of mass m and charge q are suspended from a point by strings of length L. Find the distance that each particle stands out from the vertical, assuming that the charges are weak so that the angle between the strings is small.

1-3. Given point charges of 1 C at $z = 0$, 1 C at $z = 1$, and -2 C at $z = 2$, find the field at points along the negative z axis.

1-4. In Problem 1-1, what is the field strength acting on one of the charges?

1-5. Find the field due to a uniform line charge of length L.

1-6. A uniform line charge is bent into a square of side a. Find the field at points along the axis.

1-7. A uniform line charge is bent into a semicircle lying in the xy plane with center at the origin and ends on the x axis. Find the field at points along the z axis.

1-8. A uniform semi-infinite sheet of charge lies in the xy plane with its edge along the x axis. Find the z component of the field at points along the z axis. *Hint:* Divide the sheet into strips of width dx and use the result of Example 1-5.

*E. R. Williams, J. E. Faller, and H. A. Hill, *Phys. Rev. Lett.* **26** (1971), 721.

1-9. Find the field at points along the axis of a uniformly charged circular disk of radius a.

1-10. A disk of radius a has charge density inversely proportional to the distance from the axis. Find the field at points along the axis.

1-11. Find the field at points along the axis of a uniform cylinder of charge, radius a, and length L. *Hint:* Use the result of Problem 1-9.

1-12. A line charge varies uniformly in charge density from $-\lambda_0/2$ to $\lambda_0/2$ in length L. Find the field strength at any point in the perpendicular bisector plane.

1-13. A straight line charge extending from $z = 0$ to $z = L$ has charge density proportional to z. Find the field at points along the z axis beyond L.

1-14. Find the charge distribution that gives a field of the form $C\mathbf{r}$.

1-15. Find the charge distribution that gives a field of the form $A\mathbf{r}/r^3$.

1-16. An infinitely long cylinder of radius a is filled with a uniform charge density. Find the field.

1-17. In the cylinder of Problem 1-16, an infinitely long hole is bored out parallel to the axis. The hole has radius $a/2$ and is tangent to the surface of the cylinder. Find the field at points outside in the plane containing the two axes.

1-18. Use Gauss's law to find the field due to a spherically symmetric distribution of charge $\rho(r)$.

1-19. Find the field due to a charge q distributed uniformly throughout the volume of a sphere of radius a. (Use Gauss's law.)

1-20. The field in a spherical region of radius a is $Cr^2\hat{\mathbf{a}}_r$. If there is no charge outside the region, find the field outside.

1-21. Use Gauss's law to find the field due to an infinite uniform line charge (see Example 1-4).

1-22. For the charge distribution of Problem 1-3, find the potential at an arbitrary field point. Show that this potential gives the correct answer to Problem 1-3. Be careful about the sign.

1-23. Find the potential at points on the axis of a uniform charge ring.

1-24. Find the potential at points on the axis of a uniformly charged cylindrical band of radius a, length L, and total charge q.

1-25. Derive an expression for the potential at any point due to a spherically symmetrical charge distribution.

1-26. Find the potential at any point due to a uniform solid sphere of charge.

1-27. For the charge distribution of Problem 1-7, write an explicit integral for the potential at an arbitrary field point.

1-28. Concentric spherical shells of radii a and b $(a < b)$ have uniformly distributed surface charges of Q and $-Q$, respectively. Find the potential at all points.

1-29. Repeat Problem 1-25 by means of the Poisson equation.

1-30. Find the charge distribution that gives rise to a potential of the form $U(r) = Ce^{-\alpha r}$.

1-31. Calculate the dipole moment of the charge distribution of Problem 1-3.

1-32. For Problem 1-3, find the limiting form of the field expression for very large distances and show that it corresponds with that expected from the dipole moment calculated in Problem 1-31.

1-33. Verify the correctness of the limiting forms of either the field or the potential for (a) Example 1-2, (b) Problem 1-6, (c) Problem 1-11, (d) Problem 1-12, and (e) Problem 1-18.

1-34. A hollow spherical conductor, radii a and b ($b > a$), bears charge q_1 and has charge q_2 at the center of the cavity. What is its potential?

1-35. A conductive sphere of radius 1 cm carries a charge of 10^{-8} C. It is touched to the inner surface of a larger sphere, radius 1 m, then tested for residual charge. If a meter sensitive to 10^{-8} V is available, how accurately can the inverse-square law be tested?

chapter 2

Electric Current and the Magnetostatic Field

2-1 CURRENT

It is obvious that the individual particles making up a charge distribution may in some circumstances be in motion, and that the individual velocities may or may not exactly cancel or balance each other. It is found that when unbalanced motions do exist, additional macroscopic effects may be observed, and it is therefore necessary to consider the motion of charged particles in some detail.

Charges in motion constitute *currents*. To define this concept, consider a small area A embedded in a charge distribution. Then the current through this area is defined as

$$I(A) = \text{net rate of flow of charge through area } A \text{ in a specified sense} \qquad (2\text{-}1)$$

$I(A)$ may be evaluated in terms of the particle concentrations and velocities, for an arbitrary area A. Suppose that a medium contains several species of particles, of charge q_i and average velocity $\mathbf{v}_i$ (Figure 2-1). We separately calculate the contribution of each. Taking area A as the base, we draw a cylinder of slant height $\mathbf{v}_i\, dt$. Then, starting at any instant, the particles which will get through the area A in time interval dt are just those contained within this cylinder. (*Note*: We are here neglecting

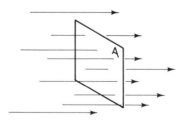

Figure 2-1 Particles streaming through area A with velocity $\mathbf{v}_i$.

the random components of the motions and treating the particles as if they were all moving with the average velocity. This is tantamount to averaging over many time intervals.) Hence for the *i*th species

$$\text{rate of penetration of particles} = n_i\mathbf{v}_i \cdot \hat{\mathbf{n}}A$$

$$\text{rate of penetration of charge due to these particles} = q_i n_i \mathbf{v}_i \cdot \hat{\mathbf{n}}A \qquad (2\text{-}2)$$

On summing over all the species present, we obtain

$$I(A) = A\hat{\mathbf{n}} \cdot \sum_i q_i n_i \mathbf{v}_i \qquad (2\text{-}3)$$

The *current density* is defined as the second factor in Eq. (2-3); since n_i and $\mathbf{v}_i$ may well vary with position and time, this constitutes a new vector point function

$$\mathbf{J}(\mathbf{r}, t) = \sum_i q_i n_i(\mathbf{r}, t)\mathbf{v}_i(\mathbf{r}, t) \qquad (2\text{-}4)$$

In the present chapter we shall be concerned only with *stationary* (i.e., time-independent) current distributions $\mathbf{J}(\mathbf{r})$.

In the MKS system, current is measured in *amperes* (coulombs per second) and current density therefore in amperes per square meter.

Example 2-1

A plasma contains 10^{20} electrons per cubic meter and an equal number of monovalent positive ions. The electrons are moving upward at 1000 m/s and the ions downward at 100 m/s. Calculate the charge and current densities.

Solution

$$\rho = -en_{\text{el}} + en_{\text{ion}} = 0$$

$$J = -en_{\text{el}}v_{\text{el}} + en_{\text{ion}}v_{\text{ion}}$$

$$= e(-n_{\text{el}}v_{\text{el}} + n_{\text{ion}}v_{\text{ion}})$$

$$= 1.6 \times 10^{-19}[-10^{20} \times 10^3 + 10^{20} \times (-10^2)]$$

$$= -1.76 \times 10^4 \text{ A/m}^2$$

Example 2-2

A copper wire 1 mm in diameter carries a current of 1000 A. Estimate the average speed of the electrons. The electron concentration in metals is roughly 10^{28} per cubic meter.

Solution $J = 10^3/(10^{-6}\pi/4) = 1.3 \times 10^9 \text{ A/m}^2$. Hence

$$|v| = \frac{J}{ne} = \frac{1.3 \times 10^9}{1.6 \times 10^{-19} \times 10^{28}} \cong 1 \text{ m/s}$$

There is a common misconception that the current travels through conductors at the speed of light. This example illustrates how wrong that is! However, what we have calculated is the *average* speed. The random velocities of electrons in metals may be vastly greater than this. Also, a *disturbance* in the distribution may travel at a high speed, as we shall discuss in Chapter 13.

Just as with charge densities, it is often convenient to deal with idealized singularities in current distributions. The possibilities are:

1. *Surface current or sheet current:* the limit of a volume current flowing in a thin layer. The MKS unit of surface current density is *amperes per meter* (of width).

2. *Line current:* the limit of a narrow strip of surface current, measured in *amperes*. It is a very useful idealization since, in practice, currents are often confined to thin wires.

3. *Point current:* the current due to a point charge in motion. It exists only at the instantaneous position of the charge.

Chargeamentum

A useful concept is the *product of current and distance* in the same direction. No name exists for this, so we have to invent one; we call it *chargeamentum* and use the symbol Ξ. For a point current, $\Xi = q\mathbf{v}$. The obvious similarity to momentum $m\mathbf{v}$ is what led to the choice of the name. For a length $\mathbf{dl}$ of a line current, $d\Xi = I\mathbf{dl}$, and for a volume element dV of a volume current, $d\Xi = \mathbf{J}\, dV$. Obviously, the MKS unit of chargeamentum is *ampere-meters*.

2-2 CURRENT CONTINUITY EQUATION

The definitions of current, Eq. (2-1), and of current density [Eq. (2-4)] lead to an important theorem which we now develop. We evaluate the rate of change of the total charge contained within a volume V. Let S be the surface of V, and consider an element dS with outward-directed normal $\hat{\mathbf{n}}$. The current through dS is, by Eq. (2-3), $\mathbf{J} \cdot \hat{\mathbf{n}}\, dS$. According to Eq. (2-1), this gives the rate of flow of charge out through dS. Since outward flow diminishes the charge remaining, we have, on integrating over the entire surface,

$$-\frac{dQ}{dt} = \oiint_S \mathbf{J} \cdot \hat{\mathbf{n}}\, ds \tag{2-5}$$

This may also be converted to a differential relation, much as we did with Gauss's law in Section 1-5. The total charge is the volume integral of the charge density; hence

$$-\frac{dQ}{dt} = -\frac{d}{dt} \iiint_V \rho\, d^3r = \iiint_V -\frac{\partial \rho}{\partial t}\, d^3r$$

The last step is valid because the limits of integration are understood not to change with time, only the function $\rho(\mathbf{r}, t)$. Thus after applying the divergence theorem to the right side, Eq. (2-5) becomes

$$\iiint_V \left(\frac{\partial \rho}{\partial t} + \operatorname{div} \mathbf{J}\right) d^3\mathbf{r} = 0$$

Since this must be true for *any* volume V whatever, it follows that the integrand must vanish, that is,

$$\frac{\partial \rho}{\partial t} + \text{div } \mathbf{J} = 0 \tag{2-6}$$

This is the usual form of the *continuity equation*. Note that it is really a statement of the *conservation of charge*, since we implicitly assumed that the only way the charge in any volume could change was by flow in or out. All observations to date are in complete agreement with this assumption.

Corollary. A *static* current distribution contained in a finite volume must flow in closed loops. In a static (i.e., steady-state) situation, all rates of change are zero. In particular, $\partial \rho / \partial t = 0$ everywhere. Thus from Eq. (2-6),

$$\text{div } \mathbf{J}_S = 0 \text{ (static)} \tag{2-7}$$

This means that in every volume element as much current flows out as flows in, in other words, that there are no sources or sinks of current. Hence it must flow in loops.

In the following example, we consider a rather more complicated problem.

Example 2-3

Consider a distribution of charges and currents contained entirely within a volume V. Derive an expression for the rate of change of the dipole moment of the charge distribution.

Solution Consider a single component of the dipole moment vector

$$p_x(t) = \iiint_V x' \rho(r', t) \, d^3\mathbf{r}'$$

$$\frac{dp_x}{dt} = \iiint_V x' \dot{\rho}(r', t) \, d^3\mathbf{r}' = \iiint_V -x' \, \text{div}' \, \mathbf{J} \, d^3\mathbf{r}'$$

where the dot over ρ denotes the time derivative. We would like to have an exact divergence in the volume integral so that we could apply the divergence theorem. Hence we use the identity

$$\text{div } (x\mathbf{J}) = J_x + x \, \text{div } \mathbf{J}$$

to obtain

$$\frac{dp_x}{dt} = \iiint_V [J_x - \text{div}' \, (x'\mathbf{J})] \, d^3\mathbf{r}'$$

$$= \iiint_V J_x \, d^3\mathbf{r}' - \oiint_{S(V)} x'\mathbf{J} \cdot \hat{\mathbf{n}} \, dS'$$

The second term vanishes since the surface lies outside all the currents so that

$J = 0$ at every point of S. Repeating for the other two components and adding vectorially, we obtain

$$\frac{d\mathbf{p}}{dt} = \iiint_V \mathbf{J}(\mathbf{r}')\, d^3\mathbf{r}'$$

2-3 MAGNETIC FIELD

It is observed that currents exert forces on each other independently of any electrostatic forces that may be present. The force law was considerably more difficult to discover for two reasons: (1) the force between two elements is *not* along the line joining them, and (2) the observations were carried out not on small isolated "point" elements but on extended circuits. The statement of the law is, however, much simpler for point elements, and is, in fact, fairly similar to Coulomb's law, Eq. (1-4), namely,

$$\mathbf{F}_{A, B} = \left(\frac{\mu_0}{4\pi}\right)\Xi_A \times \frac{\Xi_B \times (\mathbf{r}_A - \mathbf{r}_B)}{|\mathbf{r}_A - \mathbf{r}_B|^3} \tag{2-8}$$

The constant of proportionally is written in the form $\mu_0/4\pi$ for later convenience. It has the value

$$\frac{\mu_0}{4\pi} = 10^{-7}\ \text{N/A}^2$$

It is seen that the chargeamentum of each element is the significant property in this interaction, and that the force on element A involves the product of the chargeamentum of A with a quantity that depends only on the element B and the vector displacement of A from B (Figure 2-2). This quantity thus defines a *field strength* at the position of A due to the presence of element B. This is called the *magnetic field strength*. Unfortunately, this name was erroneously assigned to a somewhat different quantity in the early development of the subject, and other names, mostly meaningless and misleading (*flux density* and *induction* are two common ones), were used instead. In recent years, however, there has been a trend toward the use of the logically correct name, and we enthusiastically go along with it. Thus Eq. (2-8) is written

$$\mathbf{F}_A = \Xi_A \times \mathbf{B}(\mathbf{r}_A) \tag{2-9}$$

where $\mathbf{B}(\mathbf{r}_A)$ is the magnetic field strength at the point $\mathbf{r}_A$.* If the field is due to a single current element Ξ_B at $\mathbf{r}_B$, then

$$\mathbf{B}(\mathbf{r}) = \frac{\mu_0}{4\pi} \Xi_B \times \frac{\mathbf{r} - \mathbf{r}_B}{|\mathbf{r} - \mathbf{r}_B|^3} \tag{2-10}$$

---

*The reader is reminded that the term "chargeamentum," defined in Section 2-1, is not found in other books. More commonly, the equation is written for a single charged particle, for which $\Xi_A = q\mathbf{v}$. It then takes the form

$$\mathbf{F}_A = q\mathbf{v} \times \mathbf{B}(\mathbf{r}_A) \tag{2-9a}$$

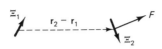

Figure 2-2   Force between two coplanar chargeamentum elements. In this case, the force vector lies in the same plane.

Equation (2-10) is known as the Biot–Savart law. The MKS unit for B may be expressed as

$$\text{newtons/ampere-meter} = \text{newton-seconds/coulomb-meter}$$

$$= \text{volt-seconds/meter}^2$$

The unit is also called the *tesla* (T) and the *weber/meter²*. To give an idea of the magnitude of this unit, the earth's field is about 5×10^{-5} T, a good iron-core electromagnet can produce a few teslas, and the best superconducting solenoids produce somewhat over 10 T. It is also quite common to measure magnetic fields in the older cgs unit, the *gauss*. The relation is

$$1 \text{ gauss} = 10^{-4} \text{ tesla}$$

For extended current distributions, the field at any point is the sum of the contributions of the individual current elements. The chargeamentum of each is written in the form appropriate to its geometry. In the case of a volume distribution, for example,

$$\mathbf{B(r)} = \frac{\mu_0}{4\pi} \iiint \frac{\mathbf{J(r')} \times (\mathbf{r} - \mathbf{r'})}{|\mathbf{r} - \mathbf{r'}|^3} d^3\mathbf{r'} \tag{2-11}$$

and for a single loop of current I,

$$\mathbf{B(r)} = \frac{\mu_0}{4\pi} \oint \frac{I \, \mathbf{dl} \times (\mathbf{r} - \mathbf{r'})}{|\mathbf{r} - \mathbf{r'}|^3} \tag{2-12}$$

Example 2-4

A long straight wire carries current I. An electron is shot parallel to the wire in the direction of the current. In what direction will it be deflected?

Solution   For the configuration drawn in Figure 2-3, the contribution of each current element in the wire to the field at the electron position is directed out of the paper [direction of the vector cross-product $I \, \mathbf{dl} \times (\mathbf{r} - \mathbf{r'})$]. The electron's chargeamentum $q\mathbf{v}$ is directed opposite to $\mathbf{v}$ (since q is negative). Thus the force is away from the wire.

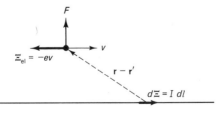

Figure 2-3   Electron moving parallel to current.

Example 2-5

Assume the wire in Example 2-4 to be infinitely long and calculate the magnetic field strength at any point around it.

Solution Choose the z axis to lie along the wire and the xy plane to pass through the field point. The contribution of the current element $I(dz')\hat{\mathbf{k}}$ at z' is

$$\frac{\mu_0}{4\pi} \frac{I\, dz'\, \hat{\mathbf{k}} \times (s\hat{\mathbf{a}}_s - z'\hat{\mathbf{k}})}{(s^2 + z'^2)^{3/2}} = \frac{\mu_0 I}{4\pi} \frac{s\, dz'\, \hat{\mathbf{a}}_\phi}{(s^2 + z'^2)^{3/2}}$$

On integrating over dz' (see Example 1-3), we obtain

$$\mathbf{B}(s, \phi, 0) = \frac{\mu_0 I}{4\pi s}\hat{\mathbf{a}}_\phi \qquad (2\text{-}13)$$

Example 2-6

An infinitely long strip of foil of width w carries a current I uniformly distributed across its width. Find the magnetic field strength at a point in the plane that perpendicularly bisects the width.

Solution Choose axes as shown in Figure 2-4.* The surface current density is $(I/w)\hat{\mathbf{k}}$ amperes per meter, hence the charageamentum of the element $dx'\, dz'$ at x', z' is $(I/w)\, dx'\, dz'\, \hat{\mathbf{k}}$, and its contribution to the field at $\mathbf{r} = y\hat{\mathbf{j}}$ is

$$\frac{\mu_0}{4\pi} \frac{I}{w} \frac{dx'\, dz'\, \hat{\mathbf{k}} \times (y\hat{\mathbf{j}} - x'\hat{\mathbf{i}} - z'\hat{\mathbf{k}})}{(y^2 + x'^2 + z'^2)^{3/2}} = \frac{\mu_0 I}{4\pi w} \frac{dx'\, dz'(-y\hat{\mathbf{i}} - x'\hat{\mathbf{j}})}{(y^2 + x'^2 + z'^2)^{3/2}}$$

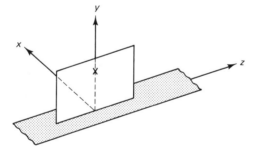

Figure 2-4 Geometry for Example 2-6.

The integration over dz' gives

$$\frac{\mu_0 I}{2\pi w} \frac{(-y\hat{\mathbf{i}} - x'\hat{\mathbf{j}})\, dx'}{y^2 + x'^2}$$

[Note that this could have been written directly from Eq. (2-13) and could, therefore, have been taken as the starting point of the problem.] The second integration from $-w/2$ to $w/2$ gives zero for the second term, so

$$\mathbf{B}(0, y, 0) = \frac{\mu_0 I}{\pi w} \tan^{-1}\frac{w}{2y}(-\hat{\mathbf{i}})$$

*Be sure to use right-hand axes when vector cross-products are involved!

Example 2-7

A circular loop of wire carries current I (Figure 2-5). Find the magnetic field strength at points on the axis.

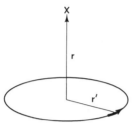

Figure 2-5 Geometry for Example 2-7.

Solution   The current element $a\,d\phi'$ located at $\mathbf{r}' = a\hat{\mathbf{a}}_{s'}$ has chargeamentum $Ia(d\phi')\hat{\mathbf{a}}_{\phi'}$. Hence its contribution to the field at $\mathbf{r} = z\hat{\mathbf{k}}$ is

$$\frac{\mu_0}{4\pi} Ia\,d\phi'\,\hat{\mathbf{a}}_{\phi'} \times \frac{z\hat{\mathbf{k}} - a\hat{\mathbf{a}}_{s'}}{(a^2 + z^2)^{3/2}} = \frac{\mu_0 I}{4\pi} a\,d\phi'\,\frac{z\hat{\mathbf{a}}_{s'} + a\hat{\mathbf{k}}}{(a^2 + z^2)^{3/2}}$$

On integrating, the first term gives zero (recall that $\hat{\mathbf{a}}_{s'}$ is a function of ϕ') and the second gives simply a factor 2π, so that

$$\mathbf{B}(0, 0, z) = \frac{\mu_0 Ia^2 \hat{\mathbf{k}}}{2(a^2 + z^2)^{3/2}}$$

Example 2-8

A solenoid of radius a and length L is wound with N total turns (Figure 2-6). Find the field strength at points along the axis.

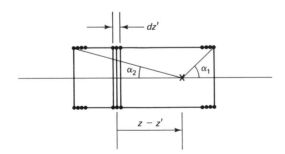

Figure 2-6 Geometry for Example 2-8.

Solution   We neglect the slight helicity of the winding and treat the current as if it flowed in planar rings. Thus the length dz' is equivalent to a ring carrying current $(N/L)\,dz'\,I$, and from Example 2-7 its field contribution at z is

$$\frac{\mu_0 NIa^2}{2L} \frac{dz'}{[a^2 + (z - z')^2]^{3/2}} \hat{\mathbf{k}}$$

The integral is essentially the same as in Example 2-5. The result may be cast in the form

$$\mathbf{B}(0, 0, z) = \frac{\mu_0 NI}{2L} (\cos \alpha_1 + \cos \alpha_2)\hat{\mathbf{k}}$$

For a point well inside a very long solenoid, the two cosines each approach unity.

2-4 MAGNETIC VECTOR POTENTIAL

We saw in Section 1-6 that the electrostatic field, with its inverse-square-law distance dependence for each source element, can be written as a derivative of a function (the potential) having an inverse-first-power dependence. The same is true for the magnetic field except that a different derivative combination, the curl, is needed. We show this by direct calculation in the same way. Let

$$\mathbf{A}_0(\mathbf{r}) \equiv \frac{\mu_0}{4\pi} \iiint \frac{\mathbf{J}(\mathbf{r}')}{|\mathbf{r} - \mathbf{r}'|} \, d^3\mathbf{r}' \tag{2-14}$$

Then

$$\operatorname{curl} \mathbf{A}_0(\mathbf{r}) = \frac{\mu_0}{4\pi} \iiint \operatorname{curl} \frac{\mathbf{J}(\mathbf{r}')}{|\mathbf{r} - \mathbf{r}'|} \, d^3\mathbf{r}'$$

$$= \frac{\mu_0}{4\pi} \iiint \left(\operatorname{grad} \frac{1}{|\mathbf{r} - \mathbf{r}'|} \right) \times \mathbf{J}(\mathbf{r}') \, d^3\mathbf{r}'$$

$$= \frac{\mu_0}{4\pi} \iiint \mathbf{J}(\mathbf{r}') \times \frac{\mathbf{r} - \mathbf{r}'}{|\mathbf{r} - \mathbf{r}'|^3} \, d^3\mathbf{r}'$$

$$= \mathbf{B}(\mathbf{r}) \tag{2-15}$$

(Here we again used the fact that the derivatives with respect to the field-point coordinates do not affect functions or integrals on the source point coordinates.)

The function $\mathbf{A}_0(\mathbf{r})$ is called a *vector potential for B*. Each source element contributes to this function a vector amount proportional to its own strength, in its own direction, and inversely proportional to its distance from the field point.

For a single filamentary loop of current, Eq. (2-14) becomes

$$\mathbf{A}_0(\mathbf{r}) = \frac{\mu_0 I}{4\pi} \oint \frac{d\mathbf{r}'}{|\mathbf{r} - \mathbf{r}'|} \tag{2-14'}$$

It should be pointed out that the vector potential is not nearly as useful a computational tool as the scalar potential of electrostatics. Attempts to evaluate $\mathbf{A}_0$ in simple-looking problems often lead to unexpected difficulties. The reader may wonder whether we could not use a scalar potential formalism for the magnetic field. The answer is that *in some circumstances* we can, and it is then a considerable simplification. The matter will be discussed in Section 12-3. However, the vector potential has the advantage of being completely general. Furthermore, it is useful in some important problems (see Section 2-5) and is vital in the development of the theory.

Divergence of B

In view of Eq. (2-15), the divergence of $\mathbf{B(r)}$ must vanish identically, that is,

$$\text{div } \mathbf{B(r)} = 0 \tag{2-16}$$

This result could also have been obtained by direct calculation on Eq. (2-11), from which the fact that $\mathbf{B}$ is a curl could have been deduced.

A function that is divergence-less at all points is called *solenoidal*. Its field lines necessarily form *closed loops*, since there is no point where they can cluster together. This is completely different from the structure of the electrostatic field, which by Eq. (1-18) is curl-free or *irrotational*. The basic structures of the two classes of vector fields are epitomized in Figure 2-7. Of course, parts of the two structures may be identical in a limited region. The global structure has to be examined to see the basic character.

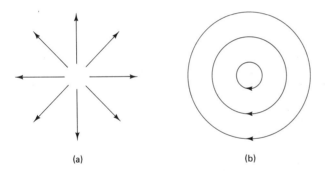

(a)   (b)

Figure 2-7 Basic structure of vector fields: (a) irrotational, (b) solenoidal.

It would seem natural to ask at this point whether the opposite kinds of fields also exist. In other words, are there also solenoidal electric fields or irrotational magnetic fields? The answer, strangely enough, is "yes and no." Solenoidal electric fields are easy to generate, as we shall see in Chapter 3. For an irrotational magnetic field, one would need to have a magnetic "charge," usually called a magnetic *monopole*. So far, no such entity has ever been detected, although there is no known fundamental reason why particles with this property should not exist. Indeed, modern particle theory seems to demand their existence, and concerted efforts have been made to detect them in many types of matter as well as in cosmic rays and among the products of very high energy collisions. But the searches to date have been fruitless and so, provisionally, Eq. (2-16) must be regarded as a fundamental property of all magnetic fields.

Divergence of A_0

As a prelude to Section 2-5, we need to evaluate div A_0. By the now familiar independence of the two sets of coordinates, we can write

$$\text{div } \mathbf{A}_0 = \frac{\mu_0}{4\pi} \iiint \mathbf{J}(\mathbf{r}') \cdot \left(\text{grad } \frac{1}{|\mathbf{r} - \mathbf{r}'|} \right) d^3\mathbf{r}'$$

The simplification of this expression is a bit tricky since, unfortunately, the straightforward procedure of evaluating the gradient leads nowhere. The trick is to note that a displacement $-\mathbf{dr}'$ produces the same change in the distance $|\mathbf{r} - \mathbf{r}'|$ as the displacement $\mathbf{dr}$. Thus for any function that depends only on this distance,

$$\text{grad } f(|\mathbf{r} - \mathbf{r}'|) = -\text{grad}' f(|\mathbf{r} - \mathbf{r}'|) \tag{2-17}$$

where grad$'$ designates differentiation with respect to the primed variables. Thus

$$\text{div } \mathbf{A}_0 = -\frac{\mu_0}{4\pi} \iiint \mathbf{J}(\mathbf{r}') \cdot \text{grad}' \, R^{-1} \, d^3\mathbf{r}'$$

where $R = |\mathbf{r} - \mathbf{r}'|$. The vector identity

$$\text{div}' \left[\frac{\mathbf{J}(\mathbf{r}')}{R} \right] = \frac{\text{div}' \, \mathbf{J}}{R} + \mathbf{J} \cdot \text{grad}' \, \frac{1}{R}$$

puts this in the form

$$\text{div } \mathbf{A}_0 = -\frac{\mu_0}{4\pi} \iiint \left[\text{div}' \left(\frac{\mathbf{J}}{R} \right) - \frac{\text{div}' \, \mathbf{J}}{R} \right] d^3\mathbf{r}'$$

Now the first term may be converted to a surface integral by means of the divergence theorem. Since the foregoing volume integrals are over all space, the surface will lie at infinity. In all real physical problems (not artificial ones like the infinite line) the currents are contained in some finite region, so that $\mathbf{J}$ is zero everywhere on the surface and the integral vanishes. Therefore, we are down to

$$\text{div } \mathbf{A}_0 = \frac{\mu_0}{4\pi} \iiint \frac{\text{div}' \, \mathbf{J}(\mathbf{r}')}{R} \, d^3\mathbf{r}' \tag{2-18}$$

Finally, in the case of *static* currents we have seen in Eq. (2-7) that the numerator vanishes. Hence

$$\text{div } \mathbf{A}_{0S} = 0 \tag{2-19}$$

Gauge Invariance

The electrostatic scalar potential is a well-defined function, flexible only to the extent of an additive constant. The magnetic vector potential is quite different. It may be modified by the addition of *any gradient* without affecting $\mathbf{B}$. Thus

$$\mathbf{B}(\mathbf{r}) = \text{curl } \mathbf{A}(\mathbf{r}) \tag{2-20}$$

where

$$\mathbf{A}(\mathbf{r}) = \mathbf{A}_0(\mathbf{r}) + \text{grad } \chi(\mathbf{r}) \tag{2-21}$$

and $\chi(\mathbf{r})$ may be any scalar function whatever.

Example 2-9

Show that $B_0 x \hat{\mathbf{j}}$, $-B_0 y \hat{\mathbf{i}}$, and $\frac{1}{2}\mathbf{B} \times \mathbf{r}$ are all suitable vector potential functions for a uniform field $B_0 \hat{\mathbf{k}}$.

Solution   Calculate the curls. They all come out to $B_0 \hat{\mathbf{k}}$.

Example 2-10

Find a suitable vector potential for the field of an infinite straight wire carrying current I.

Solution   The field is given in Eq. (2-13); hence we seek the solution of the partial differential equation

$$\frac{\partial A_s}{\partial z} - \frac{\partial A_z}{\partial s} = \frac{\mu_0 I}{2\pi} \frac{1}{s}$$

The symmetry of the problem assures us that the first term is zero, so

$$A_z = -\frac{\mu_0 I}{2\pi} \ln s + f(\phi, z)$$

The addition of a gradient is called a *gauge transformation*. Since the field is unchanged, there are no physical consequences, so the theory is said to be *gauge invariant*. This turns out to be a deeply significant property. Indeed, the entire theory can be derived by starting from the laws of mechanics, the existence of charge, and the requirement of gauge invariance.* Modern unified field theories exploit similar ideas.

For our less exalted present purposes, the flexibility allowed by gauge invariance can be very useful. In particular, it allows complete freedom for the divergence of $\mathbf{A}(\mathbf{r})$. Since

$$\text{div } \mathbf{A} = \text{div } \mathbf{A}_0 + \nabla^2 \chi$$

if we want div $\mathbf{A}(\mathbf{r})$ to be some specified function $f(\mathbf{r})$, we have only to choose χ as a solution of a "Poisson" equation

$$\nabla^2 \chi(\mathbf{r}) = f(\mathbf{r}) - \text{div } \mathbf{A}_0(\mathbf{r}) \tag{2-22}$$

2-5 AMPÈRE'S CIRCUITAL LAW

We now evaluate curl B. The calculation can be done straightforwardly but is difficult, and we now have the ingredients in hand for an easier indirect attack.

$$\text{curl } \mathbf{B} = \text{curl curl } \mathbf{A}_0$$
$$= \text{grad div } \mathbf{A}_0 - \nabla^2 \mathbf{A}_0$$

*A beautiful exposition of this approach is given in an article by D. H. Kobe, *Am. J. Phys.* **48** (1980), 348.

For the static case, Eq. (2-19) eliminates the first term, so

$$\text{curl } \mathbf{B}_S = -\nabla^2 \frac{\mu_0}{4\pi} \iiint \frac{\mathbf{J}(\mathbf{r}')}{R} \, d^3\mathbf{r}'$$

To see the form more clearly, write a single component of this vector equation, for example,

$$(\text{curl } \mathbf{B}_S)_x = -\nabla^2 \frac{\mu_0}{4\pi} \iiint \frac{J_x(\mathbf{r}')}{R} \, d^3\mathbf{r}'$$

The integral here is a *scalar*. In fact, it has exactly the same form as the electrostatic scalar potential [Eq. (1-17)] except that a different source density appears. But we know from the Poisson equation (1-22) that the Laplacian of this integral is simply a multiple of the source density (the numerator). Hence, by analogy,

$$(\text{curl } \mathbf{B}_S)_x = \mu_0 J_x$$

Adding the other two components in vectorially, we get

$$\text{curl } \mathbf{B}_S(\mathbf{r}) = \mu_0 \mathbf{J}(\mathbf{r}) \tag{2-23}$$

This is the differential form of Ampère's circuital law. To obtain the integral form, we integrate both sides over an arbitrary *open* surface and apply Stokes' theorem to the left:

$$\oint_{C(S)} \mathbf{B}_S \cdot d\mathbf{l} = \mu_0 \iint_S \mathbf{J} \cdot \hat{\mathbf{n}} \, dS$$

$$= \mu_0 I_{\text{encircled}} \tag{2-24}$$

where $I_{\text{encircled}}$ is the *net current encircled by the contour*. This same current pierces *any* surface bridging the contour, so the choice of the surface is of no importance. In the examples we have treated thus far, the currents have been confined to wires or foils, so that **J** was a highly discontinuous function. This presents some difficulties in using the differential form but not the integral form of Ampère's law.

Example 2-11

Verify Eq. (2-24) for the field of an infinitely long straight line current [Eq. (2-13)].

Solution   We may approximate any contour as closely as we wish by a series of circle arcs centered on the wire, connected by straight lines either radial or parallel to the axis (Figure 2-8). From Eq. (2-13), **B** · **dl** is zero on all the straight

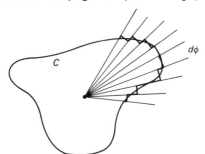

Figure 2-8   Approximation to contour C by arcs and line segments.

segments. On the circle arcs

$$\mathbf{B}_S \cdot \mathbf{dl} = \left(\frac{\mu_0 I}{2\pi s}\,\hat{\mathbf{a}}_\phi\right) \cdot (s\,d\phi\,\hat{\mathbf{a}}_\phi) = \frac{\mu_0 I}{2\pi}\,d\phi$$

Thus

$$\oint \mathbf{B}_S \cdot \mathbf{dl} = \frac{\mu_0 I}{2\pi} \times \begin{cases} 2\pi & \text{if the contour encircles the axis} \\ 0 & \text{otherwise} \end{cases}$$

$$= \mu_0 \times \text{current encircled}$$

Example 2-12

Show that near the middle of a long closely wound (i.e., ideal) solenoid the field is independent of radial position inside the winding and falls abruptly to zero outside.

Solution Choose as contour a rectangle with one side on the axis and the opposite side at distance s (Figure 2-9). Since symmetry tells us that the field must be in the axial direction, the other two sides make no contribution. Hence the integral is

$$\int_0^l B_1\,dx + \int_l^0 B_2\,dx = (B_1 - B_2)l$$

The current encircled is zero if s is less than the coil radius a, and $l(NI/L)$ if greater. Thus

$$B_2 = \begin{cases} B_1 & \text{if } s < a \\ B_1 - \dfrac{NI}{L} = 0 & \text{if } s > a \end{cases}$$

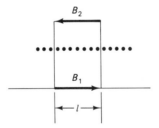

Figure 2-9 Integration contour for Example 2-12.

Ampère's circuital law can, like Gauss's law, be used to calculate fields in highly symmetrical problems. The idea is to choose a contour on which the field has uniform magnitude and known direction.

Example 2-13

A long circular cylinder carries current I uniformly distributed over its cross section. Find the magnetic field everywhere.

Solution The field must lie in the perpendicular plane since each filament of current produces a field in this plane (Figure 2-10). For the contour, choose a circle centered on the cylinder axis. Symmetry then dictates that the field could only be purely radial or purely tangential. Our previous knowledge of the form of magnetic fields rules out the former. Hence the field is everywhere tangential to the contour ($\mathbf{B} = B\hat{\mathbf{a}}_\phi$) and the integral becomes

$$\oint \mathbf{B} \cdot \mathbf{dl} = 2\pi s B = \mu_0 I_{\text{encircled}}$$

For $s > a$,

$$I_{\text{encircled}} = I \quad \text{and} \quad B = \frac{\mu_0 I}{2\pi s}$$

For $s < a$,

$$I_{\text{encircled}} = \frac{s^2}{a^2} I, \quad \text{and} \quad B = \frac{\mu_0 I s}{2\pi a^2}$$

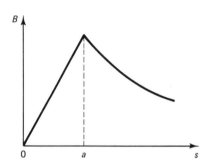

Figure 2-10 Magnetic field of a long uniform cylinder of current (Example 2-13).

Example 2-14

Verify the *differential* form of Ampère's circuital law for the field of Example 2-13.

Solution   curl $(1/s)\hat{\mathbf{a}}_\phi = 0$; curl $s\hat{\mathbf{a}}_\phi = 2\hat{\mathbf{k}}$. Thus for $s > a$,

$$\text{curl } B = 0$$

For $s < a$,

$$\text{curl } B = \frac{\mu_0 I}{\pi a^2} \hat{\mathbf{k}} = \mu_0 \mathbf{J}$$

Example 2-15

An infinite sheet of current (j amperes per meter) flows in the z direction in the xz plane. The magnetic field strength is zero on one side (due to some other sources). Find its strength on the other side.

Solution   Draw the contour *abcd* as shown in Figure 2-11 and evaluate the integral along it. Segments *ab* and *cd* make no contributions since they are perpendicular to any field there may be. Segment *bc* lies in the zero-field region.

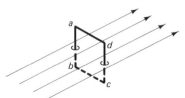

Figure 2-11 Geometry for Example 2-15.

Thus

$$\oint \mathbf{B} \cdot \mathbf{dl} \equiv Bl = \mu_0 j l$$

$$\mathbf{B} = \mu_0 j \hat{\mathbf{i}}$$

Thus a uniform sheet of current produces a magnetic field parallel to the sheet and perpendicular to the current.

Example 2-16

A toroidal solenoid (Figure 2-12) has major and minor radii b and a, respectively, and a total of N turns. The current in the wire is I. Find the field.

Solution   For a circular contour centered on the axis of the toroid the integral is $2\pi s B$. This must equal NI for contours within the toroid and zero for contours outside it.

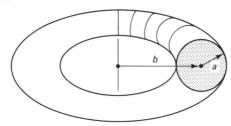

Figure 2-12   Toroidal solenoid (Example 2-16).

Example 2-17

A long cylinder of radius b has a hole of radius a running parallel to the axis but off-center. Current I is uniformly distributed over the remaining material. Find the magnetic field strength at the axis of the hole.

Solution   This problem does not seem to have the requisite symmetry for the use of Ampère's circuital law, but it may be broken into two problems each of which does. Thus consider a solid cylinder with the same *current density* (not the same total current) and a small cylinder with equal but opposite current density, and superimpose the two fields. The result is

$$\mathbf{B} = \frac{\mu_0 I d}{2\pi(b^2 - a^2)} \hat{\mathbf{a}}_\phi$$

where d is the distance between the axes.

These few examples more or less exhaust the geometries for which Ampère's circuital law is useful in simplifying field calculations. Nevertheless, the law, especially in its differential form, is of fundamental importance in the theory.

2-6 MAGNETIC DIPOLE

We saw in Section 1-8 that a very useful approximate expression for the electrostatic potential at points far outside the source distribution could be obtained by using a binomial series for the factor $1/R$ in the integral [Eq. (1-24)]. The same is, of course, true in magnetostatics. On inserting

$$\frac{1}{R} = \frac{1}{r}\left(1 + \frac{\mathbf{r} \cdot \mathbf{r'}}{r^2} + \cdots\right)$$

into Eq. (2-14) and integrating term by term, we get

$$\mathbf{A}_0(r) = \frac{\mu_0}{4\pi}\left[\frac{1}{r}\iiint \mathbf{J}(\mathbf{r'})\, d^3\mathbf{r'} + \frac{\mathbf{r}}{r^3} \cdot \iiint \mathbf{r'}\mathbf{J}(\mathbf{r'})\, d^3\mathbf{r'} + \cdots\right] \qquad (2\text{-}25)$$

For a static distribution, the first integral vanishes owing to the solenoidal (closed-loop) structure of $\mathbf{J}$; for every chargeamentum element there must somewhere be an equal opposite element, so the sum is zero.

In the second term the integral is the analog of Eq. (1-25), but this is *not* what is conventionally defined as the magnetic dipole moment. The integral as it stands is not a vector but a "dyad" (direct product of two vectors). The term can, however, be manipulated into a form containing a vector integral. Before considering the general case we work an explicit example to get a hint at how to proceed.

Example 2-18

Evaluate the second term in Eq. (2-25) for a circular loop of current, radius *a*.

Solution   Choose axes so that the loop lies in the xy plane with center at the origin and so that the field point coordinates are $(x, 0, z)$ (i.e. $\phi = 0$; Figure 2-13). The term is then

$$\frac{\mu_0}{4\pi}\frac{r\hat{\mathbf{a}}_r}{r^3} \cdot \int_0^{2\pi} (a\hat{\mathbf{a}}_{r'})(I\hat{\mathbf{a}}_{\phi'})(a\, d\phi') = \frac{\mu_0 I a^2}{4\pi r^2}\int_0^{2\pi}(\hat{\mathbf{a}}_r \cdot \hat{\mathbf{a}}_{r'})\hat{\mathbf{a}}_{\phi'}\, d\phi'$$

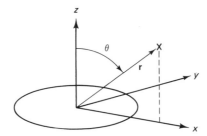

Figure 2-13   Geometry for Example 2-18.

Now $\hat{\mathbf{a}}_r \cdot \hat{\mathbf{a}}_{r'} = \sin\theta\cos\phi'$ and $\hat{\mathbf{a}}_{\phi'} = -\sin\phi'\hat{\mathbf{i}} + \cos\phi'\hat{\mathbf{j}}$, so the integral equals $\pi\sin\theta\,\hat{\mathbf{j}}$. Thus, to this approximation in the series,

$$\mathbf{A}_0(r, \theta, 0) \cong \frac{\mu_0}{4\pi}(\pi a^2 I)\frac{r\sin\theta}{r^3}\hat{\mathbf{j}}$$

This can be put in the coordinate-independent form

$$\mathbf{A}_0(\mathbf{r}) = \frac{\mu_0}{4\pi}\mathbf{m} \times \frac{\mathbf{r}}{r^3}$$

where $\mathbf{m} = \pi a^2 I\hat{\mathbf{n}}$, $\hat{\mathbf{n}}$ being the unit vector perpendicular to the plane of the current loop in the right-hand sense.

This example suggests that we try to put the second term of Eq. (2-25) in the form of a cross-product. We consider a single closed-loop filament of current. For this, the term is

$$\frac{\mu_0}{4\pi} \frac{1}{r^3} \oint (\mathbf{r} \cdot \mathbf{r}')(I \, d\mathbf{r}')$$

The combination $(\mathbf{r} \cdot \mathbf{r}') \, d\mathbf{r}'$ is recognized as one term in the vector triple product

$$(\mathbf{r}' \times d\mathbf{r}') \times \mathbf{r} = (\mathbf{r} \cdot \mathbf{r}') \, d\mathbf{r}' - \mathbf{r}'(\mathbf{r} \cdot d\mathbf{r}')$$

Also, for variation of $\mathbf{r}'$,

$$d[\mathbf{r}'(\mathbf{r} \cdot \mathbf{r}')] = (\mathbf{r} \cdot \mathbf{r}') \, d\mathbf{r}' + \mathbf{r}'(\mathbf{r} \cdot d\mathbf{r}')$$

Hence, adding and dividing by 2 gives

$$(\mathbf{r} \cdot \mathbf{r}') \, d\mathbf{r}' = \tfrac{1}{2}\{(\mathbf{r}' \times d\mathbf{r}') \times \mathbf{r}) + d[\mathbf{r}'(\mathbf{r} \cdot \mathbf{r}')]\}$$

The second term, being an exact differential, integrates to zero around a closed loop, leaving

$$\mathbf{A}_0(\mathbf{r}) \cong \frac{\mu_0}{4\pi} \mathbf{m} \times \frac{\mathbf{r}}{r^3} \tag{2-26}$$

where

$$\mathbf{m} = I \oint \frac{\mathbf{r}' \times d\mathbf{r}'}{2} \tag{2-27}$$

The integral in Eq. (2-27) has a simple geometrical interpretation (Figure 2-14). $\tfrac{1}{2}\mathbf{r}' \times d\mathbf{r}'$ is the area of the triangle formed by $d\mathbf{r}'$ and lines to the origin. Hence the components of the vector integral are the *projected areas* of the current loop on the coordinate planes. For a *planar* loop of any shape

$$\mathbf{m} = (\text{current} \times \text{area of loop})\hat{\mathbf{n}} \tag{2-28}$$

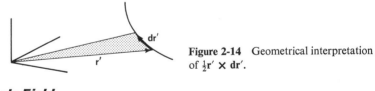

Figure 2-14 Geometrical interpretation of $\tfrac{1}{2}\mathbf{r}' \times d\mathbf{r}'$.

Magnetic Dipole Field

It remains to calculate the field corresponding to Eq. (2-26). This is perfectly straightforward.

$$\operatorname{curl} \frac{\mathbf{m} \times \mathbf{r}}{r^3} = \operatorname{grad} r^{-3} \times (\mathbf{m} \times \mathbf{r}) + r^{-3} \operatorname{curl} (\mathbf{m} \times \mathbf{r})$$

$$= -3r^{-5}\mathbf{r} \times (\mathbf{m} \times \mathbf{r}) + r^{-3}[\mathbf{m} \operatorname{div} \mathbf{r} - (\mathbf{m} \cdot \nabla)\mathbf{r}]$$

$$= -3r^{-5}[r^2\mathbf{m} - (\mathbf{m} \cdot \mathbf{r})\mathbf{r}] + r^{-3}(3\mathbf{m} - \mathbf{m})$$

$$= \frac{3(\mathbf{m} \cdot \mathbf{r})\mathbf{r} - r^2\mathbf{m}}{r^5}$$

or

$$\mathbf{B}(\mathbf{r}) = \frac{\mu_0}{4\pi} \frac{3(\mathbf{m} \cdot \hat{\mathbf{a}}_r)\hat{\mathbf{a}}_r - \mathbf{m}}{r^3} \tag{2-29}$$

This is seen to be identical in form to the electrostatic dipole field [Eq. (1-27a)]. It must be borne in mind that both of these expressions are valid only far outside the source. They cannot be identical everywhere, since an electrostatic field (irrotational) and a magnetic field (solenoidal) can never be fully identical, as we discussed in Section 2-4. The difference occurs near the sources. The electrostatic field lines start at the positive end and terminate at the negative end of the charge pair, whereas the magnetic field lines link through the current loop. In particular, within the sources the electrostatic field is directed opposite to the dipole while the magnetic field is directed along it.

2-7 MAGNETIC FLUX AND INDUCTANCE

The word *flux* is used in connection with vector fields in several different senses. For velocity fields describing particle motion or material flow, the word generally connotes amount passing through unit area in unit time (e.g., particles/cm²-s). But an entirely different usage is prevalent for electromagnetic fields: Here the connotation is simply the amount penetrating a given surface. Thus the *magnetic flux* through a surface is

$$\Phi_B(S) \equiv \iint_S \mathbf{B} \cdot \hat{\mathbf{n}} \, dS \tag{2-30}$$

This is similar to the outflow integral defined in Section 1-5, except that the domain of integration here is an *open* surface. The MKS unit of magnetic flux is the *weber*, or *tesla-meter*².

Example 2-19

Show that the total magnetic flux through any closed surface (i.e., the outflow) is zero.

Solution   For a closed surface, we may use the divergence theorem to obtain

$$\oiint_S \mathbf{B} \cdot \hat{\mathbf{n}} \, dS = \iiint_{V(S)} \operatorname{div} \mathbf{B} \, dV$$

By Eq. (2-16), the latter vanishes.

Example 2-20

Show that for any two open surfaces having the same boundary (Figure 2-15) the flux is the same.

Solution   From Example 2-19 we have

$$\iint_{S_1} \mathbf{B} \cdot \hat{\mathbf{n}}_1 \, dS + \iint_{S_2} \mathbf{B} \cdot (-\hat{\mathbf{n}}_2) \, dS = 0$$

This shows that the flux is a property of the *bounding contour*, not the particular surface.

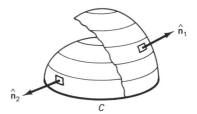

Figure 2-15 Two open surfaces having the same bounding contour C.

An expression for Φ_B that brings out its dependence on the contour alone is readily derived. Use Eq. (2-20) in Eq. (2-30), then apply Stokes' theorem to obtain

$$\Phi_B = \oint_C \mathbf{A} \cdot \mathbf{dl} \qquad (2\text{-}31)$$

Example 2-21

A rectangular loop lies in a plane containing a long straight wire carrying current I. Two sides of the rectangle are parallel to the wire and on the same side at distances a and b. Calculate the magnetic flux through the loop.

Solution 1 Use Eq. (2-13) in (2-30).

$$\Phi_B = \int_0^L dz \int_a^b \frac{\mu_0 I}{2\pi s}\, ds = \frac{\mu_0 IL}{2\pi} \ln \frac{b}{a}$$

Solution 2 Use the solution of Example 2-10 in Eq. (2-31).

$$\Phi_B = \int_0^L \left(-\frac{\mu_0 I}{2\pi} \ln a + f \right) dz + \int_L^0 \left(-\frac{\mu_0 I}{2\pi} \ln b + f \right) dz$$

$$= \frac{\mu_0 IL}{2\pi} \ln \frac{b}{a}$$

Sign Convention

Note that in order to make the signs come out the same, we must adopt a definite convention between the positive senses of the normal to the plane and the direction around the contour. The convention is that these are related via the *right-hand rule*; that is, if the fingers of the right-hand point in the positive sense around the contour, the thumb points in the positive sense of the normal.

Example 2-22

Two circular loops of radii a and b lie in the same plane with their centers a distance r apart (Figure 2-16). Assuming r to be large, calculate the flux through loop b when current I flows in loop a.

Solution The field is expressed with adequate accuracy by the dipole approxi-

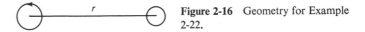

Figure 2-16 Geometry for Example 2-22.

mation [Eq. (2-29)],

$$B(r_b) = -\frac{m_a}{r^3} = -\frac{\pi a^2 I}{r^3}$$

$$\Phi_B(b) = \pi b^2 B(r_b) = -\frac{(\pi b^2)(\pi a^2)I}{r^3}$$

Example 2-23

Estimate the magnetic flux through the midplane of a long solenoid.

Solution   Since the field is approximately uniform (see Example 2-12), the flux is simply

$$\Phi_B = (\pi a^2)\mu_0 n I$$

where n is the number of turns per unit length.

Examples (2-22) and (2-23) illustrate the fact that the magnetic flux through any contour is a linear function of the currents responsible for the field. The coefficients are called *coefficients of inductance*: *self-inductance* L in the case of current along the same contour and *mutual inductance* M in the case of current along some other contour. In other words, we may write

$$\Phi_B(S_i) = L_i I_i + \sum_j' M_{ij} I_j \tag{2-32}$$

It is clear that the unit of inductance is 1 weber per ampere. This is defined as the *henry*.

$$1 \text{ henry} = 1 \text{ weber/ampere} = 1 \text{ tesla-meter}^2/\text{ampere}$$

$$= 1 \text{ newton-meter/ampere}^2 = 1 \text{ joule/ampere}^2$$

Note that μ_0 then has units of henries per meter:

$$\mu_0 = 4\pi \times 10^{-7} \text{ henry/meter}$$

Example 2-24

Calculate the mutual inductance of the two small coplanar widely separated loops of Example 2-22.

Solution   From Example 2-22, the flux through b per unit current in a is

$$M_{ab} = -\frac{(\pi a^2)(\pi b^2)}{r^3}$$

The sign is purely conventional, based on our previous choice of positive senses according to the right-hand rule. The reader should also check that $M_{ba} = M_{ab}$.

Example 2-25

If each of the foregoing loops is multiturned, calculate the mutual inductance.

Solution   The field at b is multiplied by N_a, the number of turns in loop a, which carries the current. The same field links *each* of the N_b turns of loop b, which are connected in series. Hence

$$M_{ab} = -\frac{(N_a \pi a^2)(N_b \pi b^2)}{r^3}$$

Example 2-26

Estimate the self-inductance of a long solenoid of length l with N turns.

Solution   As an approximation, we take the flux through any plane to be the same as that through the midplane. Then, from Example 2-23, the self-inductance L is

$$L = \frac{N(\pi a^2)\mu_0 N}{l} = \frac{\pi a^2 \mu_0 N^2}{l} \tag{2-33}$$

This is somewhat an overestimate, since the field gets weaker near the ends. However, the dependence on N^2 is noteworthy.

Example 2-27

Two long solenoids are wound very closely together on the same spool. Express the mutual inductance between them in terms of the individual self-inductances.

Solution   Since the flux produced by coil 1 also goes completely through coil 2,

$$M_{12} = \frac{\mu_0(\pi a^2)N_1}{l} N_2$$

$$= \sqrt{\frac{\mu_0(\pi a^2)N_1^2}{l} \frac{\mu_0(\pi a^2)N_2^2}{l}}$$

$$= (L_1 L_2)^{1/2} \tag{2-34}$$

Equation (2-34) holds in the limit of *close coupling* of the two coils. If they are somewhat separated, there is *flux leakage* between them and the mutual inductance is smaller.

We conclude this chapter by deriving a general expression for the mutual inductance between two loops. In Eq. (2-31), **A** is taken to be the vector potential at a point of loop 2 due to unit current in loop 1. Equation (2-14') gives

$$M_{12} = \oint_{C_2} \mathbf{dl}_2 \cdot \frac{\mu_0}{4\pi} \oint_{C_1} \frac{\mathbf{dl}_1}{|\mathbf{r}_2 - \mathbf{r}_1|}$$

$$= \frac{\mu_0}{4\pi} \oint_{C_2}\oint_{C_1} \frac{\mathbf{dl}_1 \cdot \mathbf{dl}_2}{|\mathbf{r}_2 - \mathbf{r}_1|} \tag{2-35}$$

This equation, known as *Neumann's formula*, brings out the essential symmetry between the two loops (Figure 2-17) that was evident in the examples above.

It might be thought that letting contour C_2 coincide with contour C_1 would give the self-inductance of a single loop. Unfortunately, the integral then diverges

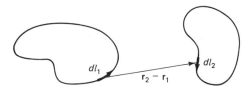

Figure 2-17   Geometry for Eq. (2-35).

because of the infinite self-field at a line current [Eq. (2-13)]. In calculating self-inductances, a more careful treatment including the finite radius of the wire is needed (see Example 2-13).

PROBLEMS

2-1. Current I is uniformly distributed over the cross section of a hollow pipe of radii a and b. Calculate the current density.

2-2. A uniform solid sphere of charge, of radius a, is spun about a central axis with angular velocity ω. Calculate the current density at all points.

2-3. The current density in a certain region is given by $\mathbf{J} = Cy\hat{\mathbf{i}}$. A square of side a has one corner at the origin, two sides parallel to z, and its normal in the xy plane at angle θ to the x axis. Calculate the current through the square.

2-4. The current density in a certain region is given by $\mathbf{J} = C\mathbf{r}$. Calculate the rate of change of the charge contained in a sphere of radius a centered at the origin.

2-5. A particle of charge q and mass m is shot into a uniform magnetic field with velocity v in a direction making angle θ with the field direction. Show that the particle moves along a helix and calculate its radius.

2-6. An infinitely long thin strip of width w carries a uniformly distributed lengthwise current I. Find the magnetic field at a point in the plane of the strip, at distance x from the centerline.

2-7. A square loop of side a carries current I. Find the magnetic field at points on the axis.

2-8. Current flows on the surface of a sphere of radius a, with nonuniform density given by $\mathbf{j}(\theta, \phi) = C \sin \theta \, \hat{\mathbf{a}}_\phi$. Calculate the magnetic field at the center.

2-9. A solenoid of radius a and length L has N turns and carries current I. Calculate the magnetic field at a point on the axis outside the solenoid at distance L from the nearer end.

2-10. It was found in Example 2-9 that two vector potential functions for the uniform magnetic field $B_0\hat{\mathbf{k}}$ are $\mathbf{A}_1 = B_0 x\hat{\mathbf{j}}$ and $\mathbf{A}_2 = -B_0 y\hat{\mathbf{i}}$. Show that these differ by the gradient of some scalar function. Find the function.

2-11. Two infinitely long straight parallel wires carry current I in opposite directions. Find the vector potential $\mathbf{A}_0$, expressed in terms of the distances r_1 and r_2 of the (arbitrary) field point from the wires. *Hint:* Start with a finite length, suitably chosen so that you get a determinate limit.

2-12. A coaxial cable carries current I in the central wire (assumed to have zero radius). The current returns uniformly through the outer sheath. Find the magnetic field everywhere.

2-13. The magnetic field in a cylindrical region of radius a is $\mathbf{B}(s) = Cs^2\hat{\mathbf{a}}_\phi$. If there is no current outside the region, calculate the field outside.

2-14. Two small circular loops of radii a and b are coaxial but separated by a large distance r. Calculate their mutual inductance.

2-15. Two small circular loops have their axes parallel but displaced. If θ is the angle between the axes and the line through the centers, find the value of θ that gives zero mutual inductance.

2-16. A toroidal solenoid of N turns has a square cross section of side a and a central radius b. Calculate the self-inductance.

2-17. A toroidal solenoid of N turns has a circular cross section of radius a and a central radius b. Calculate the self-inductance.

chapter 3

Induction

3-1 ELECTROMOTANCE

We have discussed two types of forces, electric and magnetic, that can act on charged particles. These are both due directly to the charge. There can, of course, be other forces, due to other properties of the particles. A catalog of all the possible types of forces would include:

1. Electrostatic ($q\mathbf{E}_S$)
2. Frictional (due to viscosity of the medium through which the particle moves; we shall see later (Section 5-4) that on the atomic scale frictional forces are due to random collisions)
3. Magnetic ($q\mathbf{v} \times \mathbf{B}$)
4. Chemical (due to concentration gradients)
5. Thermoelectric (due to temperature gradients)
6. Mechanical (due to contact with moving bodies)

The *electromotance* $\mathcal{E}$ is defined as the work per unit charge done by all forces other than those in categories 1 and 2. There are two cases to consider:

1. In an open segment AB of a circuit

$$\mathcal{E}_{AB} = \int_A^B \frac{1}{q} \mathbf{F}_{NE} \cdot \mathbf{dl} \tag{3-1}$$

where $\mathbf{F}_{NE}$ is the total nonelectrostatic and nonfrictional force on a particle of charge q.

2. In a complete circuit

$$\mathcal{E} = \oint \frac{1}{q} \mathbf{F}_{NE} \cdot \mathbf{dl} \tag{3-2}$$

The name *electromotance* was introduced by Lorrain and Corson.* The quantity is more commonly called *electromotive force*. This is a very poor name since, as the definition shows, it is not a force at all. Nevertheless, the abbreviation *emf* is so widespread and convenient that we too shall use it. The MKS unit of emf is the same as that of potential difference, the volt.

Example 3-1

A conveyor belt (Figure 3-1) carries particles of mass m and charge q from elevation h_A to elevation h_B in a region where there is no electric field. What is its emf?

Solution The work done on each particle is $mg(h_B - h_A)$. Thus

$$\mathcal{E}_{AB} = \frac{mg(h_B - h_A)}{q}$$

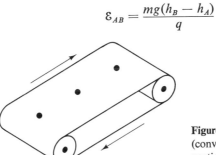

Figure 3-1 Mechanical source of emf (conveyor belt carrying charged particles).

Example 3-2

A conveyor belt carries particles of charge q from point A where the electrostatic potential is V_A to point B at potential V_B. What is the emf?

Solution The force exerted by the conveyor on each particle is $-q\mathbf{E}_s$, as required to overcome the external field. Hence the work done is $q(V_B - V_A)$. Thus

$$\mathcal{E}_{AB} = V_B - V_A$$

Note that in Example 3-1 the force had nothing to do with the charge of the particle, while in Example 3-2 it was directly proportional. Nevertheless, the emf is the work per unit charge in both cases.

The emf in a complete circuit is simply the sum of the emfs in the various segments making up the circuit (Figure 3-2). For definiteness, let us suppose that the

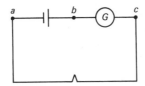

Figure 3-2 Circuit containing electrochemical (battery), electromechanical (generator), and thermoelectric (thermocouple) sources of emf in series.

*P. Lorrain and D. Corson, *Electromagnetic Fields and Waves*, 2nd ed., W. H. Freeman and Company, Publishers, San Francisco, 1970.

circuit is comprised of three segments *AB, BC, CA.* Then

$$\varepsilon_{AB} + \varepsilon_{BC} + \varepsilon_{CA} = \int_A^B \frac{1}{q} \mathbf{F}_{NE} \cdot \mathbf{dl} + \int_B^C \frac{1}{q} \mathbf{F}_{NE} \cdot \mathbf{dl} + \int_C^A \frac{1}{q} \mathbf{F}_{NE} \cdot \mathbf{dl}$$

$$= \oint \frac{1}{q} \mathbf{F}_{NE} \cdot \mathbf{dl} = \varepsilon$$

Furthermore, the integral of the total nonfrictional force per unit charge around any closed circuit is equal to the emf in the circuit.

$$\oint (q\mathbf{E}_S + \mathbf{F}_{NE}) \cdot \mathbf{dl} = q \oint \mathbf{E}_S \cdot \mathbf{dl} + q \oint \frac{1}{q} \mathbf{F}_{NE} \cdot \mathbf{dl}$$

$$= 0 + q\varepsilon$$

The vanishing of the first integral follows from the conservative nature of $\mathbf{E}_S$, as was discussed in Section 1-6.

   The foregoing result brings out the important fact that a steady current can flow in an ordinary conductive circuit only if there is a *source of emf* in the circuit. *Electrostatic forces alone cannot produce a steady current* since their line integral around any closed contour vanishes. However, if the sources of emf are confined to certain segments of the circuit (as is usually the case), the current in the *remaining* segments is driven by purely electrostatic forces.

   The *amount* of current depends not only on the emf but also on a property of the materials making up the circuit, called the *resistance*. This is related to the frictional force on the moving charges. The current takes a value such that the total work done by the sources of emf is just exactly converted to heat by the frictional forces. We shall discuss this in Chapter 5.

   Exceptions to the foregoing statements occur in materials called *superconductors*. In these, because of quantum mechanical effects, there is no frictional resistance to the motion of certain of the electrons. An emf is required to start a current but not to sustain it. We shall discuss superconductors further in Section 6-5.

   An important special case is an *open circuit*: This has infinite resistance, and the current is therefore zero no matter how much emf is supplied.

3-2 MOTIONAL ELECTROMOTANCE

One of the types of force included in $\mathbf{F}_{NE}$ is the magnetic force on a moving charged particle. The corresponding electromotance has a special significance and is called a *motional electromotance* or *motional emf.*

Example 3-3

   Calculate the emf in a straight segment *ab* of wire moving with velocity *v* (not directly along its length) in a uniform magnetic field **B** (Figure 3-3).

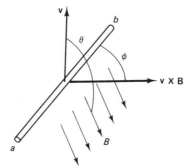

Figure 3-3   Wire segment moving in a uniform magnetic field.

Solution   The force on a charged particle q (e.g., electron or ion) in the wire is $q\mathbf{v} \times \mathbf{B}$. Thus

$$\mathcal{E}_{ab} = \int_a^b \mathbf{v} \times \mathbf{B} \cdot \mathbf{dl}$$

$$= vBl \sin \theta \cos \phi$$

where θ is the angle between $\mathbf{v}$ and $\mathbf{B}$, and ϕ is the angle between $\mathbf{v} \times \mathbf{B}$ and the wire. In the special case that the three vectors are mutually perpendicular, this reduces to vBl.

Example 3-4

Suppose that the wire in Example 3-3 is made part of a complete circuit by allowing it to slide on a pair of tracks that are connected at one end (Figure 3-4). Suppose further that the resistance of the circuit is such that current I flows. Calculate the force required to keep the wire moving at constant velocity.

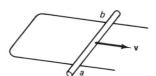

Figure 3-4   Moving segment constituting part of a closed circuit.

Solution   The magnetic force on the moving segment is, from Eq. (2-9),

$$\mathbf{F}_{\text{mag}} = \int_a^b I\, \mathbf{dl} \times \mathbf{B} = I\mathbf{l} \times \mathbf{B}$$

The applied force must just balance this to keep the wire moving at constant velocity.

Example 3-5

In Example 3-4, account for the work done by the source of the applied force.

Solution

$$\text{Rate of work} = \mathbf{F}_{\text{appl}} \cdot \mathbf{v}$$

$$= -\int_a^b I \, \mathbf{dl} \times \mathbf{B} \cdot \mathbf{v}$$

$$= I \int_a^b -\mathbf{dl} \cdot \mathbf{B} \times \mathbf{v} = I\mathcal{E}_{ab}$$

Thus the work done is used to drive the current and is converted to heat by the frictional forces (resistance) in the conductors.

Example 3-6

A metal disk is rotated about its axis at angular velocity ω in a uniform magnetic field parallel to the axis (Figure 3-5). Calculate the emf between axis and rim.

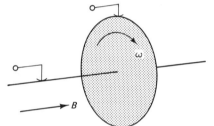

Figure 3-5   Rotating disk with sliding contacts to pick off emf (Example 3-6).

Solution   The velocity at radial distance s is $\omega s \hat{\mathbf{a}}_\phi$.

$$\mathcal{E}_{0a} = \int_0^a \omega s \hat{\mathbf{a}}_\phi \times \mathbf{B} \cdot ds \hat{\mathbf{a}}_s$$

$$= \omega B \int_0^a s \, ds = \frac{\omega B a^2}{2}$$

Example 3-7

A circular loop of radius a is rotated about one of its diameters in a uniform magnetic field perpendicular to the axis (Figure 3-6). Calculate the emf at the instant when the plane of the coil is at angle α relative to the field.

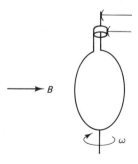

Figure 3-6   Loop rotating about axis in its plane, with sliding contacts to pick off emf (Example 3-7).

Solution   For the segment $a\,d\theta$ at angle θ, (measured from the rotation axis),

$$\mathbf{v} = \omega s\hat{\mathbf{a}}_\phi = \omega a \sin\theta\,\hat{\mathbf{a}}_\phi$$

$$\mathbf{v}\times\mathbf{B} = \omega aB\sin\theta\cos\alpha(-\hat{\mathbf{k}})$$

$$d\mathcal{E} = -\omega aB\sin\theta\cos\alpha\hat{\mathbf{k}}\cdot a\,d\theta\hat{\mathbf{a}}_\theta$$

$$= \omega a^2 B\cos\alpha\sin^2\theta\,d\theta$$

$$\mathcal{E} = \oint d\mathcal{E} = \omega a^2 B\cos\alpha\int_0^{2\pi}\sin^2\theta\,d\theta$$

$$= \pi\omega a^2 B\cos\alpha$$

This configuration forms the basic structure of practically all types of electrical *generators*. With continuous sliding contacts, as shown in Figure 3-6, the emf at the terminals reverses sense every half-turn of rotation of the coil. This results in *alternating current* (ac). By subdividing the contact rings into segments, the emf at the terminals can be kept always in the same sense, to produce *direct current* (dc).

3-3 FARADAY'S LAW OF INDUCTION

We start by reexpressing the motional emf in a moving circuit. Consider the magnetic flux through a moving circuit (not necessarily rigid) in a static magnetic field (not necessarily uniform) (Figure 3-7). At time t, the circuit lies along some contour $C(t)$,

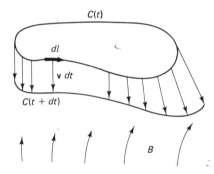

Figure 3-7   Wire loop moving deforming in a nonuniform magnetic field.

and the flux through it is then

$$\Phi_B(t) = \iint_{S(t)}\mathbf{B}\cdot\hat{\mathbf{n}}\,dS$$

where S is any open surface capping C. At time $t + dt$ the flux is

$$\Phi_B(t + dt) = \iint_{S(t+dt)}\mathbf{B}\cdot\hat{\mathbf{n}}\,dS$$

In changing from $C(t)$ to $C(t + dt)$ the loop sweeps out a closed strip in space. The change in flux is thus

$$d\Phi_B = \iint_{\text{strip}}\mathbf{B}\cdot\hat{\mathbf{n}}\,dS$$

where the integral is taken over the area of the strip swept out by the moving contour. Since each point has moved through the displacement vector $\mathbf{v}\,dt$,

$$\hat{\mathbf{n}}\,dS = \mathbf{v}\,dt \times \mathbf{dl}$$

$$d\Phi_B = \oint \mathbf{B} \cdot \mathbf{v}\,dt \times \mathbf{dl}$$

$$= -\oint (\mathbf{v} \times \mathbf{B} \cdot \mathbf{dl})\,dt$$

$$= -\varepsilon\,dt$$

Thus

$$\frac{d\Phi_B}{dt} = -\varepsilon \tag{3-3}$$

The minus sign once again follows from the conventional right-hand relation between the positive senses of $\mathbf{dl}$ and $\hat{\mathbf{n}}$.

Faraday's law of induction is a generalization of Eq. (3-3) to include *all possible changes of flux*, not only those due to motion of the circuit. It is an *independent experimental law*. However, it follows from the foregoing development with the aid of two very reasonable assumptions: (1) the effects on a circuit can be due only to *relative* motion between the circuit and the sources of the field; and (2) changes of field at the circuit due to motion of the sources cannot be distinguished from changes due to any other cause such as changes in the strength of the sources.

Example 3-8

A rigid coil having self-inductance L carries a varying current. Calculate the induced emf.

Solution $\Phi_B(t) = LI(t)$; hence

$$\varepsilon = -\frac{d\Phi_B}{dt} = -L\frac{dI}{dt}$$

Example 3-9

An infinitely long ideal solenoid of radius a with N/l turns per unit length is encircled by a loop of radius b (Figure 3-8). Calculate the induced emf in the loop when a varying current $I(t)$ flows in the solenoid.

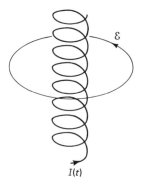

$I(t)$

Figure 3-8   Loop encircling a solenoid with a time-varying current.

Solution The only flux though the loop is that in the solenoid. Hence from Example 2-23 we have

$$\mathcal{E} = -\pi a^2 \mu_0 \frac{N}{l} \frac{dI}{dt}$$

This result seems paradoxical; there is no magnetic field at the loop at any time, yet there is an induced emf. The paradox arises entirely from the idealized nature of the problem. For a solenoid of finite length there is an external field and its lines do sweep through the loop.

Lenz's Law

It is often confusing to keep track of the sign of an induced emf by use of the right-hand rule. A simple mnemonic for this is known as *Lenz's law:* The sense of an induced emf is such that *if it were permitted to drive a current*, the field produced by that current would *tend to cancel the change* that induced the emf. In other words, induced emf's tend to *preserve the status quo.*

Example 3-10

Coil 1 is horizontal and carries a clockwise current. Coil 2 is also horizontal and lies above coil 1. Determine the sense of the emf induced in coil 2 when the current in coil 1 is interrupted.

Solution The flux through coil 2 is downward. Hence when it is diminished the induced emf must tend to produce some downward flux (to tend to cancel the change). This requires a clockwise emf.

In the case of a motional emf, Lenz's law may also be stated: If the induced emf drives a current, the force of the magnetic field on that current is such as to oppose the motion. This means, of course, that some external agency must exert a force to sustain the motion. In other words, work must be done by an external agency whenever an induced emf drives a current. We shall discuss this quantitatively in Section 9-1.

3-4 INDUCED ELECTRIC FIELDS

In the foregoing we have tried to make it clear that an emf may exist around a contour whether or not a conductive path happens to lie along the contour. If a conductive path does lie along the contour, the charged particles within it experience forces (nonelectrostatic) that make up the electomotance. It therefore follows that if there is no conducting path but only an isolated charged particle somewhere on the contour, this particle must equally well experience the force. In other words, *charged particles can experience forces due to changing magnetic fields.* Furthermore, observations on the particle alone provide *no way to distinguish these from electrostatic forces.* (In simpler terms, if one sees an isolated charged particle being accelerated, he or she cannot tell whether this is due to other charges located elsewhere or to varying magnets located elsewhere.)

We therefore generalize the concept of an electric field to include such forces:

$$\mathbf{E} = \mathbf{E}_S + \mathbf{E}_{\text{ind}} \tag{3-4}$$

where $\mathbf{E}_{\text{ind}}$ is the force per unit charge due to varying magnetic fields. Obviously, for any contour,

$$\oint_C \mathbf{E}_{\text{ind}} \cdot \mathbf{dl} = \mathcal{E} = -\frac{d}{dt} \iint_{S(C)} \mathbf{B} \cdot \hat{\mathbf{n}} \, dS \tag{3-5}$$

Stokes' theorem applied to the left side of Eq. (3-5) gives a surface integral of curl $\mathbf{E}_{\text{ind}}$. On the right side, if the contour is fixed, the only way the flux can change is through a change of $\mathbf{B}$. Thus

$$\iint_S \text{curl } \mathbf{E}_{\text{ind}} \cdot \hat{\mathbf{n}} \, dS = \iint_S -\frac{\partial \mathbf{B}}{\partial t} \cdot \hat{\mathbf{n}} \, dS$$

Since this must hold for any open surface whatever, the integrands must be equal, so that

$$\text{curl } \mathbf{E}_{\text{ind}} = -\frac{\partial \mathbf{B}}{\partial t} \tag{3-6}$$

This important relation shows that the time derivative of $\mathbf{B}$ acts as the source of $\mathbf{E}_{\text{ind}}$ in exactly the same way that a static current density $\mathbf{J}_S$ acts as the source of a magnetic field [see Eq. (2-23)]. Similarly, the integrated form of the law [Eq. (3-5)] may be written as the analog of Ampère's circuital law [Eq. (2-24)]. Furthermore, both the types of sources (right sides of the equations) have the same basic structure: Both steady-state current and magnetic flux form closed loops, or, in other words, their densities are divergenceless. Thus all the mathematical machinery for the calculation of magnetic field from current may be taken over for the calculation of induced electric field from rate of change of flux. The only difference is that the constant of proportionality is -1 instead of μ_0.

Example 3-11

A long ideal solenoid carries a steadily increasing current such that the flux through it increases at the rate of X webers per second. Find the induced electric field everywhere.

Solution   This is a direct analog of Example 2-13. The induced field is, at any point, tangential to a circle centered on the axis of the solenoid; its magnitude is, for $s > a$,

$$E_{\text{ind}} = -\frac{X}{2\pi s}$$

For $s < a$,

$$E_{\text{ind}} = -\frac{Xs}{2\pi a^2}$$

Of course, only the simple highly symmetrical problems can be handled by this approach. However, an integral of the form of Eq. (2-11) with $-\partial \mathbf{B}/\partial t$ replacing $\mu_0 \mathbf{J}$ can, in principle, always be evaluated. From the similarity in structure it follows that

$$\text{div } \mathbf{E}_{\text{ind}} = 0 \tag{3-7}$$

Thus $\mathbf{E}_{\text{ind}}$ is obviously the solenoidal electric field that was discussed in Section 2-5.

We may now write the fundamental differential equations for the total electric field defined in Eq. (3-4).

$$\text{div } \mathbf{E} = \text{div } \mathbf{E}_S = 4\pi K_f \rho \tag{3-8}$$

$$\text{curl } \mathbf{E} = \text{curl } \mathbf{E}_{\text{ind}} = -\frac{\partial \mathbf{B}}{\partial t} \tag{3-9}$$

3-5 INDUCED MAGNETIC FIELDS

It is natural to inquire whether the inverse effect to electromagnetic induction also occurs, that is, whether a changing electric field produces a magnetic field. The answer is yes. This was not discovered experimentally but was deduced by Maxwell, who showed that the other known laws would be inconsistent without it. Equation (2-23), curl $\mathbf{B} = \mu_0 \mathbf{J}$, is true for static magnetic fields. However, if it were assumed to hold also for time-varying fields the current continuity equation (2-6), div $\mathbf{J} + \partial \rho/\partial t = 0$, would be violated. To see this, simply take the divergence of both sides of the curl equation. The result is div $\mathbf{J} = 0$, which contradicts Eq. (2-6).

Thus curl $\mathbf{B} - \mu_0 \mathbf{J}$ cannot be zero in general. Following Maxwell, we deduce a value for this quantity that is consistent with the current continuity equation. Let $\mathbf{X}(\mathbf{r}, t) = \text{curl } \mathbf{B} - \mu_0 \mathbf{J}$. Then

$$\text{div } \mathbf{X} = -\mu_0 \text{ div } \mathbf{J} = \mu_0 \frac{\partial \rho}{\partial t}$$

But by Eq. (3-8), $\rho = (4\pi K_f)^{-1} \text{ div } \mathbf{E}$. Hence a solution is

$$\mathbf{X} = \mu_0 (4\pi K_f)^{-1} \frac{\partial \mathbf{E}}{\partial t}$$

At this point it is convenient to introduce a new constant,

$$\epsilon_0 = (4\pi K_f)^{-1} = 8.854 \times 10^{-12} \text{ C}^2/\text{N-m}^2$$

so

$$\mathbf{X} = \mu_0 \epsilon_0 \frac{\partial \mathbf{E}}{\partial t}$$

The complete equation is then written

$$\text{curl } \mathbf{B} = \mu_0 \left(\mathbf{J} + \epsilon_0 \frac{\partial \mathbf{E}}{\partial t} \right) \tag{3-10}$$

This is the time-varying generalization of Eq. (2-23). The added term obviously has the dimensionality of a current density, amperes per meter² in the MKS system. It is called the *displacement current density*. This term obviously produces an *induced magnetic field* whose source is a changing electric field.

Example 3-12

A battery is connected by a wire of appreciable resistance to a pair of large parallel plates of area A. A small current $I(t)$ flows temporarily while the plates are charging (Figure 3-9). Calculate the displacement current density in the space between the plates during this interval.

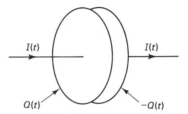

Figure 3-9   Charging of a pair of plates by a current.

Solution   Assume that the charging process is sufficiently slow that conditions are essentially static at all times. Also assume the plates to be effectively infinite in extent (i.e., neglect edge effects). Then at a time when the charges are $Q(t)$ and $-Q(t)$ the field is given to good approximation by Example 1-4 as

$$|E(t)| = \frac{Q(t)}{\epsilon_0 A} \text{ between the plates}$$

Thus

$$J_{\text{displ}} = \epsilon_0 \frac{\partial E}{\partial t} = \frac{dQ/dt}{A} = \frac{I}{A}$$

Note that the total displacement current between the plates is the same as the total real current outside the plates. In other words, the sum of real plus displacement current flows in closed loops. This holds true in general, as we see by simply taking the divergence of both sides of Eq. (3-10), that is,

$$\text{div} \ (\mathbf{J} + \mathbf{J}_{\text{displ}}) = 0 \tag{3-11}$$

3-6 MAXWELL'S EQUATIONS IN FREE SPACE

We have now completed the formulation of the basic laws of electromagnetic fields in vacuum. These have been written in various forms, but the most generally useful are the differential equations. For convenience, we collect them together here, along with their physical meanings.

Gauss's law, a direct consequence of Coulomb's law:

$$\text{div} \ \epsilon_0 \mathbf{E}(\mathbf{r}, t) = \rho(\mathbf{r}, t) \tag{3-8}$$

Faraday's law of induction:

$$\text{curl } \mathbf{E}(\mathbf{r}, t) = -\frac{\partial \mathbf{B}(\mathbf{r}, t)}{\partial t} \tag{3-9}$$

Nonexistence of isolated magnetic poles:

$$\text{div } \mathbf{B} = 0 \tag{2-16}$$

Ampère's circuital law, a direct consequence of the Biot–Savart law, plus Maxwell's term describing induction of magnetic fields:

$$\text{curl } \mathbf{B}(\mathbf{r}, t) = \mu_0 \left[\mathbf{J}(\mathbf{r}, t) + \epsilon_0 \frac{\partial \mathbf{E}(\mathbf{r}, t)}{\partial t} \right] \tag{3-10}$$

If the source densities, $\mathbf{J}$ and ρ, are regarded as *given functions*, solutions of these equations (in the form of integrals) can always be found. In fact, much of our discussion up to this point has been concerned with the form these integrals take in static situations and their evaluation in numerous simple examples. The extension to time-varying situations is not much more difficult, and we will take it up in due course (Part V).

But this is only a small part of the story. In actuality, the source densities themselves depend on the fields. Since charges and currents are commonly encountered in the interior of matter, we must first consider the nature of electromagnetic fields in matter of various types, and the ways in which the matter responds to the fields to produce the charges and currents. This is the subject of Part II.

PROBLEMS

3-1. In a chemical battery, metal is dissolved at one electrode, going into solution in the form of ions of mass m and charge q. These move to the other electrode, where they plate out. Energy of H joules per kilogram of metal transported is released. Calculate the electromotance of the battery.

3-2. A square contour lies in a plane containing a long straight wire that is carrying current I. The square has two of its sides parallel to the wire, and is moving directly away from the wire at speed v. Calculate the magnitude of the emf around the contour at the instant when the nearer side is at distance b.

3-3. A rod of length L is rotated about one of its ends at angular velocity ω in a plane perpendicular to a uniform magnetic field $\mathbf{B}$. Calculate the magnitude of the emf between its ends.

3-4. A circular loop of radius a is rotated about one of its diameters in a uniform magnetic field $\mathbf{B}$ perpendicular to the axis. Calculate the emf as function of time.

3-5. Rework Example 3-7 by means of Eq. (3-3).

3-6. Rework Example 3-6 by means of Eq. (3-3). (Think carefully about the meaning of $d\Phi/dt$ in this case.)

3-7. In Problem 3-2, if the current in the wire is flowing from west to east and the contour is moving northward, what is the sense of the induced emf?

3-8. Consider two small coplanar loops. Loop 1 (on the left) carries a counterclockwise current. Determine the sense of the emf induced in loop 2 if (a) the current is interrupted, (b) loop 1 is moved to the left, and (c) loop 1 is moved up out of the plane.

3-9. In a long closely wound solenoid of radius a with N/L turns per unit length, the current is increased uniformly at a rate of C amperes per second. Find the electric field induced at a point near the middle, at distance s from the axis ($s < a$).

3-10. Two coils, of N_1 and N_2 turns, respectively, are very closely coupled (i.e., practically coincident in space). A time-varying emf $\mathcal{E}_1(t) = \mathcal{E}_0 \cos \omega t$ is applied to coil 1. Neglecting the resistance of the coils, find the emf induced in coil 2.

3-11. Two concentric spheres have time-varying charges, $q(t)$ on the inner one and $-q(t)$ on the outer. If q is increased uniformly at the rate of C coulombs per second, find the displacement current density everywhere.

3-12. Consider a vector function $\mathbf{F}(\mathbf{r})$ such that div $\mathbf{F} = f(\mathbf{r})$ and curl $\mathbf{F} = \mathbf{G}(\mathbf{r})$, where f and $\mathbf{G}$ are given functions. Express $\mathbf{F}$ explicity in terms of integrals involving f and $\mathbf{G}$.

Part II

ELECTROMAGNETIC FIELDS IN MATTER

chapter 4

Dielectrics

4-1 CLASSIFICATION OF MATERIALS

We noted in Section 1-10 that in respect to electrical properties, materials may be divided into two broad classes: conductors and insulators (dielectrics). Conductors carry appreciable currents when subjected to even fairly weak electric fields, as we shall discuss in the next chapter. Dielectrics, on the other hand, permit only extremely small currents under quite high fields. Of course, under extremely strong fields even the best insulators "break down" and become conductive. This is often a violent process (e.g., a lightning stroke) and may lead to permanent destruction of the material. Field strengths of the order of 10^8 V/m are typically required for breakdown.

In terms of internal structure, conductors contain charged particles (electrons or ions) that can move more or less freely over distances of many atomic spacings. Dielectrics do not contain such free charges. For this reason, dielectrics are inherently simpler, and this is why we shall discuss them first.

The absence of mobile charges means that we can subdivide the material into structural units ("molecules") all of whose constituents remain relatively close together. Furthermore, this can be done in such a way that each molecule has zero net charge. Even in an ionic crystal the molecules can be taken to be neutral clusters of ions. This neutrality means that the leading term in the multipole expansion (see Section 1-8) of the external electric field produced by each molecule is the *dipole* term. Thus molecular dipole moments play a central role in the theory. The situation is often complicated, though, by the fact that the dipole moment of a molecule will in turn depend on the field acting upon it.

Dielectric materials fall into two main classes depending on the nature of the molecular dipoles. In *nonpolar* dielectrics, the molecules have no dipole moment in the absence of a field but acquire an *induced dipole moment* in a field. In *polar* dielectrics, on the other hand, each molecule has a *fixed* or *permanent* moment. In ordinary polar materials these dipoles tend to be randomly oriented because of thermal agitation, but become somewhat aligned when a field is applied. However, some polar dielectrics, called *ferroelectrics*, have their dipoles largely aligned in a definite crystallographic direction even in the absence of an applied field. We shall return to a discussion of the properties of these classes of materials in Section 4-4, after developing the basic theory of the fields.

4-2 FIELDS DUE TO DIELECTRICS

Just as in Chapter 1 we found it useful not to deal with the individual charged particles, but instead to define a mathematical *charge density* function, here we will find it useful to define a *dipole moment density* function. This is, of course, a *vector* function. It is called the *polarization* **P** and is defined as

$$\mathbf{P}(\mathbf{r}) = \text{dipole moment per unit volume in a small region around } \mathbf{r} \qquad (4\text{-}1)$$

P is regarded as a macroscopic point function, so the region must be large enough to contain many molecules [see discussion preceding Eq. (1-1), and Figure 4-1]. The MKS unit of polarization is coulombs per meter². Since the moments are those of the individual molecules, Eq. (4-1) is, explicitly,

$$\mathbf{P}(\mathbf{r}) = n(\mathbf{r})\bar{\mathbf{p}}_m(\mathbf{r}) \qquad (4\text{-}2)$$

where n is the number density (i.e., number per unit volume), $\mathbf{p}_m$ is the molecular dipole moment, and the bar denotes the average value over the surrounding region.

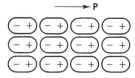

Figure 4-1 Molecules of a polarized dielectric solid.

Example 4-1

Suppose that in solid hydrogen in an electric field the average electron positions were displaced 0.01 Å from the proton positions, all in the same direction. Estimate the polarization.

Solution $p_m = (1.6 \times 10^{-19} \text{ C}) \times 10^{-12} \text{ m} = 1.6 \times 10^{-31} \text{ C-m}$. The interatomic spacing is (very roughly) 3 Å, so $n = (3 \times 10^{-10})^{-3} = 4 \times 10^{28} \text{ m}^{-3}$. Thus $P = 1.6 \times 10^{-31} \times 4 \times 10^{28} \cong 6 \times 10^{-3} \text{ C/m}^2$.

Potential at External Points

We now calculate the electrostatic potential produced by the polarization. For field points *outside* the dielectric there is no difficulty since the distance of the field point from every dipole is large, so that just the first nonvanishing term of the multipole expansion is an adequate approximation. At this time, the reader would be well advised to review this in Section 1-8.

The dipole moment of volume element $d^3\mathbf{r}'$ at point $\mathbf{r}'$ is $\mathbf{P}(\mathbf{r}')\, d^3\mathbf{r}'$. This produces at point $\mathbf{r}$ the dipole potential

$$dU(\mathbf{r}) = \frac{1}{4\pi\epsilon_0} \frac{\mathbf{P}(\mathbf{r}') \cdot \mathbf{R}}{\mathbf{R}^3} d^3\mathbf{r}'$$

where, as usual, $\mathbf{R} \equiv \mathbf{r} - \mathbf{r}'$. Thus on summing the contributions of all the volume elements in the dielectric, we obtain

$$U(\mathbf{r}) = \frac{1}{4\pi\epsilon_0} \iiint\limits_{V_d} \frac{\mathbf{P}(\mathbf{r}') \cdot \mathbf{R}}{\mathbf{R}^3} d^3\mathbf{r}' \tag{4-3}$$

The volume V_d of the integration is, of course, that of the dielectric.

Equation (4-3) is perfectly adequate to obtain the potential [and from it the field by taking $(-)$ its gradient] from any specified moment distribution. However, it is instructive and useful to transform the integral mathematically to a form in which only the first power of distance appears in the denominator. We note that the function multiplying $\mathbf{P}$ in the integrand may be written as $R^{-3}\mathbf{R} \equiv -\text{grad } R^{-1} = \text{grad}' R^{-1}$ [see Eq. (2-17)]. Thus

$$4\pi\epsilon_0 U(\mathbf{r}) = \iiint \mathbf{P}(\mathbf{r}') \cdot \text{grad}' R^{-1} d^3\mathbf{r}'$$

$$= \iiint [\text{div}'(R^{-1}\mathbf{P}) - R^{-1} \text{div}' \mathbf{P}] d^3\mathbf{r}'$$

On applying the divergence theorem to the first term, we obtain

$$U(\mathbf{r}) = \frac{1}{4\pi\epsilon_0} \iiint\limits_{V_d} \frac{-\text{div}' \mathbf{P}(\mathbf{r}')}{R} d^3\mathbf{r}' + \frac{1}{4\pi\epsilon_0} \oiint\limits_{S_d} \frac{\mathbf{P}(\mathbf{r}') \cdot \hat{\mathbf{n}}}{R} dS' \tag{4-4}$$

The integrals are, of course, taken over the volume and surface of the dielectric. It is seen that they have the same form as the potential due to certain distributions of charge. In other words, for purposes of calculating the potential (and, of course, field) the dielectric may be *replaced* by two distributions of *effective charge* (also called *bound charge* or *polarization charge*), namely,

$$\rho_P = -\text{div } \mathbf{P} \tag{4-5}$$

$$\sigma_P = \mathbf{P} \cdot \hat{\mathbf{n}} \tag{4-6}$$

Thus if **P** is a known function, the field may be calculated by the methods of Chapter 1 after calculating the effective charge densities. Alternatively, of course, Eq. (4-3) can be used directly.

Example 4-2

A dielectric cylinder is polarized uniformly in the direction of its length (Figure 4-2). Calculate the effective charge distribution.

Solution $\mathbf{P}(\mathbf{r}) = P_0\hat{\mathbf{k}}$ for **r** inside the rod, $\mathbf{P} = 0$ for **r** outside the rod. Thus div $\mathbf{P} = 0$ (i.e., $\rho_P = 0$ everywhere). $\hat{\mathbf{n}} = \hat{\mathbf{a}}_\phi$ on the cylinder surface, $\hat{\mathbf{k}}$ on the positive end face, and $-\hat{\mathbf{k}}$ on the negative end face. Thus $\sigma_P = 0$ on the cylinder surface, P_0 on the positive end face, and $-P_0$ on the negative end face. Hence the problem reduces to a pair of uniformly charged disks. The field at points on the axis can be found from Problem 1-9 by superposition.

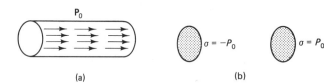

(a)   (b)   **Figure 4-2**   (a) Uniformly polarized rod; (b) equivalent charge distribution.

Example 4-3

A thin disk-shaped cavity is cut in a uniformly polarized dielectric with its axis at angle θ to the direction of polarization. Find the field in the cavity at a point not too near the rim.

Solution $\sigma_P = \pm P_0 \cos\theta$ on the two faces. Taking the radius of the disk to be effectively infinite, we have, by Example 1-4,

$$\mathbf{E}_S = \frac{P_0}{\epsilon_0}\cos\theta\,\hat{\mathbf{n}}$$

Example 4-4

A spherical cavity is carved out of a uniformly polarized dielectric. Calculate the effective surface charge density assuming that the polarization in the remaining material remains unchanged.

Solution $\sigma_P = \mathbf{P}\cdot\hat{\mathbf{n}}$, where $\hat{\mathbf{n}}$ is the unit normal directed out of the material at the point in question. Thus, at a surface point located at angle θ from the direction of **P** (see Figure 4-3),

$$\sigma_P(\theta) = \mathbf{P}\cdot\hat{\mathbf{n}}(\theta) = -P\cos\theta$$

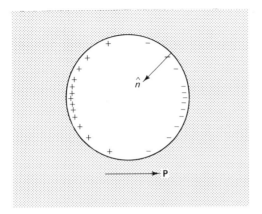

Figure 4-3 Effective surface charge distribution in a spherical cavity in a uniformly polarized dielectric.

We already have calculated the field at the center due to a surface charge distribution of this form (see Example 1-6). The result will be needed in the next section.

It is interesting to note that the *total* polarization charge is always zero. This follows simply from the divergence theorem:

$$Q_P = \iiint_V -\text{div } \mathbf{P} \, d^3r + \oiint_S \mathbf{P} \cdot \hat{\mathbf{n}} \, dS = 0 \qquad (4\text{-}7)$$

This result is intuitively obvious, since the polarization charge merely represents the effects of the molecular dipoles. Indeed, the bound surface charges can be visualized simply as the ends of the dipoles showing at each surface.

Example 4-5

A cylinder extending from $z = 0$ to $z = L$ is polarized nonuniformly (Figure 4-4) so that $\mathbf{P} = Cz^2\hat{\mathbf{k}}$. Calculate the polarization charge densities and verify Eq. (4-7) explicitly.

Solution $\rho_P = -2Cz$; $\sigma_P = 0$ on cylinder surface, 0 on left face, CL^2 on right face.

$$Q_P = \int_0^L -2Cz \, dz + 0 + CL^2 = 0$$

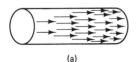

(a)

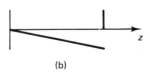

(b)

Figure 4-4 (a) Nonuniformly polarized cylinder of Example 4-5; (b) plot of effective charge density.

Field at Internal Points

We now turn to consider field points inside the dielectric. Here more care is required, since some of the molecules are very close to the field point, and their contributions therefore cannot be adequately represented by dipole terms alone. To get around this problem, we imagine a cavity surrounding the field point. We evaluate separately the contributions of the material inside and outside the cavity, and take the limit of the sum as the cavity shrinks to zero. In other words, we write

$$\mathbf{E}(\mathbf{r}) = \mathbf{E}_1(\mathbf{r}) + \mathbf{E}_2(\mathbf{r}) + \mathbf{E}_3(\mathbf{r}) \tag{4-8}$$

where

$\mathbf{E}_1$ = field due to all dielectric outside the cavity

$\mathbf{E}_2$ = field due to dielectric inside the cavity

$\mathbf{E}_3$ = field due to all real charges

For the field $\mathbf{E}_1$ there is no problem. Since the field point is exterior to the material under consideration, Eq. (4-3) or (4-4) can be used directly. If V_1 and S_1 are the volume and surface of the cavity, then $V_d - V_1$ and $S_d + S_1$ are the volume and surface of the remaining matter. Thus

$$4\pi\epsilon_0 U_1 = \iiint\limits_{V_d - V_1} \frac{-\text{div}'\,\mathbf{P}}{R}\, d^3\mathbf{r}' + \oiint\limits_{S_d + S_1} \frac{\mathbf{P}\cdot\hat{\mathbf{n}}}{R}\, dS$$

$$\mathbf{E}_1 = -\text{grad}\, U_1$$

$$= \frac{1}{4\pi\epsilon_0}\left[\iiint\limits_{V_d - V_1} \frac{(-\text{div}'\,\mathbf{P})\mathbf{R}}{R^3}\, d^3\mathbf{r}' + \oiint\limits_{S_d + S_1} \frac{(\mathbf{P}\cdot\hat{\mathbf{n}})\mathbf{R}}{R^3}\, dS'\right] \tag{4-9}$$

In the limit, the correction to the volume integral will vanish, since it will be proportional to V_1. However, the correction to the surface integral will *not*, as we shall now see. We have to calculate the contribution of a spherical cavity surface S_1 in a uniformly polarized dielectric to the field at the center of the sphere. The effective surface charge density was just worked out in Example 4-4, and by Example 1-6 the field is

$$\mathbf{E}(0, 0, 0) = \frac{\mathbf{P}}{3\epsilon_0} \tag{4-10}$$

Thus the correction due to S_1 is independent of the size of the cavity. It may, of course, depend on the *shape*. In this regard, the sphere seems a reasonable shape to use for a homogeneous isotropic (i.e., without any special directional properties) dielectric, and furthermore we may hope that when the contribution of the interior molecules is added in, the sum will be relatively shape independent. One further point: We have calculated the field only at the center. However, if the cavity is much larger than the molecules (although still small on a macroscopic scale), this will be a fair approximation to the field over most of the volume.

The field $\mathbf{E}_2$ due to the interior molecules will be a very complicated rapidly varying function of position and time. However, in keeping with the idea of a *macro-*

scopic point function, we are interested only in the *average* value. We start by calculating the average, over the volume of a sphere, of the field due to a point charge
somewhere within the sphere. Let $\mathbf{r}_0$ be the position vector of the charge. Then the
average is the integral divided by the volume, or

$$\bar{\mathbf{E}}_2 = \frac{1}{V_1} \iiint\limits_{V_1} \frac{q}{4\pi\epsilon_0} \frac{\mathbf{r} - \mathbf{r}_0}{|\mathbf{r} - \mathbf{r}_0|^3} \, d^3\mathbf{r} \qquad (4\text{-}11)$$

This expression is a familiar electrostatic result. If we rewrite it as

$$-\frac{1}{4\pi\epsilon_0} \iiint\limits_{V_1} \frac{(q/V_1)(\mathbf{r}_0 - \mathbf{r})}{|\mathbf{r}_0 - \mathbf{r}|^3} \, d^3\mathbf{r}$$

we see that it is (except for the $-$ sign) simply the field at $\mathbf{r}_0$ due to a uniform charge
density q/V_1 throughout the sphere. This was found in Problem 1-19. Thus

$$\bar{\mathbf{E}}_2 = -\frac{1}{4\pi\epsilon_0} \frac{q\mathbf{r}_0}{a^3} = -\frac{1}{3\epsilon_0} \frac{q\mathbf{r}_0}{V_1}$$

(for point charge q at $\mathbf{r}_0$). $q\mathbf{r}_0$ is the dipole moment of q with respect to the center of
the sphere.

We can now easily extend the result to an *arbitrary* charge distribution within
the sphere by simply adding the contributions of all the charges.

$$\bar{\mathbf{E}}_2 = -\frac{1}{3\epsilon_0 V_1} \sum q_i \mathbf{r}_i$$

$$= -\frac{1}{3\epsilon_0} \frac{\mathbf{p}_1}{V_1} = -\frac{\mathbf{P}}{3\epsilon_0} \qquad (4\text{-}12)$$

where $\mathbf{p}_1$ is the dipole moment of the interior charge distribution with respect to the
center of the sphere. In the last step of equating $\mathbf{p}_1/V_1$ to $\mathbf{P}$, we assume the polarization
to be homogeneous.

On comparing Eqs. (4-10) and (4-12), we see that the two nonvanishing correction terms *exactly cancel* each other. In other words, Eq. (4-4) is *valid for all field
points*, interior as well as exterior. However, the rather lengthy derivation we have
just gone through brings out two important aspects of the theory: (1) it involves a
generous amount of approximation and thus is nowhere nearly as firmly based as
vacuum electrostatics, and (2) by "the field" in a dielectric we mean the *average over
a region much larger than the molecules.*

Example 4-6

A dielectric slab with polarization P directed perpendicular to its plane is
placed between a pair of charge sheets $\pm\sigma$ (Figure 4-5). Find the field strength
everywhere.

Solution   The bound charge consists of two sheets $-P$ and P as shown. These
sheets, plus the *true* charge sheets $\pm\sigma$, give the field. Referring to the discussion
following Example 1-4 we see that in the gaps only the true charges contribute,

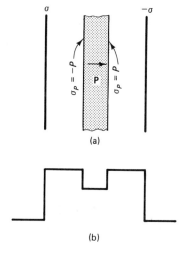

Figure 4-5 (a) True plus effective charge distributions for Example 4-6; (b) variation of field strength.

so that $\mathbf{E} = (\sigma/\epsilon_0)\hat{\mathbf{k}}$, while in the dielectric both the true and the bound charges contribute, so that $\mathbf{E} = [(\sigma - P)/\epsilon_0]\hat{\mathbf{k}}$. Outside the entire sandwich, the field is zero.

Example 4-7

The charge sheets of Example 4-6 are replaced by metal planes held at potentials U_1 and U_2. Again find the field everywhere.

Solution The results are formally the same except that we now do not know the value of σ. This is determined by the potential difference applied.

$$\int E\, dz = \frac{\sigma}{\epsilon_0}(l - d) + \frac{\sigma - P}{\epsilon_0}\, d \equiv U_1 - U_2$$

or

$$\sigma l - Pd = \epsilon_0(U_1 - U_2)$$

We simply solve for σ and substitute in the previous answers, obtaining

$$\sigma = \frac{\epsilon_0(U_1 - U_2) + Pd}{l}$$

$$\mathbf{E}_1 = \mathbf{E}_3 = \frac{\sigma}{\epsilon_0}\hat{\mathbf{k}} = \left(\frac{U_1 - U_2}{l} + \frac{P}{\epsilon_0}\frac{d}{l}\right)\hat{\mathbf{k}}$$

$$\mathbf{E}_2 = \frac{\sigma - P}{\epsilon_0}\hat{\mathbf{k}} = \left[\frac{U_1 - U_2}{l} - \frac{P}{\epsilon_0}\left(1 - \frac{d}{l}\right)\right]\hat{\mathbf{k}}$$

4-3 DISPLACEMENT VECTOR FIELD D

We have seen in Section 4-2 that $\mathbf{E}$ is determined by both the true charges and the polarization charges. In the vacuum case, the differential form of Gauss's law [Eq. (1-15)] disclosed that div $\mathbf{E}$ is given by the volume density of sources. Hence, in the presence of dielectric matter, this may be generalized to include *all* sources, that is,

$$\text{div } \mathbf{E} = \frac{1}{\epsilon_0}(\rho + \rho_P)$$

$$= \frac{1}{\epsilon_0}(\rho - \text{div } \mathbf{P}) \tag{4-13}$$

or

$$\text{div } (\epsilon_0 \mathbf{E} + \mathbf{P}) = \rho \tag{4-14}$$

In other words, the vector field whose source strength is the true charge density is the combination

$$\mathbf{D(r)} = \epsilon_0 \mathbf{E(r)} + \mathbf{P(r)} \tag{4-15}$$

D is called the *dielectric displacement*. Obviously, at any point in empty space **D** is simply $\epsilon_0 \mathbf{E}$.

Example 4-8

For Example 4-6, find **D** everywhere.

Solution In the gaps, $\mathbf{D} = \epsilon_0 \mathbf{E} = \sigma \hat{\mathbf{k}}$; in the material $\mathbf{D} = \epsilon_0 \mathbf{E} + \mathbf{P} = \sigma \hat{\mathbf{k}}$.

It must not be thought that **D** is always uniform. This result is due to the special symmetry of the example.

Example 4-9

Find **D** at the center of a uniformly polarized dielectric sphere (Figure 4-6).

Solution The calculation of **E** is exactly the same as for Eq. (4-10), except that the sign is reversed since the material is now inside rather than outside the sphere. Thus $\mathbf{E} = -\mathbf{P}/3\epsilon_0$, from which we get by the basic definition [Eq. (4-15)] $\mathbf{D} = \frac{2}{3}\mathbf{P}$.

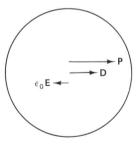

Figure 4-6 Relation of **P**, **D**, and **E** at the center of a uniformly polarized dielectric sphere.

Summary on Fields in Dielectrics

The electrostatic field at any point, either in dielectric material or in empty space, is the sum of the contributions of all true charges plus all effective charges. The latter may, if **P(r)** be a known function, be calculated directly from Eqs. (4-5) and (4-6). We have worked out their values for (1) a rod polarized lengthwise (Example 4-2), (2) a disk-shaped cavity (Example 4-3), and (3) a spherical cavity (Example 4-4). We can readily change from a cavity to a solid body, or vice versa, by simply changing sign. For the rod, if we consider it to be a very long thin needle, the effective charges are so small and so remote that their field contribution near the middle is essentially zero. Thus we can summarize the resulting contributions to $\mathbf{E}$ and $\mathbf{D}$ as in Table 4-1. In every case the value of $\mathbf{D}$ is obtained from $\mathbf{E}$ *at the same point* by Eq. (4-15), using the value of P *at that point*. If the point is in the dielectric, then $\mathbf{P} = P_0\hat{\mathbf{k}}$; if in empty space, then $\mathbf{P} = 0$.

TABLE 4-1 FIELDS IN DIELECTRIC BODIES AND CAVITIES

		Thin disk (axis along $\hat{\mathbf{k}}$)	Sphere	Long needle (along $\hat{\mathbf{k}}$)	
Solid body	σ	$\pm P_0$ on faces	$P_0\cos\theta$	$\pm P_0$ on faces	
	E	$-\dfrac{1}{\epsilon_0}P_0\hat{\mathbf{k}}$	$-\dfrac{1}{3\epsilon_0}P_0\hat{\mathbf{k}}$	0	In the body
	D	0	$\dfrac{2}{3}P_0\hat{\mathbf{k}}$	$P_0\hat{\mathbf{k}}$	
	E_{ext}	0		0	Just outside
	D_{ext}	0		0	middle of body
Cavity	σ	$\mp P_0$ on ends	$-P_0\cos\theta$	$\mp P_0\cos\theta$	
	E	$\dfrac{1}{\epsilon_0}P_0\hat{\mathbf{k}}$	$\dfrac{1}{3\epsilon_0}P_0\hat{\mathbf{k}}$	0	In the cavity
	D	$P_0\hat{\mathbf{k}}$	$\dfrac{2}{3}P_0\hat{\mathbf{k}}$	0	
	E_{ext}	0		0	Just outside
	D_{ext}	$P_0\hat{\mathbf{k}}$		$P_0\hat{\mathbf{k}}$	middle of cavity

NOTE: $\mathbf{P} = P_0\hat{\mathbf{k}}$ assumed fixed, independent of field.

It is emphasized that these values are only the contributions of the effective charges. If any true charges are present, their contributions are separately calculated, just as if the dielectric material were not present, and added in.

For the cavities, it is interesting to note the relations of $\mathbf{E}$ in the cavity to fields in the dielectric just adjacent. For the disk, $\mathbf{E}$ in the cavity is related to $\mathbf{D}$ in the material whereas for the needle, $\mathbf{E}$ in the cavity is $\mathbf{E}$ in the material. These relations are sometimes used to provide "operational" definitions of $\mathbf{E}$ and $\mathbf{D}$, since cavity fields are, in principle at least, measurable by forces on test charges.

Displacement Current

The displacement current density was introduced in Eq. (3-10) in order to preserve the current continuity relation when there is a time-varying field. Its value for fields in free space is $\epsilon_0(\partial\mathbf{E}/\partial t)$. We now have to deduce the expression for displacement current density in dielectrics. We write Eq. (3-10) as curl $\mathbf{B} = \mu_0(\mathbf{J} + \mathbf{J}_{\text{displ}})$. Then taking the divergence of both sides gives (since div curl $\equiv 0$)

$$\text{div } \mathbf{J}_{\text{displ}} = -\text{div } \mathbf{J} = \frac{\partial \rho}{\partial t} = \frac{\partial}{\partial t} \text{ div } \mathbf{D} = \text{div } \frac{\partial \mathbf{D}}{\partial t}$$

Thus a solution is

$$\mathbf{J}_{\text{displ}} = \frac{\partial \mathbf{D}}{\partial t} \tag{4-16}$$

By Eq. (4-15), this clearly reduces to the previous value [Eq. (3-10)] for empty space.

4-4 DIELECTRIC RESPONSE

We now come to the central question in the theory of dielectrics: What determines the value of $\mathbf{P}$ at any point? Stated differently, what determines the dipole moment of a given molecule? From Section 4-1 we know that this generally depends on the *field acting on the molecule*. This is called the *local field*. It is not the same as the macroscopic field $\mathbf{E}$ in the dielectric, since the latter includes a contribution from the molecule itself. What we want is the field due to *all the true charges plus all the other molecules*.

We can easily calculate the local field for a spherical molecule. We simply subtract from $\mathbf{E}$ the contribution of the molecule itself. From Eq. (4-10), this contribution is $-\mathbf{P}/3\epsilon_0$. Hence

$$\mathbf{E}_{\text{loc}} = \mathbf{E} + \frac{\mathbf{P}}{3\epsilon_0} = \frac{\mathbf{D} - \frac{2}{3}\mathbf{P}}{\epsilon_0}$$

For nonspherical molecules, the details will differ somewhat, but the important thing to note is that $\mathbf{E}_{\text{loc}}$ must be *greater* than $\mathbf{E}$ by an amount proportional to $\mathbf{P}$. Thus we write

$$\mathbf{E}_{\text{loc}} = \mathbf{E} + \frac{\nu\mathbf{P}}{\epsilon_0} \tag{4-17}$$

where ν is a constant of the order of $\frac{1}{3}$.

Linear Dielectrics

In the great majority of cases, the molecular dipole moment is *proportional* to $\mathbf{E}_{\text{loc}}$, at least for not too large values of the latter. This constitutes *linear response*. We write

$$\mathbf{p}_m = \alpha\mathbf{E}_{\text{loc}} \tag{4-18}$$

where α is a constant called the *polarizability*, characteristic of the molecule. The polarization in a field $\mathbf{E}$ is then given by

$$\mathbf{P} = n\mathbf{p}_m = n\alpha\mathbf{E}_{\text{loc}} = n\alpha\left(\mathbf{E} + \frac{\nu\mathbf{P}}{\epsilon_0}\right)$$

and solving for $\mathbf{P}$ gives $\mathbf{P} = n\alpha/(1 - \nu n\alpha/\epsilon_0)\mathbf{E}$. We write this as

$$\mathbf{P} = \chi_e \epsilon_0 \mathbf{E} \qquad (4\text{-}19)$$

where

$$\chi_e = \frac{n\alpha/\epsilon_0}{1 - \nu n\alpha/\epsilon_0} \qquad (4\text{-}20)$$

Thus $\mathbf{P}$ is directly proportional to $\mathbf{E}$. The constant of proportionality χ_e is called the *dielectric susceptibility*. It is a dimensionless number, when defined as in Eq. (4-19). The displacement is then $\mathbf{D} = \epsilon_0 \mathbf{E} + \mathbf{P} = (1 + \chi_e)\epsilon_0 \mathbf{E}$. This is usually written

$$\mathbf{D} = \epsilon \mathbf{E} \qquad \epsilon = \kappa\epsilon_0 \qquad \kappa = 1 + \chi_e \qquad (4\text{-}21)$$

which is still another way of describing the linear response of the material. ϵ is called the *permittivity* and κ the *relative dielectric constant* (often the word "relative" is omitted) of the material.

Example 4-10

Consider the slab geometry of Examples 4-6 and 4-7 but let the dielectric now be linear with dielectric constant κ. Find $\mathbf{P}$ and $\mathbf{D}$.

Solution   From the value of $\mathbf{E}$ in the dielectric in Example 4-6, we obtain

$$\sigma = \epsilon_0 E + P = D = \kappa\epsilon_0 E$$

Then

$$P = (\kappa - 1)\epsilon_0 E = \frac{\kappa - 1}{\kappa}\sigma$$

Example 4-11

A sphere of a linear dielectric material has a small spherical charge q at its center. Calculate all pertinent quantities in the dielectric.

Solution   From Eq. (4-14),

$$\mathbf{D} = \frac{q}{4\pi}\frac{\hat{\mathbf{a}}_r}{r^2} \qquad \mathbf{E} = \frac{\mathbf{D}}{\kappa\epsilon_0}$$

$$\mathbf{P} = \mathbf{D} - \epsilon_0 \mathbf{E} = \frac{q}{4\pi}\left(1 - \frac{1}{\kappa}\right)\frac{\hat{\mathbf{a}}_r}{r^2}$$

$$\rho_P = -\text{div}\,\mathbf{P} = 0$$

$$\sigma_P = \mathbf{P}\cdot\hat{\mathbf{n}} = \begin{cases} \dfrac{q}{4\pi}\left(1 - \dfrac{1}{\kappa}\right)\dfrac{1}{b^2} & \text{on outer surface} \\[2ex] -\dfrac{q}{4\pi}\left(1 - \dfrac{1}{\kappa}\right)\dfrac{1}{a^2} & \text{on inner surface} \end{cases}$$

Clausius–Mosotti Equation

On inserting Eq. (4-20) into the third part of Eq. (4-21) and solving, we obtain

$$\frac{n\alpha}{\epsilon_0} = \frac{\kappa - 1}{1 + \nu(\kappa - 1)} \qquad (4\text{-}22)$$

When v has the value $\frac{1}{3}$ appropriate to spherical molecules, this becomes

$$\frac{n\alpha}{\epsilon_0} = \frac{3(\kappa - 1)}{\kappa + 2} \tag{4-23}$$

which relates the measurable macroscopic property κ to the microscopic properties n and α.

Poisson and Laplace Equations

In a linear dielectric medium, these important equations (see Section 1-7) are only slightly modified. By Eqs. (1-16) and (4-14),

$$\nabla^2 U \equiv \text{div grad } U = -\text{div } \mathbf{E} = -\text{div } \frac{\mathbf{D}}{\kappa\epsilon_0}$$

If the dielectric is *uniform*, this becomes

$$\nabla^2 U = -\frac{1}{\kappa\epsilon_0} \text{div } \mathbf{D} = -\frac{\rho}{\kappa\epsilon_0} \tag{4-24}$$

Thus the potential (and electric field strength) are, for a given charge distribution, simply weakened by the factor $1/\kappa$. Conversely, for given potential, all charges have to be increased by the factor κ.

Some values of κ typical of various classes of materials are given in Table 4-2 later in the chapter, after we discuss (very briefly) the basic physics involved.

4-5 MECHANISMS OF DIELECTRIC RESPONSE

We first consider several mechanisms of polarization that lead to a linear polarizability.

Electronic Polarizability

The basic mechanism is that the electrons in each atom are pushed one way and the nuclei the other way by the local field.

Example 4-12

Using a simple orbit model of a hydrogen atom (Figure 4-7), estimate the displacement of the orbit center and the resulting dipole moment in a field E_{loc}, and thus obtain the polarizability. Assume E_{loc} to be much weaker than the internal field in the atom.

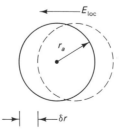

Figure 4-7   Displacement of electron orbit in an applied field (greatly exaggerated).

Solution E_{loc} will add to the field of the nucleus on one side and subtract from it on the other. Roughly, the orbit will adjust itself so as to go through regions where the total field is unchanged. Thus if the orbit radius is r_a,

$$\frac{\delta r}{r_a} = \frac{E_{loc}}{E_{int}} = \frac{E_{loc}}{e/4\pi\epsilon_0 r_a^2}$$

$$\delta r = \frac{4\pi\epsilon_0 r_a^3}{e} E_{loc}$$

Then the induced dipole moment is

$$p_m = e\,\delta r = 4\pi\epsilon_0 r_a^3 E_{loc}$$

$$\alpha_{el} = 4\pi\epsilon_0 r_a^3$$

Numerically, $r_a \cong 1$ Å, so $E_{int} = 2 \times 10^{11}$ V/m. Thus the relative distortion in any practically attainable field is minute. The value of α_{el} is seen to be about 10^{-40} C-m²/V. For other kinds of atoms and molecules it will, of course, vary, but generally not by more than about one order of magnitude.

Example 4-13

Estimate the relative dielectric constant for air at standard temperature and pressure.

Solution   From Eqs. (4-21) and (4-20) it is seen that we need estimates of both α (the polarizability) and n (the number density). α for O_2 and N_2 molecules may be "guesstimated" from Example 4-12 as about 2×10^{-40} C-m²/V, the factor of 2 coming from the two valence electrons in the molecules. For n we may use the ideal gas law, which gives a molar volume of 22.4 liters per gram-mole at STP. Since a gram-mole contains 6×10^{23} molecules (the Avogadro number),

$$n = \frac{6 \times 10^{23}}{0.0224 \text{ m}^3} \cong 3 \times 10^{25} \text{ molecules/meter}^3$$

Thus

$$\frac{n\alpha}{\epsilon_0} = \frac{3 \times 10^{25} \times 2 \times 10^{-40}}{9 \times 10^{-12}} = 6 \times 10^{-4}$$

Thus the correction term in the denominator of Eq. (4-20) is negligible and the susceptibility is ~ 0.0006. Hence $\kappa = 1 + \chi_e = 1.0006$. As can be seen in Table 4-2, this is reasonably close to the actual value of 1.00054.

Ionic Polarizability

In polyatomic molecules, the shifting of the electron orbits leads to shifts in the nuclear positions. This is a more complicated effect to calculate, but in general the resulting polarizability is comparable to the electronic term. The effect is the major one in ionic crystals.

Nonpolar Dielectrics

Electronic and ionic polarization can occur in all classes of dielectric materials. If these are the only mechanisms present, the dielectric is called *nonpolar*.

Example 4-14

Estimate the dielectric constant of a typical nonpolar solid dielectric.

Solution n is typically about 5×10^{28} atoms/meter^3, corresponding to a spacing of a few angstroms. If we take electronic and ionic polarizabilities of 10^{-40} C-m^2/V each, then

$$\frac{n\alpha}{\epsilon_0} \cong \frac{5 \times 10^{28} \times 2 \times 10^{-40}}{9 \times 10^{-12}} \cong 1.1$$

$$\chi_e \cong \frac{1.1}{1 - 1.1/3} \cong 2$$

$$\kappa = 1 + \chi_e \cong 3$$

It is interesting to note that, at the higher density prevailing in condensed matter, the correction term in the denominator is no longer negligible.

Polar Gaseous Dielectrics

We now consider molecules that have permanent dipole moments. Typical examples would be HCl, H_2O, and CO, all of which have asymmetric structures. Values of the permanent moment p_0 are of the order of charge e times a fraction of an angstrom displacement or about 10^{-30} C-m. Thus permanent moments tend to be *much greater* than induced electronic or ionic moments.

In a gas the molecules are far apart except during occasional collisions, and are thus free to rotate without interference from their neighbors. If there is no electric field, thermal agitation (through the agency of the collisions) completely randomizes the orientations, so that the *average* of any component of p_0 is zero. On the other hand, a field tends to align all the dipoles in its own direction. The conflict between these two tendencies results in a net average moment in the direction of the field. Calculation of this average is a problem in statistical mechanics and we shall not go into it here, but the result is

$$\frac{\bar{p}}{p_0} = \coth X - X^{-1} \tag{4-25}$$

where $X = p_0 E_{\text{loc}}/k_B T$, k_B being the Boltzmann constant 1.38×10^{-23} joule/kelvin (J/K).* The function on the right is called the *Langevin function* and has the form shown in Figure 4-8. For large values of X (strong field or low temperature) the molecules approach complete alignment, as we would expect. However, at ordinary temperatures X is very small even in the strongest practical fields, so we are interested in the lower end of the curve. Here it is essentially a straight line of slope $\frac{1}{3}$.

$$\frac{\bar{p}}{p_0} = \tfrac{1}{3}X$$

$$\bar{\mathbf{p}} = \frac{p_0^2 \mathbf{E}_{\text{loc}}}{3k_B T} \tag{4-26}$$

*The Boltzmann constant is the universal gas constant divided by the Avogadro number, that is, the gas constant per molecule (instead of per mole).

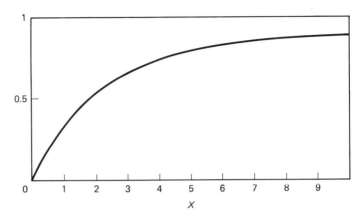

Figure 4-8 The Langevin function, defined in Eq. (4-25).

Thus the gas behaves as if the molecules have a *temperature-dependent polarizability* $p_0^2/3k_B T$. There will in addition still be electronic and ionic polarizabilities, since the same sorts of distortions occur as in nonpolar molecules. The total is then to be used in Eq. (4-20) to obtain the total susceptibility. For gases the second term in the denominator is negligible since n is very small.

Example 4-15

Show how measurements of the dielectric constant of a gas at various temparatures but constant density may be analyzed to obtain some fundamental properties of the molecules.

Solution

$$\kappa = 1 + \chi_e = 1 + \frac{n}{\epsilon_0}\left(\alpha + \frac{p_0^2}{3k_B T}\right)$$

$$\frac{\epsilon_0}{n}(\kappa - 1) = \alpha + \frac{p_0^2}{3k_B}\frac{1}{T}$$

Thus if (ϵ_0/n) $(\kappa - 1)$ is plotted against $1/T$, the intercept gives the electronic (plus ionic) polarizability and the slope is simply related to the permanent dipole moment of the molecules.

Polar Liquid and Solid Dielectrics

When the molecules are crowded close together the situation gets very complicated indeed. In solids the rotation may be limited to a number of discrete positions by the crystal structure. In liquids the rotation is still largely unhindered but each dipole so strongly affects its neighbors that the theory based on average values is inadequate. In both these cases, a polarizability proportional to $p_0^2/k_B T$ is still obtained, but the proportionality constant is different from $\frac{1}{3}$ and depends on structural details. Another complication is that the second term in the denominator of Eq. (4-20) may get quite large so that the exact value of the local field constant ν becomes critical.

When this term is close to 1, the susceptibility becomes extremely large. Liquid water, for example, has a dielectric constant of about 80 at room temperature, as seen in Table 4-2. It decreases at higher temperature, as expected from Eq. (4-26).

Nonlinear Dielectrics

It is natural to ask what happens when the correction term in the denominator gets still larger and becomes equal to 1. Then there is an infinite susceptibility, so that P can have a nonzero value even in zero field. Thus we have a *spontaneous polarization*. In terms of the Langevin function of Eq. (4-25), X has a positive value even when $E = 0$. Such materials are called *ferroelectrics*.

Since, according to Eq. (4-26), the effective polarizability is inversely temperature dependent, the spontaneous polarization disappears at sufficiently high temperature. The critical condition is obtained from Eq. (4-20) as

$$\frac{\nu n \alpha_{\text{eff}}}{\epsilon_0} = 1 \tag{4-27}$$

where

$$\alpha_{\text{eff}} = \frac{p_0^2}{3 k_B T} \tag{4-28}$$

The resulting temperature, above which the material reverts to linear dielectric behavior, is called the *Curie temperature T_c*.

The foregoing picture of ferroelectric behavior is quite oversimplified. Usually, p_0 and ν are also temperature dependent, owing to thermal expansion and to changes in crystal structure. Thus the temperature dependence of the spontaneous polarization is rather more complicated. However, there is always an upper limiting temperature, above which the material reverts to linear dielectric behavior.

4-6 FREQUENCY-DEPENDENT LINEAR RESPONSE

In the foregoing sections we have tacitly assumed that the applied field was static in time, and we now wish to remove this restriction. Since any arbitrary time variation can be expressed as a Fourier series or integral, we need consider only *harmonic* (sinusoidal) variation. Accordingly, we assume that

$$\mathbf{E} = \mathbf{E}_0 \cos \omega t \tag{4-29}$$

When such a field is first applied, there is a very complicated response as the polarization builds up and attempts to follow the field. Eventually, the initial transients die away, and P then will also vary sinusoidally at the same frequency *but not necessarily in phase* with the field. Thus in general we must write

$$P = P_1 \cos \omega t + P_2 \sin \omega t \tag{4-30}$$

Linearity has obviously been lost (unless P_2 happens to be zero) since P is no longer proportional to E. However, we can restore a generalized sort of linearity by writing

Eqs. (4-29) and (4-30) in the *complex exponential* form:

$$E = \text{Re} \, (E_0 e^{-i\omega t}) \tag{4-31}$$

$$P = \text{Re} \, (P_0 e^{-i\omega t}) \tag{4-32}$$

where the symbol Re signifies the real part of the complex number in the parentheses. P_0 will, in general, be complex, that is,

$$P_0 = P_1 + iP_2 \tag{4-33}$$

We then have

$$P = \text{Re} \, [(P_1 + iP_2) \, (\cos \omega t - i \sin \omega t)]$$
$$= \text{Re} \, [P_1 \cos \omega t + P_2 \sin \omega t + i(P_2 \cos \omega t - P_1 \sin \omega t)]$$
$$= P_1 \cos \omega t + P_2 \sin \omega t$$

which agrees with Eq. (4-30) and thus proves the correctness of the complex form.

P_1 and P_2 may also be expressed in terms of the amplitude of P_0 and the phase angle ϕ by which it lags behind the field (Figure 4-9). Since P_1 is in phase and P_2 is 90 degrees out of phase (lagging), we have

$$\tan \phi = \frac{P_2}{P_1} \quad (P_1^2 + P_2^2)^{1/2} = |P_0|$$

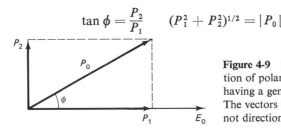

Figure 4-9 Complex-plane representation of polarization of a dielectric having a generalized linear response. The vectors represent time relations, not directions in space.

Thus

$$P_1 = |P_0| \cos \phi \qquad P_2 = |P_0| \sin \phi \tag{4-34}$$

In working in the complex exponential notation it is customary not even to write down the "Re" but merely to keep in mind that it is there and to take the real part of the final answer if it turns out to be complex. This has the further advantage of simplifying all calculus operations, since a time derivative simply multiplies the expression by $-i\omega$.

We note that the ratio of P to E is expressed as a complex number. Thus the susceptibility and the dielectric constant must, of course, also be complex.

Now, the most important aspect of the situation is that the complex response coefficients are *functions of frequency*. It can be seen that this must be the case on quite general grounds, since, whatever the mechanism of polarization may be, *inertia* and *friction* will prevent it from following the field with infinite rapidity. For this reason, κ is more properly called the *dielectric response function*.

Example 4-16

Write and solve the equation of motion for the orbit model of electronic polarizability discussed in Example 4-12 under a sinusoidally varying field.

Solution We now write x for the orbit displacement instead of δr. Under static conditions, x was seen to be proportional to the applied force, which means that there must be an internal restoring force $-Kx$. In this model

$$K = \frac{4\pi\epsilon_0 r_a^3}{e^2}$$

The equation of motion is therefore

$$m\ddot{x} = -Kx - eE_0 e^{-i\omega t}$$

Now assume that a solution exists in the form $x = x_0 e^{-i\omega t}$. For this form, $\ddot{x} = -\omega^2 x$. Thus

$$(-m\omega^2 + K)x_0 e^{-i\omega t} = -eE_0 e^{-i\omega t}$$

$$x_0 = \frac{eE_0}{m\omega^2 - K}$$

The dipole moment is given by $p_m = -ex = \alpha E$. Thus

$$\alpha = \frac{e^2/m}{(K/m) - \omega^2} = \frac{e^2/m}{\omega_r^2 - \omega^2} \tag{4-35}$$

On this simple model, the polarizability is seen to be purely real and to become infinite when $\omega = (K/m)^{1/2}$. This is called the *resonant frequency* ω_r. In actual physical systems there is always some additional force that prevents the motion from becoming infinite. Often this is a frictional effect that can be represented by a term $-\eta\dot{x}$ on the right side of the equation of motion. A vast variety of systems can be well represented by this sort of model, called the *damped harmonic oscillator*. We consider its properties in the next example.

Example 4-17

The damped harmonic oscillator: Write and solve the equation of one-dimensional motion of a particle of charge q, bound to a center by a restoring force $-Kx$ and subject to a frictional restraining force $-\eta\dot{x}$, under the action of an alternating applied field. From the result, obtain an expression for the dielectric response function $\kappa(\omega)$ of a medium consisting of n such particles per unit volume.

Solution The equation of motion can be written as

$$\ddot{x} = -\omega_r^2 x - \gamma\dot{x} + \frac{qE_0}{m} e^{-i\omega t} \tag{4-36}$$

where $\omega_r^2 = K/m$, $\gamma = \eta/m$. On assuming a solution of the form $x = x_0 e^{-i\omega t}$, we obtain

$$x_0 = \frac{qE_0/m}{\omega_r^2 - \omega^2 - i\gamma\omega}$$

Since the dipole moment is qx, the polarizability is

$$\alpha(\omega) = \frac{q^2/m}{\omega_r^2 - \omega^2 - i\gamma\omega} \tag{4-37}$$

To keep things simple, we shall assume n to be small enough so that the local field correction is negligible. Then

$$\kappa(\omega) = 1 + \frac{n\alpha(\omega)}{\epsilon_0}$$

$$= 1 + \frac{nq^2}{m\epsilon_0} \cdot \frac{1}{\omega_r^2 - \omega^2 - i\gamma\omega} \qquad (4\text{-}38)$$

As expected, κ turns out to be a complex function of the frequency. We can separate it into its real and imaginary parts by the usual method of multiplying numerator and denominator by the complex conjugate of the latter, obtaining

$$\kappa_{\mathrm{re}}(\omega) = 1 + \frac{nq^2}{m\epsilon_0} \frac{\omega_r^2 - \omega^2}{(\omega_r^2 - \omega^2)^2 + \gamma^2\omega^2} \qquad (4\text{-}39)$$

$$\kappa_{\mathrm{im}}(\omega) = \frac{nq^2}{m\epsilon_0} \frac{\gamma\omega}{(\omega_r^2 - \omega^2)^2 + \gamma^2\omega^2} \qquad (4\text{-}40)$$

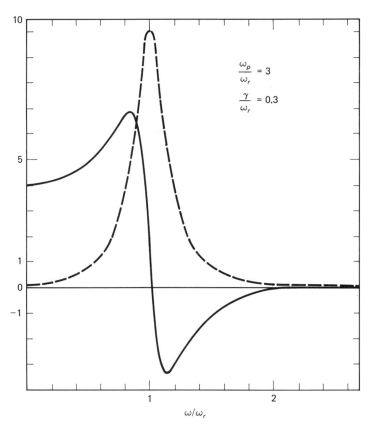

$$\frac{\omega_p}{\omega_r} = 3$$

$$\frac{\gamma}{\omega_r} = 0.3$$

ω/ω_r

Figure 4-10 Real part (solid line) and imaginary part (dashed line) of relative dielectric response function, Eqs. (4-38) to (4-40).

The form of these functions has the typical *resonance* shape, as shown in Figure 4-10. The peaks get narrower and higher as γ is made smaller, approaching Eq. (4-35) as $\gamma \to 0$. Thus κ_{im} is very small except in the vicinity of the resonance. We shall see later [Eq. (8-44)] that κ_{im} is associated with absorption of energy from the field.

In a real material there are a variety of resonance and relaxation processes. The electronic resonances are the fastest and occur in the visible, ultraviolet, and x-ray frequency ranges. Far beyond the highest of them, the material behaves essentially like empty space. Processes involving atomic and molecular motions typically occur in the infrared, and those involving large groups of molecules are very much slower.

Kramers–Kronig Relations

Even though the expressions for κ_{re} and κ_{im} resulting from the damped oscillator model look rather complicated, the "real-life" situation is far worse. There are always so many resonances over such a wide range of frequencies that a single expression cannot come close to an accurate description. Only direct measurement will suffice. But because of the so-called *Kramers–Kronig relations*, it is not necessary to measure both functions. If *either one* is known at all frequencies the other can be obtained at any desired frequency.

One of the Kramers–Kronig relations is

$$\kappa_{re}(\omega) = 1 + \frac{2}{\pi} P \int_0^\infty \frac{\omega' \kappa_{im}(\omega')}{\omega'^2 - \omega^2} \, d\omega' \tag{4-41}$$

where the symbol P designates the *principal part* of the integral: The integral is evaluated in two pieces with a gap around the singularity at $\omega' = \omega$, then the limit is taken as the gap shrinks to zero. Usually, the integrals have to be evaluated numerically, but the following (unphysical) example illustrates the general nature of the results.

Example 4-18

Evaluate $\kappa_{re}(\omega)$ for a hypothetical material having a single uniform absorption band, $\kappa_{im} = C$ for $\omega_1 \leq \omega \leq \omega_2$, 0 otherwise.

Solution

$$\kappa_{re}(\omega) = 1 + \frac{2}{\pi} P \int_{\omega_1}^{\omega_2} \frac{C\omega' \, d\omega'}{\omega'^2 - \omega^2}$$

$$= 1 + \frac{C}{\pi} \ln \left| \frac{\omega_2^2 - \omega^2}{\omega_1^2 - \omega^2} \right|$$

This illustrates the general result that an absorption band (peak in κ_{im}) is always accompanied by a "wiggle" in κ_{re}.

The Kramers–Kronig relations are independent of any specific model, and are based directly on the *linearity* and *causality* (an effect cannot precede its cause) of the dielectric response. The proof relies on certain properties of analytic complex function and cannot be given here, but the relations are of great practical as well as fundamental importance. For example, as stated, they make it unnecessary to measure both parts of the complex function $\kappa(\omega)$ in investigating properties of materials.

Values of Dielectric Response Function κ

Table 4-2 lists a few values of κ, in two frequency ranges, for some typical examples of important classes of materials. The frequencies cited are *static* or very low (well below any resonances) and *optical*, or roughly 3×10^{14} Hz.

TABLE 4-2   DIELECTRIC CONSTANTS OF TYPICAL MATERIALS

	Low frequency	High frequency[a]
Diamond	5.5	5.8
NaCl	5.62	2.34
Polystyrene	2.5	2.5
Water	78.5 (25°C)	1.77
	34.5 (200°C)	
Benzene	2.28	2.35
Air	1.00054	1.0006
Rochelle salt	4000[b]	2.25

[a]Frequency in red or near-infrared.
[b]Along crystallographic *a* axis.

A number of interesting comparisons emerge. Looking first at the low-frequency column, we see that gases have very low values; this is because of the low density. Liquid or solid organics (benzene and polystyrene) range around 2 to 3. Ionic crystals (NaCl) are generally a bit higher since there is ionic as well as electronic polarization. Diamond is a covalently bonded crystal from the fourth column of the periodic table and has high electronic polarizability because of the very high concentration of bonding electrons. Water is a polar molecule and thus has an even higher, although temperature-dependent, value. Finally, the huge value for Rochelle salt is characteristic of a ferroelectric crystal slightly above its Curie temperature.

Now looking at the high-frequency column, we see that in those cases where the polarization is dominantly electronic, little change with frequency has occurred. In these materials, the resonances all lie at even higher frequency, in the ultraviolet or x-ray regions. For ionic crystals, there is typically a strong resonance in the infrared owing to the ion motions. This is passed once the frequency reaches the visible range, so κ is somewhat lower. Similarly in the case of water; the molecular rotations cannot follow such high frequencies and only the electronic part of the polarizability remains. Also, of course, the large-scale organized displacements in the ferroelectric crystal are damped out at even lower frequency.

4-7 CAPACITORS

An extremely useful circuit element based on dielectrics is the *capacitor*, which consists in its most common form of a sandwich of dielectric between two conductive sheets. The *capacitance* is defined as the amount of charge *transferred* from the negative to the positive plate per unit applied potential difference. The capacitance of a given structure depends only on the geometrical arrangement and the dielectric response functions of the materials. It is independent of the applied potential difference, as long as the dielectrics are linear.

To calculate the capacitance of a given structure we may either:

1. Assume a potential difference, obtain the field and from it the charge distribution, or
2. Assume a charge displacement, obtain the field and from it the potential difference.

Example 4-19

Calculate the capacitance of a dielectric sandwich, of area A and thickness d. Neglect "fringing" of the field around the edges.

Solution 1 For a potential difference V, the field strength is V/d. From Example 4-6, this corresponds to surface charge densities $\pm\sigma$ given by $\sigma = \kappa\epsilon_0 E = \kappa\epsilon_0 V/d$. Hence the displaced charge is $Q = \sigma A = (\kappa\epsilon_0 A/d)V$, and

$$C \equiv \frac{Q}{V} = \frac{\kappa\epsilon_0 A}{d} \tag{4-42}$$

Solution 2 For displaced charge Q, the surface charge densities are $\pm\sigma = \pm Q/A$. From Example 4-6, the field strength is $\sigma/\kappa\epsilon_0 = Q/\kappa\epsilon_0 A$. The potential difference is $V = Ed = Qd/\kappa\epsilon_0 A$. Thus

$$C \equiv \frac{Q}{V} = \frac{\kappa\epsilon_0 A}{d}$$

Example 4-20

Find the capacitance of a sandwich in which the dielectric fills only a part d' of the total thickness d.

Solution From Example 4-8, we find $D = \sigma = Q/A$ *everywhere*. Thus $E_{\text{vac}} = Q/\epsilon_0 A$ while $E_{\text{diel}} = Q/\kappa\epsilon_0 A$. Thus

$$V = \frac{Q}{\epsilon_0 A}(d - d') + \frac{Q}{\kappa\epsilon_0 A}d'$$

$$= \frac{Q}{\epsilon_0 A}\left(d - d' + \frac{d'}{\kappa}\right)$$

and

$$C \equiv \frac{Q}{V} = \frac{\epsilon_0 A}{d - d' + d'/\kappa}$$

This illustrates once again that the sources of D are the true charges only, while the sources of E are the true plus effective charges. In linear dielectrics, the latter always tend to weaken the field.

Example 4-21

Now consider a sandwich in which the dielectric fills the thickness completely but covers only a part A' of the total area A.

Solution   What is constant here is the field strength, since the same potential difference exists between all parts of the electrodes. That is, $E = V/d$ everywhere. Hence σ must be different: $\sigma = \epsilon_0 E$ in the empty regions and $\kappa\epsilon_0 E$ in the filled regions. Thus

$$Q = \epsilon_0 \frac{V}{d}(A - A') + \kappa\epsilon_0 \frac{V}{d}A'$$

$$= \frac{\epsilon_0 V}{d}(A - A' + \kappa A')$$

$$C \equiv \frac{Q}{V} = \frac{\epsilon_0}{d}(A - A' + \kappa A')$$

A few more simple geometries are treated in the problems at the end of the chapter.

The usefulness of capacitors in electric circuits arises from the flow of *displacement current* [Eq. (4-16)] under a time-varying applied potential difference. Considering the simple sandwich configuration of Example 4-19, we find for the displacement current density

$$J_d \equiv \frac{dD}{dt} = \kappa\epsilon_0 \frac{dE}{dt} = \kappa\epsilon_0 \frac{d}{dt}\left(\frac{V}{d}\right)$$

Hence the total displacement current is

$$I_d \equiv J_d A = \frac{\kappa\epsilon_0 A}{d}\frac{dV}{dt} = C\frac{dV}{dt} = \frac{dQ}{dt}$$

Thus the displacement current in the capacitor exactly "bridges the gap" in the true current flowing in the wires: I_d exactly equals I. Thus the current-voltage relation for a capacitor is

$$I = C\frac{dV}{dt} \tag{4-43}$$

It is evident that high-frequency currents are passed easily, whereas steady (zero-frequency) currents are blocked.

FURTHER REFERENCES

In this chapter we have been able to give only a sketchy introduction to the complicated and fascinating subject of dielectric behavior. The reader will find deeper-going treatments in the following works:

R. S. ELLIOTT, *Electromagnetics*, McGraw-Hill Book Company, New York, 1966, Chap. 6.

A. NUSSBAUM, *Electromagnetic and Quantum Properties of Materials*, Prentice-Hall, Inc., Englewood Cliffs, N.J., 1966, Chap. 5.

C. KITTEL, *Introduction to Solid State Physics*, 3rd ed., John Wiley & Sons, Inc., New York, 1966, Chaps. 12 and 13.

A. VON HIPPEL, *Dielectric Materials and Applications*, John Wiley & Sons, Inc., New York, 1954.

A. J. DEKKER, *Solid State Physics*, Prentice-Hall, Inc., Englewood Cliffs, N.J., 1957, Chaps. 6 and 8.

N. F. MOTT and R.W. GURNEY, *Electronic Processes in Ionic Crystals*, Oxford University Press, London, 1940, Chap. 1.

PROBLEMS

4-1. A cylindrical cavity of length L and radius a is cut parallel to **P** in a uniformly polarized dielectric. Find the field at the center of the cavity.

4-2. A dielectric cylinder of length L and radius a with center at the origin is polarized uniformly along its length. Find the field at points on the axis inside the cylinder.

4-3. For Problem 4-2, find **D** at points on the axis both inside and outside the cylinder.

4-4. A dielectric sphere of radius a is polarized nonuniformly so that $\mathbf{P} = C\mathbf{r}$. Find the bound charge densities.

4-5. A point charge q is located at the center of a sphere of linear dielectric, of radius a. Find **D**, **E**, and **P** everywhere.

4-6. Derive the Clausius–Mosotti formula (4-23).

4-7. Gases at standard temperature and pressure contain about 3×10^{25} molecules per cubic meter. Air under these conditions has a relative dielectric constant of 1.00054. Determine the average polarizability of air molecules.

4-8. Using the result of Problem 4-7, estimate the relative dielectric constant of air compressed to 1000 times its standard density.

4-9. A dielectric material has a complex dielectric constant given by $\kappa = C(1 + i\delta)$. Calculate the phase angle between **D** and **E** in an alternating field.

4-10. For Problem 4-9, calculate the phase angle between **P** and **E**.

4-11. The polarization in a dielectric has a magnitude of $2\epsilon_0 E_0$ and lags by 30 phase degrees behind the (alternating) electric field. Write **P** in the complex exponential form.

4-12. For Problem 4-11, write the complex susceptibility and dielectric constant.

4-13. For the damped linear oscillator model, write the real and imaginary parts of the dielectric constant if the spring constant is reduced to zero.

4-14. For the damped linear oscillator model, find approximations valid for frequencies near resonance when the damping is small.

4-15. A parallel-plate capacitor is initially empty. When a dielectric slab covering the entire area but only half the thickness is inserted, the capacitance increases by one-third of the initial value. What is the relative dielectric constant of the material inserted?

4-16. Two long coaxial cylindrical conductors have radii a and b. Calculate the capacitance per unit length.

4-17. Calculate the capacitance of two concentric conductive spherical shells of radii a and b.

chapter 5

Conductors

5-1 LINEAR CONDUCTIVITY

We mentioned in Section 4-1 that conductors are distinguished by the presence of *free carriers*, that is, charged particles that can move large distances through the material. The current density $\mathbf{J}$ was given in Eq. (2-4) in terms of the number densities and average velocities of all the species of carriers present. In most common conductors at any fixed temperature it is found that:

1. All the n_i are independent of the electric field (if the field is not too strong).
2. All the v_i are proportional to the electric field.

The latter relation is written

$$|\mathbf{v}_i| = \mu_i |\mathbf{E}| \tag{5-1}$$

where μ_i is called the *mobility* of the ith species of carrier. Under these conditions, it follows that the current density is proportional to the field,

$$\mathbf{J} = g\mathbf{E} \tag{5-2}$$

where g is called the *conductivity* of the material. Inserting Eq. (5-1) into (2-4), we obtain

$$\mathbf{J} = \sum_i n_i |q_i| \mu_i \mathbf{E} \tag{5-3}$$

so that the conductivity is

$$g = \sum_i n_i |q_i| \mu_i \tag{5-4}$$

Note that carriers of both signs contribute *additively* to the conductivity. Negative charges move opposite to the field but their current is along the field.

Example 5-1

Find the relation between total current and potential difference along a cylinder of a linear conductor, of length l and area A.

Solution   The current is JA and the electric field strength is V/l, hence $I/A = gV/l$ or

$$V = \frac{l}{gA} I \qquad (5\text{-}5)$$

This relation is known as *Ohm's law*. The coefficient of I is the *resistance* of the cylinder. The MKS unit of resistance is the *Ohm* (1 ohm = 1 volt per ampere) and its reciprocal is called the *mho* (1 mho = 1 ampere per volt). Thus conductivity is measured in mhos per meter. More commonly quoted is the reciproal of conductivity, called *resistivity*, measured in ohm-meters.

Example 5-2

Copper at room temperature has a conductivity of 6×10^7 mhos per meter (mho/m). Find the resistance of 1 km of 1-mm-diameter wire.

Solution

$$R = \frac{l}{gA} = \frac{10^3}{6 \times 10^7 \times (\pi/4) \times (10^{-3})^2} \cong 20 \text{ ohms}$$

Example 5-3

In metals the carriers are electrons. Assuming one free electron per atom, estimate the mobility in copper at room temperature.

Solution   From Eq. (5-4) and the conductivity value just given,

$$\mu = \frac{g}{ne} = \frac{6 \times 10^7}{10^{28} \times 1.6 \times 10^{-19}}$$

$$\cong 4 \times 10^{-2} \text{ m}^2/V\text{-s}$$

For example, a field of 1 V/m produces an average speed of roughly 0.04 m/s.

The steady velocity comes about because frictionlike forces (actually due to collisions) tend to counteract the acceleration by the field. The behavior is analogous to an object falling through the air with a parachute; a steady *terminal velocity* is quickly attained. The energy imparted to the carriers is *dissipated* (converted to heat) by the frictional forces. Thus each carrier dissipates $|q|V$ of energy in "falling" through potential difference V, and the total rate of dissipation is

$$\text{power consumed} = VI \qquad (5\text{-}6)$$

For linear conductivity, this can also be written as

$$VI = I^2R = \frac{V^2}{R} \qquad (5\text{-}7)$$

Example 5-4

Calculate the *power density* in a conductive material.

Solution   Assuming a cylindrical shape, the volume is lA. Thus the power per unit volume is

$$\frac{VI}{lA} = \frac{V}{l}\frac{I}{A} = \mathbf{E} \cdot \mathbf{J} \tag{5-8}$$

Example 5-5

A heating element for a toaster designed to operate on 115 V is found to work too slowly. How should the wire size be changed?

Solution   Since the voltage is to be held constant, we use the form V^2/R for the power. Thus R must be decreased; that is the wire diameter must be increased.

Steady Currents in Circuits

In any circuit segment *ab*, the current is driven by the *algebraic sum of potential differences and emfs*, and is limited by the resistance, so that Ohm's law is generalized to

$$I_{a\rightarrow b}R_{ab} = U_a - U_b + \mathcal{E}_{ab} \tag{5-9a}$$

$$= \int_a^b \mathbf{E}_S \cdot \mathbf{dl} + \int_a^b \frac{1}{q}\mathbf{F}_{NE} \cdot \mathbf{dl} \tag{5-9b}$$

where $\mathbf{F}_{NE}$ is the total nonelectrostatic nonfrictional force per particle, as discussed in Section 3-1.

Case 1: Open Segment.   If the segment is not part of a complete circuit, there can be no steady current. In this case, Eq. (5-9a) gives

$$U_b - U_a = \mathcal{E}_{ab} \text{ (open segment)} \tag{5-10}$$

For example, the potential difference between the terminals of an *isolated* battery is equal to the emf of the battery.

Case 2: Complete Circuit.   If points *a* and *b* are identical, then $U_a - U_b = 0$ (in other words, $\oint \mathbf{E}_S \cdot \mathbf{dl} = 0$), so Eq. (5-9a) applied to the entire circuit gives

$$IR_{\text{total}} = \mathcal{E}_{\text{total}} \tag{5-11}$$

where R_{total} is the sum of all resistances in the circuit, *including internal resistances* in the sources of emf. When applied to only a segment, Eq. (5-9a) gives

$$U_b - U_a = \mathcal{E}_{ab} - I_{a\rightarrow b}R_{ab} \tag{5-12}$$

Thus the potential difference is altered by the ohmic voltage drop in the resistance of the segment. If the current flow is in the direction of $\mathcal{E}_{ab}$, the potential difference is diminished. If it is in the opposite direction (driven by other sources of emf in the circuit but not within segment *ab*), the potential difference is *increased*.

Example 5-6

In the circuit shown in Figure 5-1, find the potential difference between points *a* and *b*.

Solution

$$I = \frac{5.0}{20 + 0.1} = 0.249 \text{ A}$$

(1) $U_a - U_b = 5.0 - 0.249 \times 0.1 = 4.975$ V

or

(2) $U_a - U_b = 0.249 \times 20 = 4.975$ V

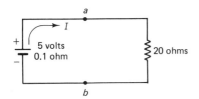

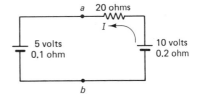

Figure 5-1   Circuit for Example 5-6.   **Figure 5-2**   Circuit for Example 5-7.

Example 5-7

In the circuit shown in Figure 5-2, find the potential difference between points *a* and *b*.

Solution

$$I = \frac{10 - 5}{20 + 0.1 + 0.2} = 0.2463 \text{ A}$$

(1) $U_a - U_b = 5 - (-0.2463 \times 0.1) = 5.0246$ V

or

(2) $U_a - U_b = 10 - (0.2463 \times 20.2) = 5.0246$ V

These simple examples will repay careful study, as questions of voltages in current-carrying circuit elements are often perplexing.

5-2 ELECTRONS IN SOLIDS

The carriers in most solid (and many liquid and gaseous) conductors are electrons. Their motions are governed, as in atoms and molecules, by the laws of quantum mechanics. This subject is beyond our present scope, but we can give a rough idea of the salient features so as to glean some understanding of conductive processes.

In isolated atoms the electrons can have only certain *states of motion* (roughly speaking, *orbits*), each of which has a definite discrete energy. The electrons seek the lowest possible energy levels, but the *Pauli exclusion principle* prevents more than one from occupying any single state. Thus the levels are filled from the lowest energy up, with vacant levels generally present above the occupied ones. When many atoms are brought close together in a crystal, two important effects take place: (1) the states no longer are confined to single atoms but spread over the entire crystal, and (2) the energies are "perturbed" to various degrees by the neighboring atoms so that they are no longer discrete but spread out into an *energy band*. The situation is depicted schematically in Figure 5-3. It is seen that the higher energy levels, which generally correspond to larger orbits, tend to split sooner and more. Thus the highest bands will always overlap. The filling of the levels follows the same principles as in isolated atoms except that now, since the levels in a band are so close together, thermal agitation may also play a significant role.

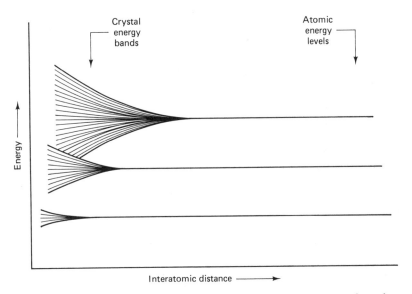

Figure 5-3 Spreading of atomic energy levels into bands as atoms are brought together to form a crystal (schematic).

Since the electron states are extended throughout the crystal, it might be thought that all electrons could move freely from atom to atom, so that all crystals would be conductors. This is not the case. The velocity vectors of the individual electrons, although large in magnitude (speed), are distributed over all directions, so that the average velocity is zero in the absence of an applied field. In order to conduct, the electrons must be able to *change their states of motion* in response to the field. This can occur only if there are adjacent empty levels for them to move into. In other words, a completely filled band widely separated in energy from the other bands cannot conduct. It can only respond as a dielectric (i.e., polarize).

Thus the essential condition for conductivity is the presence of one or more *partially filled bands*. This can come about in three essentially different ways:

1. There may not be enough electrons to fill the highest occupied band completely. This is the case in the monovalent metals such as sodium, copper, and many others. The band made up of the outermost filled atomic orbits contains two states per atom, but there is only one electron per atom available to occupy it.

2. The highest occupied band may be overlapped by higher bands. This is the case in polyvalent metals. If the overlap is very small, the material is called a *semimetal*; bismuth is a familiar example.

3. There are just enough electrons to fill the highest occupied band but the next higher band is only slightly separated in energy. Then thermal agitation may excite some electrons into the next band, leaving empty levels available in both bands. Such a material is called an *intrinsic semiconductor*. The lower (nearly full) band is called the *valence band* since the electrons in it form the chemical bonds that hold the atoms together; the upper is called the *conduction band*. However, conduction occurs in *both bands*. In the valence band, the empty states ("holes") act like *positive* charges. Thus there are two kinds of carriers present in equal number, and this number is a rapidly increasing function of temperature.

It remains to consider how the electrons are distributed over the levels of the partially filled bands. This is somewhat analogous to the filling of an empty container with water. If just a small amount is put in, it will vaporize and the available energies will be only fractionally occupied, although with more at the lower energies (lower velocities in the gas) than at the higher. As more water is added, eventually some of it will remain condensed and fill the container from the bottom. Of course, the liquid surface is not precisely sharp; there is a bit of "fuzziness" due to evaporation and condensation but its extent is minuscule compared with the total depth. The latter situation corresponds to the filling of the energy levels in metals, and is obviously simpler than the "vaporlike" occupancy in semiconductors.

Quantitatively, the probability that any given level is occupied is given by the *Fermi–Dirac distribution* function

$$f(\mathsf{E}) = \left(1 + \exp\frac{\mathsf{E} - \mathsf{E}_F}{k_B T}\right)^{-1} \tag{5-13}$$

where E is the energy of the level and E_F is called the *Fermi energy* or *Fermi level*. The shape of this function is shown in Figure 5-4. It is seen that states well below the Fermi level are essentially fully occupied, whereas those well above it are essentially empty. The width of the step is about $k_B T$. The value of E_F depends on the distribution of the levels in energy (i.e., the quantum mechanics of the crystal), the number of electrons present, and the temperature. Thus it is a rather complicated quantity to calculate. In metals, however, it is largely determined by the very high number-density of electrons, and a fair estimate of its value may be obtained by treating the electrons in the partially occupied band as a "gas" of free particles. It is a consequence of quantum mechanics that a free particle of mass m confined to a cube of side L must have a kinetic energy of at least $3h^2/8mL^2$, where h is *Planck's*

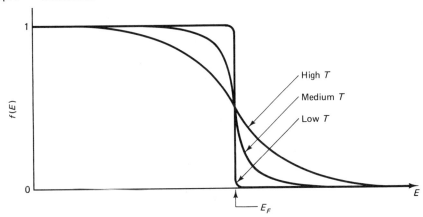

Figure 5-4   Fermi–Dirac distribution function, giving occupation probabilities of energy levels in equilibrium.

constant, 6.7×10^{-34} J-s. From this we can readily estimate the average kinetic energy of the free electrons in a typical monovalent metal. Take $L^3 = 1/n$ as the volume available to the average electron, where n is the electron density, typically about 10^{28} per cubic meter. Then the foregoing formula gives

$$E \cong \frac{3 \times (6.7 \times 10^{-34})^2}{8 \times (9 \times 10^{-31}) \times (10^{28})^{-2/3}}$$

$$\cong 8 \times 10^{-19} \text{ J}$$

$$\cong 5 \text{ electron-volts}$$

This is roughly the Fermi energy in typical metals. The value is much greater than $k_B T$ at all accessible temperatures, so the step width at the top of the distribution is, for most considerations, inconsequential.

In semiconductors the problem is more complex, and we can only quote some results. If the material is pure, the Fermi level lies about halfway between the top of the valence band and the bottom of the conduction band. Thus the occupation of the latter and the emptying of the former are due to the "tails" on the step of the distribution function. By suitable *doping* (deliberate addition of certain impurities) the Fermi level can be moved toward or even across one or the other band edge. One thus obtains *n-type* (more electrons than holes) or *p-type* (more holes than electrons) semiconductors. The various situations are illustrated in Figure 5-5.

5-3 ELECTROSTATICS OF CONDUCTORS

Now that we have some idea of the nature of conductive materials, we are in a position to start considering fields within them. Normally, every (macroscopic) region is electrically neutral, owing to an exact balance between the densities of positive and negative charge. However, in a field the mobile charges may move macroscopic

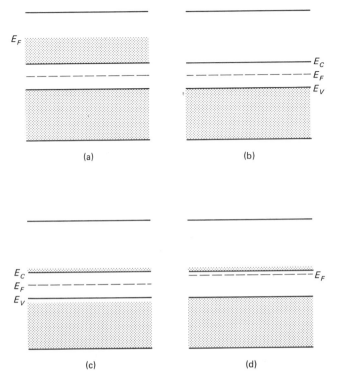

Figure 5-5 Illustrations of electron energy band occupation in various types of materials: (a) metal; (b) insulator or intrinsic semiconductor at very low temperature; (c) intrinsic semiconductor at higher temperature; (d) n-type semiconductor.

distances, and this may lead to unbalance of the densities in some regions. For example, in a metal the regions (if any) where electrons accumulate will acquire a net negative charge density, leaving other regions with a deficiency of electrons and hence a net positive charge density.

A general relation between charge density and field strength is given by Gauss's law or, in more convenient form, Poisson's equation (4-24). A second relation describes the response of the material, above and beyond the polarization of the electrons in the fully occupied lower bands. The essential effect is that all energy levels are shifted by any electrostatic potential that may be present. The reason is simply that the potential energy adds onto all other energy contributions involved. Since an energy level is an energy of an electron, which has a charge of $-e$,

$$E(\mathbf{r}) = E(\mathbf{r}_0) - e[U(\mathbf{r}) - U(\mathbf{r}_0)]$$
$$= \text{const.} - eU(\mathbf{r}) \tag{5-14}$$

The Fermi level, on the other hand, is everywhere *the same when the system is in equilibrium.* The reason is that two levels at the same energy must have equal occupation probabilities; otherwise, there would be a flow of electrons from the more heavily occupied one to the other. The resulting situation is illustrated in Figure 5-6; there, E represents any specified energy level, such as the edge of a particular band. Clearly, the net negative charge density will depend on the amount of lowering of the bands. Since the levels are essentially filled up to the Fermi level, the increase in concentration will be proportional to the energy shift, that is,

$$\frac{\Delta n}{n} \cong \frac{eU}{E_F} \tag{5-15}$$

where U is the potential difference relative to a neutral field-free region of the metal. Then the charge density is

$$\rho = -e\,\Delta n \cong -\frac{e^2 n}{E_F}\,U \tag{5-16}$$

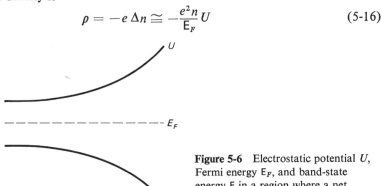

Figure 5-6 Electrostatic potential U, Fermi energy E_F, and band-state energy E in a region where a net negative charge density exists, Eq. (5-14).

Example 5-8

Evaluate the coefficient in Eq. (5-16) for a typical metal.

Solution From our estimate of the Fermi energy this is

$$-\frac{(1.6 \times 10^{-19})^2 \times 10^{28}}{8 \times 10^{-19}} = 3 \times 10^8 \ \text{C/m}^3\text{-V}$$

This is an extremely large number; it means that only *very small* potential differences can occur statically in metals. The basic reason, of course, is the very large value of n. In conductors with fewer carriers, correspondingly larger potential differences can occur. However, the calculations are more complicated, and we shall not go into them.

By inserting Eq. (5-16) into the Poisson equation (4-24) we derive the differential equation obeyed by the electrostatic potential in a metal.

$$\nabla^2 U = -\frac{1}{\kappa_e \epsilon_0}\frac{-e^2 n}{E_F}\,U$$

This is often written

$$\nabla^2 U = \frac{U}{L_{TF}^2} \tag{5-17}$$

where L_{TF} is called the *Thomas–Fermi length* or, more descriptively, the *screening length*. Its value is

$$L_{TF} = \left(\frac{\kappa_e \epsilon_0 E_F}{ne^2}\right)^{1/2} \tag{5-18}$$

Here κ_e is the dielectric constant of the metal. It arises from the polarization of the electrons in the lower (filled) energy bands (i.e., the inner atomic shells). Equation (5-17) is readily solved in the case that the potential varies in one dimension only. Let z be the direction of variation. Then the differential equation is

$$\frac{d^2U}{dz^2} = \frac{U}{L_{TF}^2}$$

If the metal has its surface at $z = 0$ and extends toward the positive z direction, the solution that remains finite everywhere in the material is

$$U(z) = U_0 \exp\left(\frac{-z}{L_{TF}}\right) \tag{5-19}$$

Thus any field applied to the surface of a metal is *screened out* exponentially with a decay length L_{TF}.

Example 5-9

Estimate the screening length in a typical metal.

Solution   Since the dielectric constant arises from inner electron shells, it will not be too much greater than 1. Thus, roughly,

$$L_{TF} = \left[\frac{9 \times 10^{-12} \times 8 \times 10^{-19}}{10^{28} \times (1.6 \times 10^{-19})^2}\right]^{1/2} \cong 10^{-10} \text{ m}$$

Thus the screening length in a metal is very small, well under an interatomic distance. This is a consequence of the very high electron concentration, and goes hand in hand with the result of Example 5-8. In other types of conductors, the screening length and the potential difference across the charged layer may both be much greater.

Example 5-10

A fixed point charge is inserted into a metal. Find the potential in the surrounding region.

Solution   In vacuum the potential would vary as r^{-1}, where r is the distance from the point charge. This suggests that we seek a solution of Eq. (5-17) in the form $r^{-1}F(r)$. By substituting in the spherical-coordinate form of the equation, it is readily verified that

$$U(r) = \frac{q}{4\pi\kappa_e\epsilon_0} r^{-1} \exp\left(\frac{-r}{L_{TF}}\right)$$

This is called the *screened Coulomb potential*; it plays an important role in the theory of metals.

The foregoing results hold when the conductor is *in equilibrium*. We now consider the rate of approach to equilibrium following a disturbance. Suppose that some mobile charge be abruptly injected into a conductor (e.g., some electrons may be shot in from an external source). From Eqs. (2-6), (5-2), and (4-24) we obtain

$$\frac{\partial \rho}{\partial t} = -\text{div } J = g \nabla^2 U = -\left(\frac{g}{\kappa_e \epsilon_0}\right)\rho$$

The first and last terms constitute a differential equation for ρ. The solution is

$$\rho(\mathbf{r}, t) = \rho(\mathbf{r}, 0) \exp\left(\frac{-g}{\kappa_e \epsilon_0}\right)t \tag{5-20}$$

Thus the injected charge density at any interior point decays away exponentially with a time constant of $\kappa_e \epsilon_0/g$. This is called the *dielectric relaxation time*. Physically, the repulsion of the injected charge causes all the carriers to adjust their positions slightly so that all the net charge ends up in the thin surface layer described by Eq. (5-19).

Example 5-11

Estimate the dielectric relaxation time in a typical metal.

Solution   Taking the conductivity of copper from Example 5-2 and a dielectric constant of about 1, we obtain

$$\frac{9 \times 10^{-12} \text{ C/V-m}}{6 \times 10^7 \text{ A/V-m}} \simeq 10^{-19} \text{ s}$$

Actually, as we shall see in Section 5-6, this time is so short that the conductivity will not have its ordinary (low-frequency) value. Nevertheless, the result shows that in a metal the equilibrium is reached very rapidly indeed. In poorer conductors and near-insulators the times may be very much longer.

Ideal Conductors

For most calculations involving fields near metals it is a good approximation to treat the metal as an *ideal conductor*, which, as we discussed in Section 1-10, is a *fictitious* material in which the *screening length is zero*. From Eq. (5-18) it is seen that this would require an *infinite carrier concentration*. This in turn would lead to an *infinite conductivity*, hence *zero relaxation time*. It therefore follows that, as was assumed in Section 1-10, $\mathbf{E} = 0$ in an ideal conductor at all times and positions.

5-4 CURRENTS IN CONDUCTORS

Drift Currents

The carriers in a conductor are in constant random motion due to either thermal agitation or to the quantum-mechanical kinetic energy discussed in Section 5-2, or both. Each particle moves freely for a while, then suffers a "collision" which abruptly changes its velocity. Since the motion is random, the average *velocity* of any species is zero. The average *speed* (magnitude of the velocity vector) may, of course, be very large.

If an electric field is present in the material, each particle is *accelerated* during the intervals between its collisions. Thus at time t after a collision

$$\mathbf{v}(t) = \mathbf{v}(0) + \frac{q\mathbf{E}}{m} t \qquad (5\text{-}21)$$

(Actually, m is not the true mass of the particle but a quantity called the *effective mass* that characterizes the response to the macroscopic field $\mathbf{E}$ in the presence of the strong microscopic fields in the material. Thus it depends on the type of particle and on the quantum mechanics of the material.) From Eq. (5-21) we can calculate the mobility of a given species of particle in terms of its charge, effective mass, and mean free time between collisions. According to Eq. (5-1), we need the average velocity in the field. Accordingly, we average Eq. (5-21) over time. The average of the first term is zero, since this is the random velocity. The average of the second term gives

$$\mathbf{v}_i = \frac{q_i \tau_i}{m_i} \mathbf{E} \qquad (5\text{-}22)$$

Hence,

$$\mu_i = \frac{|q_i| \tau_i}{m_i} \qquad (5\text{-}23)$$

where τ_i is the mean free time.

It is seen the collisions provide a mechanism whereby the energy imparted to the particles by the field is delivered to the material as heat. In terms of the average velocity, this amounts to a *frictional* effect that prevents the continued acceleration of the swarm of particles.

Example 5-12

Calculate the effective frictional force on a given species of carriers in a field.

Solution   The frictional force must just balance the applied force in order to give a steady average velocity. Thus

$$\mathbf{F}_i = -q_i \mathbf{E} = -\frac{m_i}{\tau_i} \mathbf{v}_i$$

The nature of the collision processes requires a bit of discussion. If the atoms of the materials were at rest in a perfect crystal lattice, the carriers would simply move smoothly through it and there would be no collisions. In other words, collisions are due to *departures from perfection* of the lattice. Such departures come about in two ways: (1) due to thermal agitation of the lattice, which increases with temperature; and (2) due to various fixed imperfections (impurities, displaced atoms, etc.) inevitably present in any real material. Thus mobilities tend to *increase with decreasing temperature* and reach a limiting maximum value at very low temperatures. If the carrier concentrations are temperature independent (as in metals), the conductivity behaves in the same way. In intrinsic semiconductors, the carrier concentrations increase strongly with increasing temperature, and this overpowers the effects of decreasing mobilities.

Diffusion Currents

If the carrier concentrations are *nonuniform*, the random motion leads to a *net flow* from regions of high to low concentration, as suggested in Figure 5-7. For each species the resulting current is given by *Fick's law* as

$$\mathbf{J}_{\text{diff}} = -Dq \text{ grad } n \tag{5-24}$$

where D is the *diffusion constant*. The total current due to both drift and diffusion for a material containing several species of free carriers is simply the sum of all the current contributions.

$$\mathbf{J} = \sum_i q_i[\mu_i n_i(\pm\mathbf{E}) - D_i \text{ grad } n_i] \tag{5-25}$$

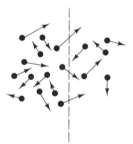

Figure 5-7   Illustration of diffusion current flow from a region of high concentration (left) to low concentration (right).

We can find a *generalized potential* function that acts as the driving force for the current contribution of a given species of carrier. Equation (5-25) is

$$\mathbf{J} = \sum_i q_i(\mp n_i\mu_i \text{ grad } U - D_i \text{ grad } n_i) \tag{5-26}$$

The quantity in parentheses is called the *electrochemical potential* for the ith species. It expresses the combined effects of the electric force and the chemical concentration gradients in tending to move the carriers. The condition for equilibrium in a conductor is that the electrochemical potentials of all the carrier species be uniform throughout. There may well be an electric field present but the drift currents it evokes are exactly cancelled by the diffusion currents.

Two important classes of devices in which the foregoing considerations are of primary importance are electrochemical cells (batteries) and *p-n* junctions in semiconductors. In cells, the concentration gradients are produced by the unequal "dissolving tendencies" of the two electrode materials. On open-circuit, a potential difference builds up to just stop the flow of current. When an external connection is provided, this potential is somewhat diminished, and a current flows. This current is a *diffusion current within the cell* and a *drift current in the external connection*. A *p-n* junction is a piece of semiconductor material with different types of doping (see Section 5-2) in two regions. Thus concentration gradients of both electrons and holes exist near the boundary. This again causes the buildup of a potential difference to stop the diffusion currents. An applied potential difference upsets the balance and allows a net current to flow. The amount of current depends very much on the *direction* of the applied field. If it is such as to drive both kinds of carriers toward the junction (positive on the *p*-type end) the copious supply of carriers leads to a large current, which flows through the junction by *diffusion*. But the opposite sign of applied field pulls both kinds of carriers away from the junction and thus can, at most, reduce the diffusion currents to zero. As a result, a *p-n* junction is a *rectifier*, having a current-voltage characteristic of the form shown in Figure 5-8. Clearly, diffusion currents can lead to *highly nonlinear* conduction.

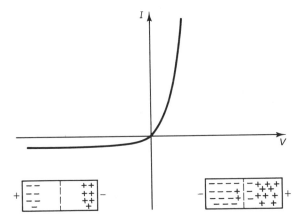

Figure 5-8 Current–voltage curve for a *p-n* junction. The inserts indicate the carrier distributions under the two polarities of applied voltage.

Space-Charge-Limited Currents

In some cases the charge of the moving carriers themselves makes a significant contribution to the field strength. This greatly complicates matters, since all the pertinent quantities are then variables subject to a set of simultaneous equations. We illustrate this for the simplest possible case, namely, a single species of carrier moving in a vacuum. This is the situation in most types of vacuum tubes: The carriers are electrons and they are delivered into the vacuum by *thermionic emission* from a heated *cathode*.

We consider a planar cathode and a parallel electrode (the *anode*) held at a positive potential so as to attract the electrons. The equations (differential or algebraic) that relate the current density, field strength, electron concentration, and electron velocity at any distance x from the cathode are the following:

1. Since there is only one species of carrier, Eq. (2-4) is

$$J(x) = qn(x)v(x) \tag{5-27}$$

2. For steady-state conditions the current continuity equation gives

$$\frac{dJ}{dx} = 0 \qquad J = \text{const.} \tag{5-28}$$

3. Gauss's law [Eq. (1-15)] is

$$\frac{dE}{dx} = \frac{q}{\epsilon_0} n(x) \tag{5-29}$$

4. Since there is no frictional medium, the carriers are freely accelerated by the field; hence Newton's law $a = F/m$ becomes

$$v \frac{dv}{dx}\left(= \frac{dv}{dt}\right) = \frac{q}{m} E(x) \tag{5-30}$$

We now seek to combine these into an equation for a single one of the variables. In view of Eq. (5-28), Eq. (5-27) may be solved for either n or v to substitute in the others. It works a bit better to retain v, hence Eq. (5-29) becomes

$$\frac{dE}{dx} = \frac{J}{\epsilon_0} v^{-1} \tag{5-31}$$

Then by differentiating (5-30), we obtain

$$\frac{dE}{dx} = \frac{m}{q}\left[v \frac{d^2v}{dx^2} + \left(\frac{dv}{dx}\right)^2\right] \tag{5-32}$$

and equating the right sides of (5-31) and (5-32) gives the desired equation.

A solution can be found in the form of a power function of x, and from it all the pertinent variables can be evaluated. If we try $v = Ax^h$, the equation becomes

$$Ax^h(Ah(h-1)x^{h-2} + A^2h^2x^{2h-2} = \frac{Jq}{m\epsilon_0 A} x^{-h}$$

The equality can hold only if all powers of x are equal, so $2h - 2 = -h$; $h = \frac{2}{3}$. Then $A = (9Jq/2m\epsilon_0)^{1/3}$. From this, $n = J/qv = (J/qA)x^{-2/3}$, and Eq. (5-31) gives

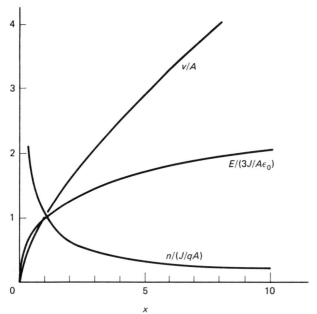

Figure 5-9 Variation of velocity, field strength, and electron concentration with distance from cathode in space-charge-limited current flow in vacuum.

$$E = \frac{J}{A\epsilon_0} \int_0^x (x')^{-2/3} \, dx' = \frac{3J}{A\epsilon_0} x^{1/3}$$

The various functions are sketched in Figure 5-9.

Example 5-13

From the foregoing solution, calculate the current–voltage characteristic.

Solution   The potential difference is

$$V = \int_0^L -E \, dx = -\frac{3J}{\epsilon_0 A}\left(\frac{3}{4} L^{4/3}\right)$$

$$= \text{const.} \cdot J^{2/3} L^{4/3}$$

or

$$J = \text{const.} \cdot \frac{V^{3/2}}{L^2} \tag{5-33}$$

Equation (5-33) is known as *Child's law*. It is valid as long as the supply of electrons from the cathode is copious enough to keep the field strength close to zero there. If V is made too large, the supply of electrons becomes inadequate, and the current levels off at a *saturation* value, governed by the rate of emission from the cathode.

Space-charge-limited currents can also occur in semiconductors, but the theory is rather more complicated, owing to the presence of several species of charged particles.

5-5 MAGNETIC FIELD EFFECTS

Hall Effect

If a magnetic field is present, the moving carriers are subject to the magnetic force $q\mathbf{v} \times \mathbf{B}$, Eq. (2-9), in addition to the electrical and frictional forces. Thus the time-averaged equation of motion becomes, instead of Eq. (5-22),

$$\frac{m_i}{\tau_i} \mathbf{v}_i = q_i(\mathbf{E} + \mathbf{v}_i \times \mathbf{B}) \tag{5-34}$$

Obviously, if $\mathbf{B}$ is not parallel to $\mathbf{v}$, the magnetic force tends to deflect the carriers sideways. In a finite sample such as a bar, the accumulation of charges at the side faces produces an additional electric field which can, in principle, be measured. This is known as the *Hall effect* (see Figure 5-10).

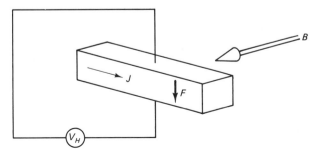

Figure 5-10   Geometry of the Hall effect.

Example 5-14

Consider a conductor having a single species of carrier. Let a current-carrying bar be oriented along the x axis and let a magnetic field B be applied parallel to the z axis. Calculate the *Hall field* E_y.

Solution   In steady state the current is constrained to flow along the bar, hence $v_y = 0$. The x and y components of Eq. (5-34) are

$$\frac{m}{\tau} \mathbf{v}_x = qE_x \tag{5-35}$$

$$0 = q(E_y - v_x B) \tag{5-36}$$

Thus

$$E_y = v_x B = \frac{J}{nq} B$$

The quotient E_y/JB is called the *Hall coefficient*. It is seen to give a measure of the carrier concentration in a one-carrier material. The geometry of the situation is shown in Figure 5-10.

Example 5-15

A centimeter-square bar of a typical metal carries a current of 1000 A in a transverse magnetic field of 3 T. Calculate the Hall voltage developed across the bar.

Solution

$$V_y = LE_y = L \frac{I}{L^2 nq} B = \frac{IB}{nqL}$$

$$= \frac{10^3 \times 3}{10^{28} \times (-1.6 \times 10^{-19}) \times 10^{-2}} = -2 \times 10^{-4} \text{ V}$$

The Hall effect in metals is thus seen to be quite small. In semiconductors, having much lower carrier concentrations, it is generally large enough to be easily measurable and is widely used to determine the concentration and sign of the majority carriers.

The carrier mobility can also be measured via the conductivity and Hall coefficient R_H. Since $g = nq\mu$ and $R_H = (nq)^{-1}$, we get

$$\mu = R_H g \tag{5-37}$$

The mobility measured in this way is called the *Hall mobility*. Owing to the averaging over the carrier speed distribution, it is slightly different from the "conductivity mobility."

The Hall coefficient in a semiconductor containing both electrons and holes is somewhat more complicated. We seek the transverse electric field E_y such that the *total current* flows along the bar. For the electrons the y component of Eq. (5-34) is, instead of (5-36),

$$\frac{m_n}{\tau_n} v_{ny} = -e(E_y - v_{nx}B)$$

By Eqs. (5-22) and (5-23) this may be written

$$v_{ny} = -\mu_n(E_y + \mu_n E_x B)$$

Similarly, for the (positive) holes,

$$v_{py} = \mu_p(E_y - \mu_p E_x B)$$

The y component of the total current density vanishes, that is,

$$-env_{ny} + epv_{py} = 0$$

where n and p are the electron and hole concentrations, respectively. Hence, on multiplying the two previous equations by n and p, respectively, subtracting, and solving, we obtain

$$E_y = \frac{p\mu_p^2 - n\mu_n^2}{p\mu_p + n\mu_n} E_x B$$

But the x component of total current density is, neglecting small terms of order B^2,

$$J = e(n\mu_n + p\mu_p)E_x$$

Thus

$$R_H \equiv \frac{E_y}{JB} = \frac{p\mu_p^2 - n\mu_n^2}{(p\mu_p + n\mu_n)^2} \frac{1}{e} \tag{5-38}$$

It is seen that the two types of carriers tend to cancel each other in the Hall effect. Physically, this comes about because they drift in *opposite* directions along the bar, hence are deflected toward the *same* side, so that the buildup of charge is alleviated.

Magnetoresistance

Another magnetic field effect is the *magnetoresistance*, an increase in the resistivity of the conductor when a magnetic field is applied. For a single-carrier material, Eq. (5-35) seems to imply that there is no such effect, but this is not quite correct, since we have neglected there the random motion of the carriers. Actually, Eq. (5-34) applies only to the *average* carrier. Those moving either faster or slower will be deflected in one direction or the other, and will thus make less headway along the bar between collisions.

Since the Hall field tends to minimize deflection, the magnetoresistance can be enhanced by minimizing this field. One approach is to use a very short bar with highly conductive electrodes to act as "short circuits" for the transverse voltage. Better yet is to use a cylindrical geometry with concentric circular electrodes, so that the current flow is radial and there are no side faces. This is called a *Corbino disk*.

Example 5-16

Calculate the current density in a single-carrier material when the Hall field is completely shorted out.

Solution   In Eq. (5-34) $E_y = 0$, so instead of Eqs. (5-35) and (5-36) we have

$$v_x = \frac{q\tau}{m}(E_x + v_y B)$$

$$v_y = -\frac{q\tau}{m}v_x B$$

Solving for v_x gives

$$J_x = nqv_x = \frac{nq^2\tau/m}{1 + (q\tau B/m)^2} E_x \tag{5-39}$$

Since the second term in the denominator is always positive, the conductivity is reduced.

This simple treatment is valid only for rather weak fields such that $q\tau B/m \ll 1$. For the opposite limit, the carriers make many complete orbits before colliding, so that the entire picture of acceleration to a terminal drift velocity breaks down. The effects are too complicated to discuss here but provide valuable probes of the detailed structure of the electron quantum states in solids.

5-6 CONDUCTIVITY AT HIGH FREQUENCY

When the period of the field alternations becomes comparable with or shorter than the mean free time, it is certainly no longer correct to assume that the carriers are subject to a constant acceleration between their collisions, as was done in deriving Eqs. (5-22) and (5-34). Instead, we must write the full equation of motion for each species of carrier as

$$m\frac{d\mathbf{v}}{dt} + \frac{m}{\tau}\mathbf{v} = q(\mathbf{E} + \mathbf{v} \times \mathbf{B}) \tag{5-40}$$

As in Section 4-6, we consider sinusoidal time variation and work in the complex exponential notation. For simplicity we consider only a single-carrier material. Thus we let $\mathbf{E}(t) = \mathbf{E}_0 e^{-i\omega t}$ and assume that $\mathbf{v}(t)$ has the same time dependence. Then Eq. (5-40) with $\mathbf{B} = 0$ gives

$$-i\omega m\mathbf{v}_0 + \frac{m}{\tau}\mathbf{v}_0 = q\mathbf{E}_0$$

$$\mathbf{v}_0 = \frac{q\mathbf{E}_0}{-i\omega m + m/\tau} = \frac{q\tau/m}{1 - i\omega\tau}\mathbf{E}_0$$

$$\mathbf{J} = nq\mathbf{v} = \frac{nq^2\tau}{m}\frac{1}{1 - i\omega\tau}\mathbf{E}$$

$$g \equiv \frac{J}{E} = \frac{nq^2\tau/m}{1 - i\omega\tau} = \frac{nq^2\tau/m}{1 + \omega^2\tau^2}(1 + i\omega\tau) \tag{5-41}$$

The conductivity is seen to become complex, and to approach a purely imaginary (and very small) value as $\omega\tau$ gets much greater than 1 (Figure 5-11).

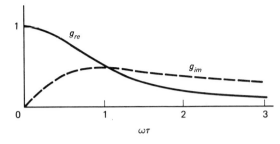

Figure 5-11   Real and imaginary parts of the conductivity of a single-carrier conductor at frequencies much less than the plasma frequency.

The meaning of the complex conductivity is made evident by consideration of both the conduction and displacement currents. The total current density is

$$J_{\text{tot}} = J + J_{\text{displ}} = gE + \epsilon \frac{\partial E}{\partial t} = (g - i\omega\epsilon)E$$

$$= \left[g_{\text{re}} + \omega\epsilon_{\text{im}} - i\omega\left(\epsilon_{\text{re}} - \frac{g_{\text{im}}}{\omega}\right) \right]E$$

$$= -i\omega\left[\left(\epsilon_{\text{re}} - \frac{g_{\text{im}}}{\omega}\right) + i\left(\epsilon_{\text{im}} + \frac{g_{\text{re}}}{\omega}\right) \right]E \qquad (5\text{-}42)$$

Thus g_{im}, the imaginary part of the conductivity, makes a *negative* contribution to the effective dielectric constant of the material.

Plasma Frequency

For a single-carrier conductor, we now calculate the frequency at which the effective dielectric constant changes sign. From Eq. (5-42) the effective dielectric constant is

$$\kappa_{\text{eff}} = \kappa - \frac{g_{\text{im}}}{\omega\epsilon_0} \qquad (5\text{-}43)$$

where g_{im} is the imaginary part of the right side of Eq. (5-41). Thus, assuming that $\omega\tau \gg 1$,

$$\kappa_{\text{eff}} = \kappa - \frac{nq^2}{m\omega} \frac{1}{\omega\epsilon_0}$$

(see Figure 5-12). For real κ this vanishes at a frequency ω_p, called the *plasma frequency*, given by

$$\omega_p^2 = \frac{nq^2}{m\kappa\epsilon_0} \qquad (5\text{-}44)$$

In terms of this frequency the preceding equation may be written

$$\kappa_{\text{eff}} = \kappa\left(1 - \frac{\omega_p^2}{\omega^2}\right) \qquad (5\text{-}45)$$

It is seen that the effective dielectric constant is *negative* for frequencies below the plasma frequency (as long as the assumption $\omega\tau \gg 1$ remains valid). We shall see later that this has profound effects on the propagation of electromagnetic waves through the material.

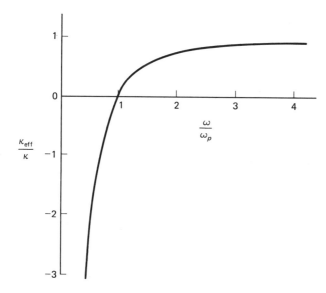

Figure 5-12 Effective dielectric response function for collisionless plasma.

Magnetoplasma Effects

We now consider the effects of a static magnetic field at right angles to the varying electric field. It works out best to deal with a *rotating* electric field of the form

$$\mathbf{E}(t) = E_0(\hat{\mathbf{i}} + i\hat{\mathbf{j}})e^{-i\omega t} \tag{5-46}$$

Example 5-17

Show that Eq. (5-46) does represent a rotating field and give the direction of rotation.

Solution Remember that the physical field is represented by the real part of the complex function. Thus

$$\text{Re}\,[\mathbf{E}(t)] = E_0(\hat{\mathbf{i}}\cos\omega t + \hat{\mathbf{j}}\sin\omega t)$$

The amplitude is always E_0 and the direction rotates in the clockwise sense as viewed in the $\hat{\mathbf{k}}$ direction.

To calculate the conductivity in the combination of fields just described, we assume that the velocity vector will also be a rotating vector of the form

$$\mathbf{v}(t) = v_0(\hat{\mathbf{i}} + i\hat{\mathbf{j}})e^{-i\omega t}$$

Then if the magnetic field is $B\hat{\mathbf{k}}$,

$$\mathbf{v} \times \mathbf{B} = v_0 B(-\hat{\mathbf{j}} + i\hat{\mathbf{i}})e^{-i\omega t}$$
$$= iv_0 B(\hat{\mathbf{i}} + i\hat{\mathbf{j}})e^{-i\omega t}$$

On substituting this into Eq. (5-40) and solving, we find

$$v_0 = \frac{q\tau/m}{1 - i(\omega \pm \omega_c)\tau}E_0 \tag{5-47}$$

where

$$\omega_c = \frac{|q|B}{m} \tag{5-48}$$

The conductivity is then

$$g = \frac{J}{E} = \frac{nqv}{E} = \frac{nq^2\tau/m}{1 + (\omega \pm \omega_c)^2\tau^2}[1 + i(\omega \pm \omega_c)\tau] \tag{5-49}$$

where the $\pm$ sign corresponds to the sign of q. The imaginary part of the conductivity in Eq. (5-49) may be written

$$g_{im} = \frac{g_0 X}{1 + X^2} \tag{5-50}$$

where $g_0 = nq^2\tau/m = dc$ conductivity, and $X = (\omega - \omega_c)\tau$. This has a negative maximum at $X = -1$. Hence, in view of Eq. (5-43), the effective dielectric constant may be very large at the corresponding frequency (Figure 5-13).

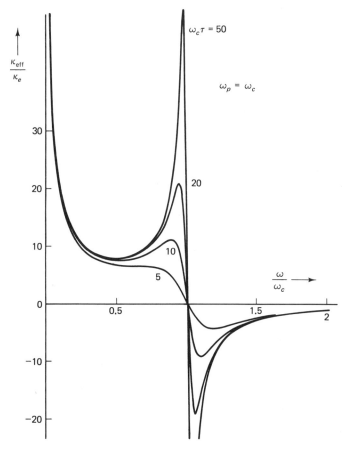

Figure 5-13   Dielectric response function for a plasma in a magnetic field from Eqs. (5-43) and (5-49).

It is also seen in Eq. (5-49) that the real part of the conductivity has a maximum (a *resonance*) if the carriers are negatively charged; a similar resonance would occur for positive carriers if the electric field rotation or the magnetic field were reversed. This is called the *cyclotron resonance*, and the angular frequency ω_c at which it occurs is called the *cyclotron frequency*.

The physical interpretation of cyclotron resonance is as follows: The particles in the magnetic field alone would tend to move in circular orbits since the force is always perpendicular to the velocity. The angular velocity is obtained by equating the applied magnetic force to the centripetal force required by the orbit radius (i.e., $|q|vB = mv^2/r$). The angular velocity is thus $\omega_c = v/r = \dfrac{|q|B}{m}$. For a magnetic field along the positive z direction, the motion is clockwise (as viewed along $\hat{\mathbf{k}}$) for negative particles, and vice versa. Thus cyclotron resonance occurs when the rotating electric field is applied at the same frequency and in the same sense as the "free" motion of the particles.

FURTHER REFERENCES

In this chapter we have touched on a few aspects of the complex subject of electrical conduction. A glaring apparent omission is *superconductivity* (resistanceless current flow), but this is more appropriately regarded as a magnetic effect, and so will be discussed in Chapter 6.

The following references have been selected for readability without a detailed working knowledge of quantum mechanics. There are, of course, dozens of other excellent works at the same and higher levels for those who wish to delve more deeply into this important and elegant subject.

R. S. ELLIOTT, *Electromagnetics*, McGraw-Hill Book Company, New York, 1966, Chap. 8.

C. KITTEL, *Introduction to Solid State Physics*, 3rd ed., John Wiley & Sons, Inc., New York, 1966, Chaps. 7 and 8.

A. J. DEKKER, *Solid State Physics*, Prentice-Hall, Inc., Englewood Cliffs, N.J., 1957, Chaps. 9 and 11.

F. SEITZ, *Modern Theory of Solids*, McGraw-Hill Book Company, New York, 1940, Chap. 4.

PROBLEMS

5-1. A sample of germanium has electron and hole concentrations of 10^{14} and 10^{12} per cubic centimeter, respectively, with mobilities of 2000 and 900 cm²/V-s, respectively. Calculate the resistivity.

5-2. What will be the resistance of a 1-cm cube of the material in Problem 5-1 with electrodes fully covering a pair of opposite faces?

5-3. If 100 V is applied across the two electrodes of the preceding problem, what will be the drift velocities of the two kinds of carriers?

5-4. In a solid, at what energy is an electron energy level whose probability of being occupied is $\frac{1}{2}$?

5-5. At room temperature (300 K), what is the energy relative to the Fermi level of an electron energy level whose occupation probability is 0.01?

5-6. Estimate the average speed of the free electrons in a typical monovalent metal.

5-7. From Eq. (5-19), calculate how the field strength varies from the surface in a metal.

5-8. A field of 10^7 V/m (very strong) is applied to the surface of a typical metal. Estimate the potential difference developed across the space-charge layer at the surface.

5-9. For Problem 5-8, calculate the effective surface charge density due to the space-charge layer and the average volume charge density within the layer.

5-10. An imperfect insulator has a dielectric constant of 3 and a dielectric relaxation time of 1 s. What is its resistivity?

5-11. From the mobility of electrons in copper (Example 5-3) estimate the mean free time and the mean free path length between collisions.

5-12. Why does Child's law [Eq. (5-28)] not apply in ordinary conductors?

5-13. Calculate the Hall coefficient for the germanium sample of Problem 5-1.

5-14. Given that the product of the electron and hole concentrations is a constant (at a given temperature), what concentrations in germanium would give a Hall coefficient of zero?

5-15. For copper, at what frequency would the real part of the conductivity be reduced to one-half of the low-frequency value?

5-16. At this frequency, what would the imaginary part of the conductivity be?

5-17. Estimate the plasma frequency for copper.

chapter 6

Magnetic Materials

6-1 FIELDS DUE TO MAGNETIZED MATTER

Before embarking on this chapter, the reader would be well advised to review Chapter 4, since much of the material is fairly (although by no means exactly) similar. Once again we consider the molecules of matter as sources of a field, but this time we are concerned with the intramolecular *currents* rather than the charges. There are currents due to both the *orbital motion* and the *spin* of each electron, so that each electron contributes a certain amount of magnetic dipole moment. There are also contributions from the nuclei, but they are very small. These contributions add vectorially to make up the molecular magnetic dipole moment $\mathbf{m}_m$. In other words, $\mathbf{m}_m$ is the magnetic moment of a single molecule, the analog of $\mathbf{p}_m$. The *magnetization* is defined as the analog of the polarization, that is,

$\mathbf{M}(\mathbf{r})$ = magnetic dipole moment per unit volume in a small region around $\mathbf{r}$

$$= n(\mathbf{r})\bar{\mathbf{m}}_m(\mathbf{r}) \tag{6-1}$$

Again, the bar over $\mathbf{m}_m$ designates the average value in the region considered.

Example 6-1

The electron in the lowest-energy state in a hydrogen atom moves in such a way that its angular momentum about the nucleus is $h/2\pi$. Calculate the magnetic moment. (h is Planck's constant.)

Solution Regard the orbit as circular with radius r, velocity v, and frequency $v/2\pi r$. Then $m_e v r = h/2\pi$, where m_e is the electron mass. The current is the charge times the frequency, so that the magnitude of the magnetic moment (current times area) is

$$|\mathbf{m}_m| = e\frac{v}{2\pi r}(\pi r^2) = \frac{eh}{4\pi m_e}$$

$$= 9.3 \times 10^{-24} \text{ A-m}^2$$

This is a convenient unit for atomic and molecular dipole moments, and is called the *Bohr magneton*.

Proceeding as in Chapter 4, we write the vector potential at an external point due to a piece of magnetized matter. Since the field point is distant from all the molecules, the dipole approximation [Eq. (2-26)] can be used for the contribution of each volume element.

$$\mathbf{A}_0(\mathbf{r}) = \frac{\mu_0}{4\pi} \iiint \frac{\mathbf{M}(\mathbf{r}') \times (\mathbf{r} - \mathbf{r}')}{|\mathbf{r} - \mathbf{r}'|^3} \, d^3\mathbf{r}' \tag{6-2}$$

By a manipulation similar to that used for Eq. (4-4), the foregoing integral can be transformed so that only the first power of the distance appears in the denominator. The integrand may be written $\mathbf{M} \times \text{grad}' \, R^{-1}$, which is equal to $R^{-1} \, \text{curl}' \, \mathbf{M} - \text{curl}' \, (R^{-1}\mathbf{M})$. By the cross-product analog of the divergence theorem,

$$\iiint \text{curl}' \, (R^{-1}\mathbf{M}) \, d^3\mathbf{r}' = \oiint \hat{\mathbf{n}} \times (R^{-1}\mathbf{M}) \, dS'$$

we get

$$\mathbf{A}_0(\mathbf{r}) = \frac{\mu_0}{4\pi} \iiint_V \frac{\text{curl}' \, \mathbf{M}}{R} \, d^3r' + \frac{\mu_0}{4\pi} \oiint_S \frac{\mathbf{M} \times \hat{\mathbf{n}}}{R} \, dS' \tag{6-3}$$

The integrals are, of course, taken over the volume and surface of the magnetized matter.

It is seen that the integrals have the same form as the vector potential due to certain distributions of current. In other words, for purposes of calculating the vector potential (and the magnetic field), the matter may be *replaced* by two distributions of *effective current* (also called *magnetization current*), namely

$$\mathbf{J}_M = \text{curl } \mathbf{M} \quad \text{(volume density)} \tag{6-4}$$

$$\mathbf{j}_M = \mathbf{M} \times \hat{\mathbf{n}} \quad \text{(surface density)} \tag{6-5}$$

Thus, if $\mathbf{M}$ is a known function, the field may be calculated by the methods of Chapter 2 after calculating the effective current densities. Alternatively, of course, Eq. (6-2) can be used directly.

Example 6-2

A cylinder has uniform magnetization $\mathbf{M}$ in the direction of its length (a bar magnet). Calculate the effective current distribution.

Solution $\mathbf{M} = M\hat{\mathbf{k}}$; $\mathbf{J}_M = 0$; $\mathbf{j}_M = 0$ on end faces; $\mathbf{J}_M = M\hat{\mathbf{a}}_\phi$ on cylindrical surface. Thus the effective current distribution is the same as an ideal solenoid, with M replacing NI/L (see Example 2-8).

Example 6-3

The bar magnet of Example 6-2 has a hole drilled through along its axis. Find the field at a point on the axis.

Solution   The effective current distribution is that of two concentric solenoids with equal and opposite currents. Therefore, the field is

$$\mathbf{B}(0, 0, z) = \mu_0 M \hat{\mathbf{k}}(\cos \alpha_1 + \cos \alpha_2 - \cos \alpha'_1 - \cos \alpha'_2)$$

where the α's are the angles subtended by the end radii at the field point.

Example 6-4

Find the effective current distributions of a uniformly magnetized sphere, with magnetization M_0.

Solution $\mathbf{M} = M_0 \hat{\mathbf{k}}$; $\mathbf{J}_M = \text{curl } \mathbf{M} = 0$;

$$\mathbf{j}_M(\theta) = M_0 \hat{\mathbf{k}} \times (\cos \theta \, \hat{\mathbf{k}} + \sin \theta \, \hat{\mathbf{a}}_s)$$
$$= M_0 \sin \theta \, \hat{\mathbf{a}}_\phi$$

Example 6-5

A sphere of radius a is magnetized *nonuniformly* so that the magnetization is parallel to the z axis and proportional to the distance from the axis with proportionality constant C (Figure 6-1). Find the effective current distributions.

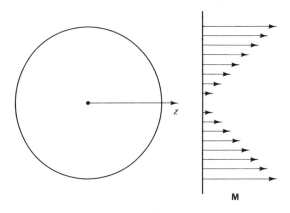

Figure 6-1   Magnetization distribution for Example 6-5.

Solution $\mathbf{M} = Cs \hat{\mathbf{k}}$; $\mathbf{J}_M = \text{curl } \mathbf{M} = C \hat{\mathbf{a}}_\phi$;

$$\mathbf{j}_M = Cs \hat{\mathbf{k}} \times (\cos \theta \, \hat{\mathbf{k}} + \sin \theta \, \hat{\mathbf{a}}_s)$$
$$= C(a \sin \theta) \sin \theta \, \hat{\mathbf{a}}_\phi$$

As was true of the polarization charges, the magnetization currents are not mere mathematical fiction. Figure 6-2 illustrates a distribution of molecular dipoles such that $\mathbf{M}$ has a curl (a transverse derivative). In this example, $\partial M_z/\partial x$ is negative, so curl $\mathbf{M}$ lies in the positive y direction. It is seen that on any intermolecular plane there is, in fact, more current flowing up than down. Also, on the left face there is a net surface current downward, and on the right face a smaller one upward.

We now turn to consider field points inside the magnetized matter. As in the dielectric case, some of the molecules are close to the field point. Accordingly, we

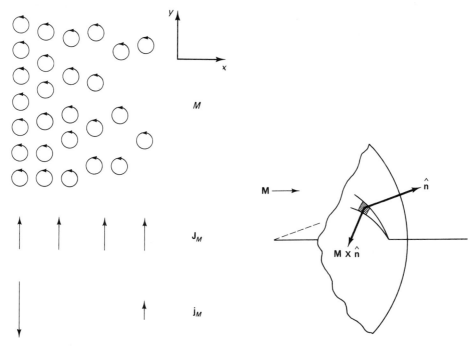

Figure 6-2 Nonuniform distribution of magnetization such that curl $\mathbf{M} \neq 0$. The effective current densities are indicated.

Figure 6-3 Effective surface current density on a uniformly magnetized sphere.

again imagine a cavity around the field point and evaluate separately the contributions of the matter outside and inside (see Figure 6-3).

For the field $\mathbf{B}_1$ in the cavity due to the magnetized matter outside, the potential is

$$\mathbf{A}_1 = \frac{\mu_0}{4\pi} \iiint\limits_{V - V_1} \frac{\text{curl}' \, \mathbf{M}}{R} \, d^3r' + \frac{\mu_0}{4\pi} \oiint\limits_{S + S_1} \frac{\mathbf{M} \times \hat{\mathbf{n}}}{R} \, dS'$$

Then

$$\mathbf{B}_1 = \text{curl} \, \mathbf{A}_1$$

and since $\mathbf{M}$ is a function of $\mathbf{r}'$, not $\mathbf{r}$, this is

$$\mathbf{B}_1 \cong \frac{\mu_0}{4\pi} \iiint\limits_{V} \frac{(\text{curl}' \, \mathbf{M}) \times \mathbf{R}}{R^3} \, d^3r' + \frac{\mu_0}{4\pi} \oiint\limits_{S + S_1} \frac{(\mathbf{M} \times \hat{\mathbf{n}}) \times \mathbf{R}}{R^3} \, dS' \qquad (6\text{-}6)$$

In Eq. (6-6) we have taken account of the vanishing of the correction to the volume integral in the limit as the cavity shrinks. For the spherical cavity surface S_1 the effective current distribution is the same as in Example 6-4 except that $\hat{\mathbf{n}}$ now points inward. Thus a ring of width $a \, d\theta'$ carries an effective current

$$-M \sin \theta' (a \, d\theta') \hat{\mathbf{a}}_{\phi'}$$

The radius of the ring is $a \sin \theta'$ and the distance from its plane to the field point is $-a \cos \theta'$. Thus from Example 2-7 the field at the center due to the surface effective current on S_1 is

$$\mathbf{B}(0) = -\frac{\mu_0 \mathbf{M}}{2} \int_0^\pi \frac{\sin \theta'(a \, d\theta')(a^2 \sin^2 \theta')}{(a^2 \sin^2 \theta' + a^2 \cos^2 \theta')^{3/2}}$$

$$= -\frac{\mu_0 \mathbf{M}}{2} \int_0^\pi \sin^3 \theta' \, d\theta'$$

$$= -\tfrac{2}{3}\mu_0 \mathbf{M} \tag{6-7}$$

For the field B_2 due to the interior molecules we again consider only the average value over the volume of the cavity. The calculation is rather more complicated than the electrostatic one, and for simplicity we first treat a special case.

Example 6-6

Calculate the average, over the volume of a sphere, of the field due to a magnetic dipole located at the center.

Solution

$$\bar{\mathbf{B}}_2 = \frac{1}{V_1} \iiint \mathbf{B}_2 \, d^3\mathbf{r} = \frac{1}{V_1} \iiint \text{curl } \mathbf{A}_2 \, d^3\mathbf{r}$$

$$= \frac{1}{V_1} \oiint \hat{\mathbf{n}} \times \mathbf{A}_2 \, dS = \frac{\mu_0}{4\pi V_1} \oiint \hat{\mathbf{n}} \times \frac{\mathbf{m} \times \mathbf{R}}{R^3} \, dS$$

(In the next-to-last step we have used the cross-product analog of the divergence theorem and in the last step have inserted the "distant" vector potential of the dipole.) The integrand works out to

$$\frac{m}{a^2} \sin \theta (\hat{\mathbf{a}}_r \times \hat{\mathbf{a}}_\phi) = \frac{m}{a^2} \sin \theta (-\cos \theta \, \hat{\mathbf{a}}_s + \sin \theta \, \hat{\mathbf{k}})$$

The first term integrates to zero over $d\phi$, leaving

$$\bar{\mathbf{B}}_2 = \frac{\mu_0 \mathbf{m}}{4\pi V_1} \int_0^\pi 2\pi \sin^3 \theta \, d\theta$$

$$= \frac{2\mu_0 \mathbf{m}}{3V_1} \tag{6-8}$$

For a dipole located at an arbitrary point within the sphere, the calculation gets quite complicated, and we shall merely indicate the procedure. Choose the z axis in the direction from the center to the dipole and the xz plane parallel to the dipole. Then the vectors in the integrand are

$$\mathbf{m} = m_1 \hat{\mathbf{i}} + m_3 \hat{\mathbf{k}}$$

$$\mathbf{R} = (a \cos \theta - r_0)\hat{\mathbf{k}} + a \sin \theta (\hat{\mathbf{i}} \cos \phi + \hat{\mathbf{j}} \sin \phi)$$

The integrals are straightforward but tedious, and the results work out to be the same as Eq. (6-8).

For the total contents of the sphere we simply sum all the interior dipoles to get the total moment MV_1. Thus the desired average value is

$$\mathbf{B}_2 = \tfrac{2}{3}\mu_0\mathbf{M} \tag{6-9}$$

On comparing with Eq. (6-7), we see that the correction terms again cancel, as in the electrostatic case. However, the individual terms are different in the two cases, $\mathbf{B}_2$ being in the same direction as $\mathbf{M}$, whereas $\mathbf{E}_2$ is opposite to $\mathbf{P}$. This reflects the complete difference in the form of the near fields of the two kinds of dipoles, as was emphasized at the end of Section 2-6.

Finally, we note that the interior field, although given correctly by the effective currents, involves some approximations and designates an *average over a region much larger than the molecules*.

6-2 MAGNETIC INTENSITY FIELD **H**

We have seen in Section 6-1 that **B** is determined by the true currents plus the magnetization currents. In the static vacuum case, Ampère's circuital law [Eq. (2-23)] discloses that curl **B** is given by the volume density of sources. Hence, in the presence of magnetized matter this must be generalized to include *all* sources, that is,

$$\operatorname{curl}\mathbf{B}_S = \mu_0(\mathbf{J} + \mathbf{J}_M) = \mu_0(\mathbf{J} + \operatorname{curl}\mathbf{M}) \tag{6-10}$$

or

$$\operatorname{curl}\mathbf{H} = \mathbf{J} \tag{6-11}$$

where

$$\mathbf{H}(\mathbf{r}) \equiv \frac{\mathbf{B}(\mathbf{r})}{\mu_0} - \mathbf{M}(\mathbf{r}) \tag{6-12}$$

H is called the *magnetic intensity*. Obviously, at any point in vacuum **H** reduces to $\mathbf{B}/\mu_0$.

The MKS unit of **H** is *amperes per meter*, the same as magnetization **M** or surface current density. The quotient B/H obviously has the same units as μ_0, namely *newtons per ampere²*, or *joules per ampere²-meter*, or *henries per meter*. In the cgs system, the unit of **H** is the *oersted*, which is essentially the same as the gauss (unit of **B**) since in this system **B** and **H** are dimensionally equal. Units and dimensions will be discussed more thoroughly in Chapter 7.

Example 6-7

Calculate **B** and **H** at points on the axis of a bar magnet with uniform magnetization **M**.

Solution   Since we now know that the effective currents give the field inside the matter as well as outside, Example 6-3 holds at all points of the axis. Thus, since $\alpha_1' = \alpha_2' = 0$ in the present example,

$$\mathbf{B}(0, 0, z) = \frac{\mu_0 M}{2} (\cos \alpha_1 + \cos \alpha_2)\hat{\mathbf{k}}$$

$$\mathbf{H}(0, 0, z) = \begin{cases} \dfrac{M}{2} (\cos \alpha_1 + \cos \alpha_2 - 2)\hat{\mathbf{k}} & \text{in the bar} \\[2mm] \dfrac{M}{2} (\cos \alpha_1 + \cos \alpha_2)\hat{\mathbf{k}} & \text{outside} \end{cases}$$

Note that $\mathbf{B}$ is continuous, while $\mathbf{H}$ has discontinuities at the ends of the bar, as shown in Figure 6-4, such that $\mathbf{H}$ is directed *opposite* to $\mathbf{B}$ and $\mathbf{M}$ within the bar.

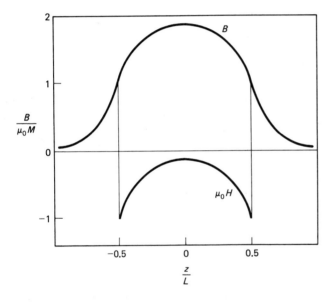

Figure 6-4   Magnetic field at points on axis of bar magnet with length/radius ratio of 5. Inside the magnet, $\mu_0\mathbf{H} = \mathbf{B} - \mu_0\mathbf{M}$; outside, $\mu_0\mathbf{H} = \mathbf{B}$.

Example 6-8

Calculate $\mathbf{B}$ and $\mathbf{H}$ at the center of a uniformly magnetized sphere.

Solution   Equation (6-9) gives $\mathbf{B}$. Then from Eq. (6-12)

$$\mathbf{H} = -\tfrac{1}{3}\mathbf{M}$$

Example 6-9

A thin circular disk is uniformly magnetized perpendicular to its plane. Calculate $\mathbf{H}$ at the center.

Solution   The disk may be regarded as a very short solenoid. Thus

$$\mathbf{B} \cong 0 \qquad \mathbf{H} \cong -\mathbf{M}$$

It is instructive to compare the last three geometries as we go from a long rod to a sphere to a disk. Table 6-1, which should be compared with Table 4-1, summarizes the results.

TABLE 6-1 FIELDS IN UNIFORMLY MAGNETIZED BODIES

	Thin disk	Sphere	Long needle
B	0	$\frac{2}{3}\mu_0\mathbf{M}$	$\mu_0\mathbf{M}$
H	$-\mathbf{M}$	$-\frac{1}{3}\mathbf{M}$	0

NOTE: These are the fields due to the magnetization. Any external fields present must be added. $\mathbf{M} = M_0\hat{\mathbf{k}}$, assumed fixed.

6-3 MAGNETIC RESPONSE

This section is the analog of Section 4-4, but there are some differences and it would be advisable to reread the latter carefully at this point. We again want to calculate a *local field*, that is, the field acting on a given molecule due to all the true currents plus all the *other* molecules. We again assume a spherical molecule, and subtract from **B** the contribution of the molecule itself. From Eq. (6-9) this is $\frac{2}{3}\mu_0\mathbf{M}$. Hence

$$\mathbf{B}_{\text{loc}} = \mathbf{B} - \tfrac{2}{3}\mu_0\mathbf{M} \tag{6-13}$$

We now come to a somewhat subtle point. To obtain the local value of the magnetic intensity **H**, we must keep in mind that $\mathbf{B}_{\text{loc}}$ is a field *in vacuum* (i.e., in the *cavity* obtained by removing the molecule in question). Thus

$$\mathbf{H}_{\text{loc}} = \frac{\mathbf{B}_{\text{loc}}}{\mu_0}$$

$$= \frac{1}{\mu_0}[\mu_0(\mathbf{H} + \mathbf{M}) - \tfrac{2}{3}\mu_0\mathbf{M}]$$

$$= \mathbf{H} + \tfrac{1}{3}\mathbf{M} \tag{6-14}$$

Linear Magnetic Response

As with dielectrics, linear response occurs when the molecular moment is *proportional* to the local field, that is,

$$\mathbf{m}_m = \alpha_m\mathbf{B}_{\text{loc}} = \alpha_m\mu_0\mathbf{H}_{\text{loc}} \tag{6-15}$$

The magnetic moment per unit volume is

$$\mathbf{M} = n\mathbf{m}_m = n\alpha_m(\mathbf{B} - \tfrac{2}{3}\mu_0\mathbf{M}) = n\alpha_m\mu_0(\mathbf{H} + \tfrac{1}{3}\mathbf{M})$$

On solving, we obtain

$$\mathbf{M} = \frac{n\alpha_m}{1 + \tfrac{2}{3}n\alpha_m\mu_0}\mathbf{B} = \frac{n\alpha_m\mu_0}{1 - \tfrac{1}{3}n\alpha_m\mu_0}\mathbf{H} \tag{6-16}$$

The relation between **B** and **H** for a linear magnetic material then follows from Eq. (6-12) as

$$\mathbf{H} = \frac{\mathbf{B}}{\mu_0} - \mathbf{M} = \left(\frac{1}{\mu_0} - \frac{n\alpha_m}{1 + \frac{2}{3}n\alpha_m\mu_0}\right)\mathbf{B} \tag{6-17}$$

or, alternatively,

$$\mathbf{B} = \mu_0(\mathbf{H} + \mathbf{M}) = \mu_0\left(1 + \frac{n\alpha_m\mu_0}{1 - \frac{1}{3}n\alpha_m\mu_0}\right)\mathbf{H} \tag{6-18}$$

We see that in a linear material (but not otherwise) **B** and **H** are proportional to each other. Logically, the relation should be written in the form of Eq. (6-17) with **B** appearing in the role of independent variable, since **B** is the true field quantity that gives rise to forces on bodies. However, owing to historical development, it has become customary to write the relation with **H** in the "independent" role. Accordingly, Eqs. (6-16) and (6-18) are written

$$\mathbf{M} = \chi_m\mathbf{H} \qquad \chi_m = \frac{n\alpha_m\mu_0}{1 - \frac{1}{3}n\alpha_m\mu_0} \tag{6-19}$$

$$\mathbf{B} = \mu\mathbf{H} = \kappa_m\mu_0\mathbf{H} \qquad \kappa_m = 1 + \chi_m \tag{6-20}$$

χ_m is called the *magnetic susceptibility*, μ the *permeability*, and κ_m the *relative permeability*. μ has the same dimensionality as μ_0 while the other two are dimensionless.

Example 6-10

A long solenoid is filled with a linear magnetic material. Find **B** and **H** at points on the axis near the middle.

Solution   For simplicity, we assume the solenoid to be long enough so that the subtended angles α_1 and α_2 are very small. **B** arises from both the true and the effective currents so that, from Example 6-7 and Example 2-8,

$$B = \mu_0\left(M + \frac{NI}{L}\right)$$

But $M = \chi_m H = \frac{\chi_m B}{\mu_0(1 + \chi_m)}$, hence

$$B = \left(\frac{\chi_m}{1 + \chi_m}\right)B + \frac{\mu_0 NI}{L}$$

Solving gives

$$B = \frac{(1 + \chi_m)\mu_0 NI}{L} = \frac{\mu NI}{L}$$

$$H = \frac{B}{\mu} = \frac{NI}{L}$$

Note that the magnetic flux, and thus the inductance of the coils, is changed by the factor κ_m.

The foregoing example is a bit deceptive in that **H** appears to be determined directly by the true current alone. The reason for this simplicity is that the true and effective currents happen to coincide in space. The *magnetic circuit* concept, often used in discussing devices such as transformer cores, is based on this type of approximation. We shall discuss it briefly in Section 12-5. More general geometries involve boundary value problems, which we shall also discuss in Chapter 12.

6-4 MECHANISMS OF MAGNETIC RESPONSE

Diamagnetism

We saw in Example 6-1 that an atomic electron will generally have a magnetic moment due to its orbital motion. Each electron also has an intrinsic *spin*, which produces a magnetic moment one Bohr magneton (9.3×10^{-24} A-m²) in magnitude. The atomic nuclei may have additional magnetic moments, but these are smaller than the electronic moments by factors of 1000 or more, and can thus usually be disregarded. The molecular dipole moment is thus, very nearly, the vector sum of the moments (orbital and spin) of the electrons in the molecule. Diamagnetic materials are those in which the molecular moment is *zero in the absence of a field*. By far the majority of substances fall into this class, since electrons generally tend to "pair off" so that both their spin and orbital moments cancel.

Diamagnetism results from *changes in the orbital motions* resulting from the force of the local magnetic field on the moving electrons. To get an idea of its magnitude, we again consider a highly simplified model, namely an electron in a circular orbit around a nucleus, with a magnetic field applied parallel to the orbital magnetic moment. From the geometry shown in Figure 6-5, it is seen that the magnetic force is radially outward. Hence the centrifugal acceleration equation is

$$m_e \omega^2 r = \frac{Ze^2}{4\pi\epsilon_0 r^2} - e\omega r B$$

where m_e is the electron mass and Ze is the nuclear charge. As a first approximation, we take r to have its original value r_0 and put the original value of the angular velocity ω_0 in the small correction term. The equation thus becomes

$$\omega^2 r_0 = \omega_0^2 r_0 - \frac{e\omega_0 r_0}{m_e} B$$

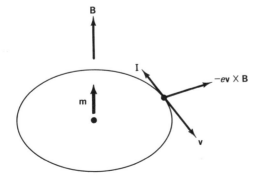

Figure 6-5 Magnetic force on an electron in a circular orbit around a nucleus.

Thus

$$\omega^2 - \omega_0^2 \equiv (\omega + \omega_0)(\omega - \omega_0) = -\frac{eB}{m_e}\omega_0$$

$$\Delta\omega \equiv \omega - \omega_0 \cong -\frac{eB}{2m_e} \tag{6-21}$$

(*Note*: To verify the validity of the approximation used, the result may be put back into the equation of motion and the latter then solved for r; it turns out that the change in r is of *second order* in the small quantity $\Delta\omega/\omega_0$.)

Equation (6-21) shows that the motion is *slowed down* by the field. In other words, the orbital moment is *reduced* by a parallel field. Since the moment (see Example 6-1) is

$$m = \frac{ev}{2\pi r}(\pi r^2) = \frac{erv}{2} = \frac{er^2\omega}{2}$$

the change in the moment is

$$\Delta\mathbf{m} = \frac{er^2}{2}\Delta\omega = -\frac{e^2}{4m_e}r^2\mathbf{B} \tag{6-22}$$

Example 6-11

Derive the preceding result from Lenz's law (Section 3-3).

Solution   Imagine the field to be built up from zero to its final value. During this process, there is an induced electric field given by

$$2\pi r E_\phi = -\pi r^2 \frac{dB}{dt}$$

This electric field accelerates the electron according to

$$m_e \frac{dv_\phi}{dt} = -eE_\phi = \frac{er}{2}\frac{dB}{dt}$$

The change in v is thus in the positive ϕ direction (i.e., opposite to v) and the change in ω comes out the same as Eq. (6-21).

In the foregoing example, a parallel field decreased the magnitude of **m**. Obviously, an antiparallel field would increase it. In both extreme cases, therefore, the *change* in **m** is opposite to the field. Thus for a pair of electrons whose moments initially cancel, the result is a net *induced moment* directed opposite to the applied field. Each pair contributes twice the amount of Eq. (6-22). Thus for a molecule containing several electron pairs with various orbit radii,

$$\mathbf{m}_{\text{ind}} = \sum_i er_i\,\Delta v_i = \sum_i er_i^2\,\Delta\omega_i$$

$$= -\left(\frac{e^2}{2m_e}\sum_i r_i^2\right)\mathbf{B} \tag{6-23}$$

The quantity in parentheses is seen to be the coefficient α_m of Eq. (6-15).

Example 6-12

Estimate a typical value for the coefficient α_m in Eq. (6-23).

Solution   Typical outer orbit radii are of the order of 1 Å, and there are generally only a few electron pairs with orbits this large. Hence

$$\alpha_m \cong -\frac{e^2}{2m_e}r^2 = \frac{(1.6 \times 10^{-19})^2}{2 \times 10^{-30}} \times (10^{-10})^2 \cong -10^{-28} \text{ A-m}^2/\text{T}$$

Example 6-13

Estimate a typical value of the diamagnetic susceptibility of a solid.

Solution   In Eq. (6-19) the correction term in the denominator will be negligible, so

$$\chi_m = n\alpha_m\mu_0 \cong -10^{-6}$$

Thus, although some diamagnetism is always present, since all atoms and molecules have orbiting paired electrons, it is usually very small and may easily be overshadowed. It is a *negative* magnetic effect as indicated by the sign of the result. As we shall see in Chapter 9, diamagnetic materials are *repelled* by magnets.

Paramagnetism

Certain atoms and ions have permanent magnetic dipole moments. When this occurs due to unpaired outer electrons, there is a strong tendency to form chemical bonds so that the electrons can pair off. However, in elements of either a transition series or a rare earth series in the periodic table, atoms may have *incompletely filled inner shells* of electrons. According to quantum mechanics, such electrons will tend to align their spin and orbital moments as nearly parallel as possible, rather than pairing off as do the outer ones. Net moments as large as about 5 Bohr magnetons then occur.

The behavior of the permanent dipoles in a field is much the same as in the electric dipole case (Section 4-4). The local field tends to align them and thermal agitation tends to randomize them. There is a slight difference in that the magnetic dipole, being associated with an angular momentum, is restricted to certain discrete orientations relative to the field (again a quantum effect). As a result, the average moment component in the field direction is

$$\bar{m} = m_0 \mathcal{B}(X) \tag{6-24}$$

where $X = m_0 B_{\text{loc}}/k_B T$, and $\mathcal{B}$ is one of a family of functions called *Brillouin functions*, all very similar in shape to the Langevin function (4-25). In particular, for small values of X, $\mathcal{B}$ reduces to $\frac{1}{3}X$, just like the Langevin function. Thus there is a positive susceptibility inversely proportional to temperature, a relation known as *Curie's law*. In analogy with Eq. (4-26),

$$\bar{\mathbf{m}} = \frac{m_0^2}{3k_B T}\mathbf{B}_{\text{loc}} \tag{6-25}$$

The coefficient of $\mathbf{B}_{loc}$ is the effective value of α_m, which is then substituted in Eq. (6-19).

Example 6-14

Estimate a rough maximum value for the paramagnetic susceptibility of a solid at room temperature.

Solution   Taking 5 Bohr magnetons as the permanent magnetic moment, we obtain

$$\frac{m_0^2}{3k_B T} = \frac{(5 \times 9 \times 10^{-24})^2}{3 \times 1.4 \times 10^{-23} \times 300} = 10^{-25}$$

$$\chi_m \cong 10^{28} \times 10^{-25} \times 4\pi \times 10^{-7} \cong 10^{-3}$$

Thus paramagnetism, when it occurs, may exceed the diamagnetism by about 1000-fold at room temperature, and even more so at lower temperatures. Of course, weaker paramagnetism can also occur if the permanent moments are present only in small concentration, as often occurs in paramagnetic salts.

6-5 FERROMAGNETISM

Example 6-14 shows that when permanent moments are present, the denominator of Eq. (6-19) could go to zero at a sufficiently low temperature. This means that the moments would be aligned spontaneously, even in the absence of an applied field. The cause of the alignment is the mutual magnetic interaction of the dipoles. From the example it is evident that the temperature would have to be of the order of 0.1 K or lower. In other words, the magnetic interaction is so weak that very slight thermal agitation is able to overcome it.

There exist in nature, however, many materials that are spontaneously magnetized even at temperatures of several hundred degrees. To explain this, it was proposed (by Pierre Weiss in 1907) that some *nonmagnetic* interaction of much greater strength is present and tends to align a given dipole in the same direction as its neighbors. To incorporate this idea into the theory, Eq. (6-14) is replaced by

$$\mathbf{H}_{loc} = \mathbf{H} + \gamma \mathbf{M} \tag{6-26}$$

where γ (called the *Weiss molecular field constant*) is assumed to $\gg 1$. The general nature of the Weiss interaction was explained some 20 years later by Heisenberg as a quantum-mechanical effect; it is basically electrostatic in nature, which is the reason for its great strength. However, many aspects of the interaction are still not well understood and are the subject of active research today.

As the temperature increases, the spontaneous magnetization decreases since a fraction of the moments are disaligned by thermal agitation. It disappears altogether at the *Curie temperature* T_C.

Example 6-15

The Curie temperature of iron is about 1000 K. Estimate the Weiss molecular field constant.

Solution   In analogy with Eqs. (4-27) and (4-28), T_C is the temperature at which $\gamma n \mu_0 m_0 / 3 k_B T = 1$. On inserting reasonable values for n and m_0 (see Example 6-14), we obtain

$$\gamma = \frac{3 k_B T_C}{\mu_0 n m_0^2} = \frac{3 \times 1.4 \times 10^{-23} \times 10^3}{4\pi \times 10^{-7} \times 10^{-28} \times (5 \times 9 \times 10^{-24})^2} \cong 2 \times 10^3$$

This is greatly in excess of the value of order $\frac{1}{3}$ given by the magnetic interaction.

The relation between Curie temperature and molecular field constant may also be written

$$k_B T_C = \frac{\gamma \mu_0 n m_0^2}{3} = \gamma \mu_0 n (\alpha_m k_B T) \tag{6-27}$$

Above T_C there is no spontaneous magnetization, which means that the local field correction term is < 1. The material is then paramagnetic with a susceptibility given by Eq. (6-19) except that the $\frac{1}{3}$ is replaced by γ. Equation (6-25) gives the effective value of α_m, and from Eq. (6-27) this can be written as $T_C / n \gamma \mu_0 T$. Hence

$$\chi_m = \frac{T_C}{\gamma (T - T_C)} \qquad \text{for } T > T_C \tag{6-28}$$

This is known as the *Curie–Weiss law*.

Ferromagnetic Domains

Many ferromagnetic materials (e.g., soft iron) do not behave as permanent magnets on a large scale. Weiss also proposed an explanation for this. As we shall see in Chapter 9, the external field of a magnet contains a considerable amount of energy. By subdividing the magnetization into *domains* (i.e., small regions) the external energy may be reduced. The domain walls are disordered thin layers separating the differently oriented domains. They have, of course, a higher energy density than the perfectly ordered regions, and the subdivision continues until the *total* energy is minimized. The situation is illustrated schematically in Figure 6-6. Domain walls may be rendered visible by allowing suspensions of fine magnetic powders to settle on the material. Typical domain sizes are 10 to 100 μm, while the wall thicknesses are of the order of 100 Å.

When an external field is applied, those domains that happen to be oriented parallel to it are energetically more favored than the others, and the walls tend to move so that these domains grow at the expense of the others. Thus the material as a whole acquires a net magnetic moment (i.e., a large-scale magnetization). However, the wall motion is not entirely free, but is impeded by crystalline defects such as impurities, strains, and so on. As a result, the net magnetization is, at first, roughly proportional to the applied field. In other words, the material behaves in a *quasi-paramagnetic* manner, but with a *very large apparent susceptibility*. In carefully prepared pure iron, for example, κ_m may reach several hundred thousand, and even more in certain special alloys. Of course, once all the domains have been "swallowed up" by the ones initially in the field direction, the susceptibility returns to normal; the material is said to be

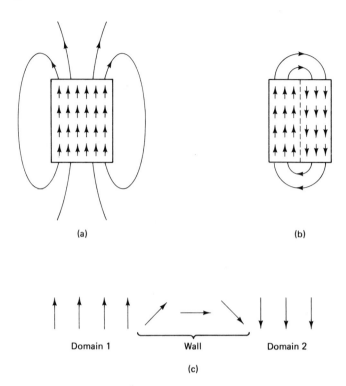

Figure 6-6   (a) External field of a fully aligned bar magnet; (b) same after sub-dividing the magnet into two domains; (c) detail of the domain wall (schematic).

saturated. Finally, on removing the applied field the impediments prevent the domain walls from returning fully to their original positions, which leads to the *hysteresis* effect. The sequence is illustrated in Figure 6-7.

Antiferromagnetism and Ferrimagnetism

In certain materials (usually oxides or other insulating compounds) the interaction between neighboring dipoles has the opposite sign, so that adjacent moments tend to align antiparallel to each other. Again there is a critical temperature (the *Neel temperature*) T_N, above which the thermal agitation overcomes the interaction. The material behaves in a paramagnetic manner, but the susceptibility decreases below T_N, as the moments tend to lock into alignment and resist rotation by the field.

If two unequal kinds of moments are present, there will be a net magnetization even under complete antiparallel ordering. Such materials are called *ferrites.* The best known material of this type is the magnetic oxide of iron, $FeO \cdot Fe_2O_3$, the famous *lodestone* known even in antiquity. The combination of spontaneous magnetization along with very low conductivity gives these materials considerable technological importance in high-frequency circuitry.

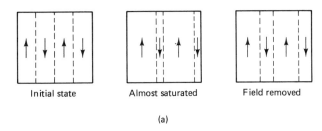

| Initial state | Almost saturated | Field removed |

(a)

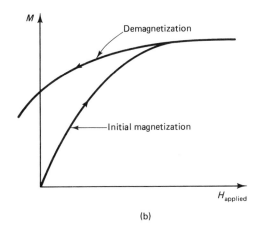

(b)

Figure 6-7 (a) Domain configurations; (b) net magnetization during magnetization and demagnetization of a soft ferromagnet.

Lest it be thought that these few types exhaust the subject, we mention that there are a myriad of other complex and beautiful magnetic alignment patterns known, and that magnetism is still an active and vital field of research.

6-6 MAGNETIC EFFECTS OF FREE ELECTRONS

Pauli Paramagnetism

Due to the spin magnetic moment, the free electrons in a metal constitute a dense collection of permanent magnetic dipoles. According to quantum mechanics, these cannot assume arbitrary orientations but are restricted to one of two possibilities: The component in the direction of any magnetic field can be only ± 1 Bohr magneton. These result in a temperature-independent paramagnetism (called *Pauli paramagnetism*). However, the effect is very small and is often masked by the diamagnetism of the inner electron shells.

Diamagnetism

A free electron moving perpendicular to a magnetic field will be deflected into a circular orbit and thus will have a magnetic moment. In accordance with Lenz's law, this will be opposite to the field, and thus at first glance we might expect metals (with a great many free electrons) to be strongly diamagnetic. However, the situation is really much more complicated. When the quantum mechanics of the motion in a finite-size sample is properly taken into account, it turns out that there is only a very small magnetization which *alternates in direction* as the field is increased. This is called the *DeHaas–VanAlphen effect*. It is a powerful tool for the study of the detailed behavior of electrons in real metals.

Superconductivity

This remarkable and important effect was first discovered as an anomaly in conductivity but is basically magnetic in nature. In 1911, K. Onnes observed that the resistance of mercury went abruptly to zero when the temperature was reduced below 4.15 K. In a weak magnetic field the transition was repressed to a lower temperature, and in a field of more than about 0.05 T it did not occur at all. Since then about two dozen elements and many alloys have been found to display similar behavior with transition temperatures ranging from a few hundredths of a degree to more than 20 K. The general form of the effect is illustrated in Figure 6-8.

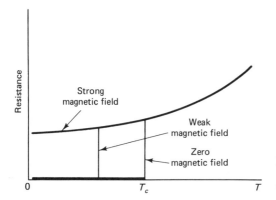

Figure 6-8   Resistance of a superconductor as function of absolute temperature and magnetic field.

The central magnetic aspect was discovered in 1933 by Meissner and Ochsenfeld: On cooling a sphere through the transition temperature in a weak field, the magnetic field was *repelled out of the sample*, as shown in Figure 6-9. This is called the *Meissner*

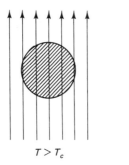
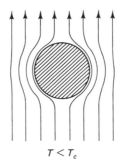

Figure 6-9   Field lines of a weak externally applied magnetic field in the vicinity of a superconductive sphere.

effect. Above some critical value H_c of the applied field, the Meissner effect, and with it the superconductivity, is destroyed as the magnetic field again penetrates the material. In some superconductors (called type II) there is only a partial penetration, so that superconductivity persists. We shall discuss this later.

Example 6-16

A long solenoid of radius b is wound coaxially around a type I superconducting cylinder of radius a (Figure 6-10a). Find the magnetic field at all points both (a) above and (b) below the transition temperature T_c (Figure 6-10b and c).

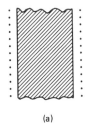

(a)

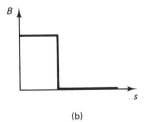

(b)

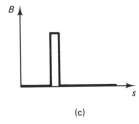

(c)

Figure 6-10 (a) Geometry for Example 6-16: bar of type I superconductor in a solenoid; (b) magnetic field distribution for $T > T_c$; (c) magnetic field distribution for $T < T_c$.

Solution

(a) Above T_c, the material is a normal metal. We may neglect the very small diamagnetic and Pauli paramagnetic susceptibilities, so that from Example 2-12,

$$B = \begin{cases} \mu_0 \dfrac{N}{L} I & \text{for } 0 < s < b \\ 0 & \text{for } s > b \end{cases}$$

(b) Below T_c, B must be zero in the material. However, the field in the empty space must be unchanged, as can be seen from Ampère's circuital law [Eq. (2-24)] applied to a contour encircling part of the winding. Thus

$$B = \begin{cases} \mu_0 \dfrac{N}{L} I & \text{for } a < s < b \\ 0 & \text{for } s < a \text{ and } s > b \end{cases}$$

Example 6-17

Find a distribution of magnetization that would account for the preceding results.

Solution Suppose that $M = -(N/L)I$ for $0 < s < a$ (i.e., in the superconductor) and $H = (N/L)I$ for $0 < s < b$ (i.e., everywhere within the coil). Then **H** satisfies curl $\mathbf{H} = \mathbf{J}$, as required by Eq. (6-11) and $\mathbf{B} = \mu_0(\mathbf{H} + \mathbf{M})$ everywhere as required by Eq. (6-12).

Thus type I superconductors may be described as *perfect diamagnets* ($\mathbf{M} = -\mathbf{H}$). However, it is only an *interpretation*. There is no way of ascertaining the value of **H** inside the material.

Example 6-18

Find a distribution of currents that would also account for the results of Example 6-16.

Solution The **M** field given in Example 6-17 has an effective surface current density [Eq. (6-5)] of $-(NI/L)\hat{\mathbf{a}}_\phi$. The same **B** field would be produced by a *real* surface current of the same density. In this case, **H** would be zero inside the material, but **B** would be the same as before everywhere.

Thus an alternative interpretation of the Meissner effect is that the field in the material is *canceled by surface currents*. In either interpretation, the magnetic behavior is illustrated in Figure 6-11. As the magnetizing current is increased from zero, the magnetic field in the material remains zero until **H** reaches the critical value H_c. At this point, B jumps to the normal-state value, essentially $\mu_0 H_c$, then continues to rise lineally as H is further increased. H_c is, of course, a function of the temperature, increasing to a finite limit as $T \rightarrow 0$.

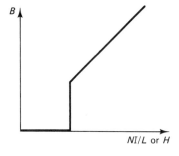

Figure 6-11 Magnetic field B in a long bar of type I superconductor as a function of the current in a surrounding solenoid.

Example 6-19

The cylinder of Example 6-16 is hollow and the following sequence of operations is carried out: (1) Current is started in the solenoid while the superconductor is in the normal state at a temperature above T_c, (2) the material is cooled below T_c, and (3) current in the solenoid is stopped. Calculate the field and surface current distribution at each stage.

Solution

1. The field is uniform everywhere within the solenoid.
2. The field is excluded from the material by sheets of surface current, flowing opposite to the solenoid current on the outer surface and in same sense as the solenoid on the inner surface of the superconductor.
3. The field goes to zero outside the cylinder as well as in the material. The outer surface current also goes to zero; otherwise, it would produce a field in the material. However, the inner surface current *continues to flow* and sustains the original field in the inner space.

Persistent currents of this type have been observed to flow for several years with no measurable decay. This provides evidence that the resistance in the superconducting state is very small indeed and may be truly zero!

The foregoing examples show that "supercurrents" are entirely different from ordinary conduction currents in that they are caused primarily by *magnetic* rather than electric fields. In effect, **H** acts as a *source density* for **J**, which flows in loops around it. To describe this relationship, F. and H. London proposed the equation

$$\text{curl } \mathbf{J} = -\frac{1}{L_L^2}\mathbf{H} \tag{6-29}$$

where L_L is a characteristic length called the *London penetration depth*. Essentially, the equation says that curl **J** is proportional to the source strength **H**. The constant of proportionality is written as an inverse area to make the dimensionality come out right. Note that the equation describes the *response of the material*. It does not replace or vitiate any of the Maxwell equations. Other than this, the material is assumed to have essentially no magnetic response, so that $\mathbf{B} = \mu_0\mathbf{H}$ at every point.

To show that these relations lead to superconductive behavior, we take the time-derivative of the London equation and insert Faraday's law to obtain

$$\text{curl } \dot{\mathbf{J}} = -L_L^{-2}\dot{\mathbf{H}} = -\mu_0^{-1}L_L^{-2}\dot{\mathbf{B}}$$

$$= \mu_0^{-1}L_L^{-2} \text{ curl } \mathbf{E}$$

An obvious solution is

$$\dot{\mathbf{J}} = \mu_0^{-1}L_L^{-2}\mathbf{E} \tag{6-30}$$

Thus the *rate of increase* of the current density is proportional to the field, as if the carriers were continuously accelerated without the frictional effects that lead to finite conductivity in normal metals.

To show that the London equation also predicts the Meissner effect (exclusion of magnetic flux), we take its curl and insert the Maxwell equation (6-11),

$$\text{curl curl } \mathbf{J} = -L_L^{-2} \text{ curl } \mathbf{H} = -L_L^{-2} \mathbf{J}$$

Since div $\mathbf{J} = 0$ (for static conditions), this is

$$\nabla^2 \mathbf{J} = L_L^{-2} \mathbf{J} \tag{6-31}$$

The nature of the behavior is seen most easily in a one-dimensional case, where the equation reduces to

$$\frac{d^2 \mathbf{J}}{dz^2} = L_L^{-2} \mathbf{J}$$

so that

$$\mathbf{J}(z) = \mathbf{J}(0) \exp\left(-\frac{z}{L_L}\right) \tag{6-32}$$

Thus, from (6-29) again,

$$\mathbf{H} = -L_L^2 \text{ curl } \mathbf{J}$$

$$= \mathbf{H}(0) \exp\left(-\frac{z}{L_L}\right) \tag{6-33}$$

where $\mathbf{H}(0) = \mathbf{J}(0) \times \hat{\mathbf{n}}$. Thus both the current density and the magnetic field decay exponentially with distance into the material, and the characteristic decay length is just the London penetration depth L_L. If this be small enough, the field is excluded from practically the entire volume of the material.

To estimate the magnitude of L_L, we consider the dynamics of the superconducting electrons. Since these move without friction their equation of motion is simply

$$m_e \dot{\mathbf{v}} = -e\mathbf{E}$$

If there are n_s such electrons per unit volume, the equation of motion for the supercurrent is

$$\dot{\mathbf{J}} = -en_s \dot{\mathbf{v}} = \frac{n_s e^2}{m_e} \mathbf{E} \tag{6-34}$$

On comparing with Eq. (6-30), it is seen that

$$L_L = \left(\frac{m_e}{\mu_0 n_s e^2}\right)^{1/2} \tag{6-35}$$

The number of electrons n_s in the superconducting state increases rapidly as the temperature is lowered below T_c and approaches the total electron concentration as $T \rightarrow 0$.

Example 6-20

Estimate the value of the London penetration depth in a typical superconducting metal at $T = 0$.

Solution

$$L_L = \left[\frac{9 \times 10^{-31}}{4\pi \times 10^{-7} \times 10^{28} \times (1.6 \times 10^{-19})^2} \right]^{1/2} \cong 5 \times 10^{-8} \text{ m}$$

This is so small that it is not discernible in measurements on bulk samples, so the Meissner effect is essentially complete.

The underlying mechanism of superconductivity was explained in 1957 by Bardeen, Cooper, and Schrieffer. The central feature of the so-called *BCS theory* is that the electrons tend to associate into *pairs* with a binding energy of the order of $k_B T_c$. The ordinary collision processes do not involve nearly enough energy to disrupt the pairs, so the supercurrents are not subject to the usual frictional processes that limit ordinary conduction currents.

The pairing introduces another important length parameter into the problem, namely the *correlation length* ξ_0. This is, roughly speaking, the distance over which the pair partners influence each others' motions. Its value depends on the electron concentration and the pair binding energy, and may range from some tens of angstroms (in transition metals and certain alloys) to several thousand angstroms (in aluminum). A major effect of the correlation is that no significant variation of any quantity (e.g., field or current density) can occur on a scale smaller than ξ_0. This means that the London equation can be valid only if $\xi_0 \ll L_L$. Materials in which this is true are called *type II superconductors*. The results derived in the last few examples are valid for these materials. For the *type I superconductors* (discovered earlier), similar field and current patterns pertain, but the length scale of variation is ξ_0 rather than L_L.

One overwhelmingly important effect of type II behavior is that the initial breakdown of the Meissner effect is *incomplete*. This means that the resistanceless current flow is much more stable and not nearly so easily destroyed by magnetic fields. Thus type II superconductors have been used to build huge magnets producing fields of several tens of teslas while consuming no power whatever! The magnetic response of type II superconductors is illustrated in Figure 6-12. Magnetic flux is completely excluded (for cylindrical geometry as in our previous examples) up to a *lower critical field* H_{c1}. For $H > H_{c1}$, there is flux penetration, but the field is extremely nonuni-

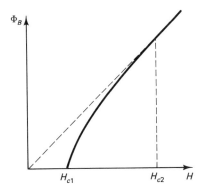

Figure 6-12 Magnetic flux in a long bar of type II superconductor as a function of the current in a surrounding solenoid.

form, and all the flux is concentrated in tiny filaments, each surrounded by a *current vortex*. The material between the vortices remains in the superconductive state, so the electrical resistance remains zero. As *H* is increased further, the vortices get more and more crowded until at the *upper critical* field H_{c2} they occupy the entire volume and bulk superconductivity is destroyed. Even then, however, there is still a superconductive surface sheath, which finally disappears at a still higher field H_{c3}.

6-7 MAGNETIC EFFECTS AT HIGH FREQUENCIES

The time-dependent behavior of a magnetic dipole in a magnetic field is considerably more complicated than that of an electric dipole in an electric field. In both cases, the field exerts a torque on the dipole. But the magnetic dipole is always associated with an angular momentum, and its response to the torque is profoundly affected thereby. Instead of simply rotating toward the field direction it *precesses*, which means that its direction moves on a cone about the field direction. This motion obviously leaves the component along the field direction completely unchanged. However, there are also frictional processes which dissipate the magnetic energy and allow the dipoles to "relax" toward the field direction. The time lag again leads to a "generalized linear" response, with a complex susceptibility, quite analogous to the dielectric behavior in Section 4-4. A variety of *magnetic resonance* effects can also occur when an alternating field is applied at or near the precessional frequency. These have many important applications, but we cannot go into the details here.

The main point we wish to make is that at sufficiently high frequency (e.g., infrared or higher) *none* of the *strong* magnetic mechanisms are able to follow the field. In other words, the susceptibility is essentially zero, and μ is essentially equal to μ_0.

FURTHER REFERENCES

R. S. ELLIOTT, *Electromagnetics*, McGraw-Hill Book Company, New York, 1966, Chap. 7.

C. KITTEL, *Introduction to Solid State Physics*, 3rd ed., John Wiley & Sons, Inc., New York, 1966, Chaps. 11, 14, and 15.

A. J. DEKKER, *Solid State Physics*, Prentice-Hall, Inc., Englewood Cliffs, N.J., 1957, Chaps. 18 and 19.

A. NUSSBAUM, *Electromagnetic and Quantum Properties of Materials*, Prentice-Hall, Inc., Englewood Cliffs, N.J., 1966, Chap. 6.

F. LONDON, *Superfluids*, Vol. 1, Dover Publications, Inc., New York, 1959.

P. G. DE GENNES, *Superconductivity of Metals and Alloys*, W. A. Benjamin, Inc., New York, 1966.

PROBLEMS

6-1. A sphere of radius a has a uniform magnetization $\mathbf{M}_0$. Calculate the dipole moment of the sphere by use of the effective current distribution.

6-2. A semi-infinite block with surface in the xy plane has a uniform magnetization $M_0(\hat{\mathbf{k}} \cos \theta + \hat{\mathbf{i}} \sin \theta)$, where θ is any fixed angle. Find the effective current distribution.

6-3. A sphere of radius a centered at the origin has a nonuniform magnetization $M = Cx^2\hat{\mathbf{k}}$. Find the effective current distribution.

6-4. A spherical shell of radii a and b has a uniform magnetization $M_0\hat{\mathbf{k}}$. Find $\mathbf{B}$ and $\mathbf{H}$ at the center.

6-5. An infinitely long cylinder of radius a with axis along z is magnetized uniformly perpendicular to its length, so $M = M_0\hat{\mathbf{i}}$. Find $\mathbf{B}$ and $\mathbf{H}$ at points on the axis.

6-6. A linear magnetic material of permeability μ has a uniform field $\mathbf{B}_0$ within it. Calculate the magnetization.

6-7. For the solenoid of Example 6-10, calculate the effective current distribution.

6-8. An infinitely long thin wire carrying current I is surrounded coaxially by a cylindrical shell, radii a and b, of a linear magnetic material of susceptibility χ. Find $\mathbf{B}$ and $\mathbf{H}$ everywhere, and also find the distribution of effective currents in the shell.

6-9. Starting from Eqs. (6-24) and (6-26), write a pair of simultaneous equations satisfied by the spontaneous magnetization at any temperature below T_C.

6-10. Estimate the value of the spontaneous magnetization in a ferromagnetic metal at low temperature. Take 2 Bohr magnetons as a molecular moment.

6-11. In an iron sample, after $\mathbf{B}$ is saturated in the upward direction it requires an $\mathbf{H}$ of 2×10^{-3} A/m downward to bring $\mathbf{B}$ back to zero. What is M then?

chapter 7

Maxwell's Equations

7-1 FIELD EQUATIONS IN THE PRESENCE OF MATTER

In Section 3-6 we wrote down the basic laws of electromagnetic fields in vacuum. These are experimental laws but are believed to be exact (for macroscopic phenomena). In Chapters 4, 5, and 6 we discussed fields in various kinds of matter. Some of the laws had to be modified and replaced by equations that were only approximately correct. For convenience, we recapitulate them here. We shall have frequent occasion to write partial derivatives with respect to time, and for conciseness will use the dot notation, that is,

$$\dot{F}(\mathbf{r}, t) \equiv \frac{\partial}{\partial t} F(\mathbf{r}, t)$$

The equations are

$$(\text{I}) \qquad \text{div } \mathbf{D}(\mathbf{r}, t) = \rho(\mathbf{r}, t) \qquad\qquad (4\text{-}14)\ (7\text{-}1)$$

$$(\text{II}) \qquad \text{curl } \mathbf{E}(\mathbf{r}, t) = -\dot{\mathbf{B}}(\mathbf{r}, t) \qquad\qquad (3\text{-}9)\ (7\text{-}2)$$

$$(\text{III}) \qquad\qquad \text{div } \mathbf{B} = 0 \qquad\qquad\qquad (2\text{-}16)\ (7\text{-}3)$$

$$(\text{IV}) \qquad \text{curl } \mathbf{H}(\mathbf{r}, t) = \mathbf{J}(\mathbf{r}, t) + \dot{\mathbf{D}}(\mathbf{r}, t) \qquad\qquad (6\text{-}11)\ (7\text{-}4)$$

(II) and (III) are the same as in vacuum and are still exact; (I) and (IV) are approximate because in their derivations the molecules of the matter were treated as ideal dipoles (i.e., their higher multipole moments were ignored in calculating the fields they produced).

There are a total of 16 variables: the 15 components of the five vectors $\mathbf{E}$, $\mathbf{D}$, $\mathbf{B}$, $\mathbf{H}$, and $\mathbf{J}$, and the scalar ρ. If the source densities ρ and $\mathbf{J}$ are regarded as "given," there still remain 12 "unknown" functions. There are, however, only eight equations, the second and fourth being vector equations with three components each.

Obviously, additional relations are needed to make the system determinate. These are the *constitutive* relations describing the properties of the media in which the fields exist. The general form is to specify **P**, **M**, and **J**, and therefore also **D** and **H**, as functions of **E** and **B**.

Example 7-1

What form do the Maxwell equations take in a *completely linear* medium?

Solution   The constitutive relations take the form $\mathbf{D} = \epsilon\mathbf{E}$, $\mathbf{B} = \mu\mathbf{H}$, $\mathbf{J} = g\mathbf{E}$. Hence the Maxwell equations become

$$\text{div } \mathbf{E} = \frac{\rho}{\epsilon} \tag{7-5}$$

$$\text{curl } \mathbf{E} = -\dot{\mathbf{B}} \tag{7-6}$$

$$\text{div } \mathbf{B} = 0 \tag{7-7}$$

$$\text{curl } \mathbf{H} = g\mathbf{E} + \epsilon\dot{\mathbf{E}} \tag{7-8}$$

Example 7-2

Show that the electric and magnetic fields are "decoupled" under static conditions.

Solution   If all time derivatives are zero, the Maxwell equations fall apart into two pairs:

$$\text{curl } \mathbf{E} = 0 \qquad\qquad \text{div } \mathbf{B} = 0$$
$$\text{div } \mathbf{D} = \rho \qquad\qquad \text{curl } \mathbf{H} = \mathbf{J}$$

A general solution of either of these pairs of equations can be found only under quite restricted conditions. The medium must be *linear* and *homogeneous*. The two fields (either **E** and **D** or **B** and **H**) are then proportional (linearity) and the proportionality factor (either ϵ or μ) is the same at every point (homogeneity). The first pair, for example, becomes

$$\text{curl } \mathbf{E} = 0 \qquad \text{div } \mathbf{E}(\mathbf{r}) = \frac{\rho(\mathbf{r})}{\epsilon}$$

and the general solution is, from Chapter 1,

$$\mathbf{E}(\mathbf{r}) = \frac{1}{4\pi\epsilon} \iiint \frac{\rho(\mathbf{r}')(\mathbf{r} - \mathbf{r}')}{|\mathbf{r} - \mathbf{r}'|^3} \, d^3\mathbf{r}'$$

If ϵ were not a constant (e.g., if the medium were vacuum in some places and dielectric material in other places), it could not be factored out of div **D**. Thus we would not know both the divergence and curl of either **E** or **D** and could not write down an integral solution for either. Obviously, the most interesting cases fall into this category, and we shall explore various methods of finding particular solutions in Chapters 11, 12, and 14.

Example 7-3

Derive the current continuity equation (2-6) from the Maxwell equations.

Solution   Take the divergence of the fourth Maxwell equation, then substitute the first, to obtain

$$0 = \text{div } \mathbf{J} + \text{div } \dot{\mathbf{D}} = \text{div } \mathbf{J} + \dot{\rho}$$

Thus current continuity (charge conservation) is built into the system and does not constitute a separate law.

Example 7-4

Consider a medium that is completely linear but also contains some externally controlled distribution of charges and currents, ρ_0 and $\mathbf{J}_0$ (an example would be a pair of metal plates connected to an ac generator and embedded in the medium). Write the Maxwell equations for this situation.

Solution   The total current density at any point is the sum of the current density $g\mathbf{E}$ produced by the field plus the value of $\mathbf{J}_0$ at that point. Thus the fourth equation becomes

$$\text{curl } \mathbf{H} = g\mathbf{E} + \epsilon\dot{\mathbf{E}} + \mathbf{J}_0$$

The others are unchanged.

7-2 VECTOR AND SCALAR POTENTIALS

Since div $\mathbf{B} = 0$ in general (not only for static fields), Eq. (2-20) still holds, that is,

$$\mathbf{B} = \text{curl } \mathbf{A} \qquad \text{(2-20) (7-9)}$$

However, as was indicated in Eq. (2-21) the vector function $\mathbf{A}$ is not completely specified by this relation. In fact, according to Eq. (2-23), div $\mathbf{A}$ is completely arbitrary. The relation is

$$\mathbf{A}(\mathbf{r}, t) = \mathbf{A}_0(\mathbf{r}, t) + \text{grad } \chi(\mathbf{r}, t) \qquad \text{(2-21) (7-10)}$$

Example 7-5

For Example 2-9, find the functions χ that convert the first vector potential into the second and third.

Solution 1

$$-B_0 y\hat{\mathbf{i}} = B_0 x\hat{\mathbf{j}} + \text{grad } \chi$$

$$\chi = -B_0 xy + \text{const.}$$

Solution 2

$$\frac{B_0}{2}(x\hat{\mathbf{j}} - y\hat{\mathbf{i}}) = B_0 x\hat{\mathbf{j}} + \text{grad } \chi$$

$$\chi = \tfrac{1}{2}B_0 xy + \text{const.}$$

$\mathbf{E}$ can also be expressed in terms of potential functions. On inserting Eq. (7-9) into the second Maxwell equation we obtain

$$\text{curl } (\mathbf{E} + \dot{\mathbf{A}}) = 0$$

Hence $(\mathbf{E} + \dot{\mathbf{A}})$ is a gradient, which is conventionally written

$$\mathbf{E} + \dot{\mathbf{A}} = -\text{grad } U$$

or

$$\mathbf{E} = -\dot{\mathbf{A}} - \text{grad } U \tag{7-11}$$

$U(\mathbf{r})$ is called the *scalar potential*. Note that if $\mathbf{A}$ is varied according to Eq. (7-10), U must, in general, also be varied. To keep $\mathbf{E}$ unchanged we require that

$$-(\dot{\mathbf{A}}_0 + \text{grad } \dot{\chi}) - \text{grad } U = -\dot{\mathbf{A}}_0 - \text{grad } U_0$$

$$U = U_0 - \dot{\chi} \tag{7-12}$$

The combination of transformations of the potential functions given in Eqs. (7-10) and (7-12) is the general form of a gauge transformation. Since the fields are left unchanged by such a transformation, all physical effects must also be left unchanged. In other words, all physical laws are gauge invariant. This provides a check on any proposed physical law written in terms of the potentials.

For a linear medium, the equations satisfied by the potentials are readily found. We have already used the second and third Maxwell equations. The resulting expressions may be substituted into the other two to obtain

$$\text{div } (-\dot{\mathbf{A}} - \text{grad } U) = \frac{\rho}{\epsilon} \tag{7-13}$$

and

$$\text{curl curl } \mathbf{A} = \mu(\mathbf{J} - \epsilon\ddot{\mathbf{A}} - \epsilon \text{ grad } \dot{U})$$

or

$$\text{grad div } \mathbf{A} - \nabla^2\mathbf{A} + \mu\epsilon\ddot{\mathbf{A}} + \mu\epsilon \text{ grad } \dot{U} = \mu\mathbf{J} \tag{7-14}$$

This is quite a messy pair of equations. However, by taking advantage of the flexibility allowed by gauge transformations, we can achieve considerable simplification.

Example 7-6

Find the equations that result from the gauge choice

$$\text{div } \mathbf{A} = 0 \tag{7-15}$$

Solution

$$\nabla^2 U = -\frac{\rho}{\epsilon} \tag{7-16}$$

$$\nabla^2\mathbf{A} - \mu\epsilon\ddot{\mathbf{A}} - \mu\epsilon \text{ grad } \dot{U} = -\mu\mathbf{J} \tag{7-17}$$

This choice is called the *Coulomb gauge*. It has the advantage that the equation for the scalar potential has the same form as in electrostatics. However, one must bear in mind that (1) the electric field is *not* given by this potential alone, and (2) Eq. (7-17) is valid only in combination with Eq. (7-15).

Example 7-7

Find the equations that result from the gauge choice

$$\text{div } \mathbf{A} = -\mu\epsilon \dot{U} \tag{7-18}$$

Solution

$$\nabla^2 \mathbf{A} - \mu\epsilon\ddot{\mathbf{A}} = -\mu\mathbf{J} \tag{7-19}$$

$$\nabla^2 U - \mu\epsilon\ddot{U} = -\frac{\rho}{\epsilon} \tag{7-20}$$

This choice is called the *Lorentz gauge*, and Eq. (7-18) is the *Lorentz condition*. Its main advantage is the symmetry of form between the two equations. This is particularly useful in radiation problems, which we shall take up in Chapter 15.

7-3 SUPERPOTENTIALS

Since the source densities ρ and $\mathbf{J}$ are not independent but are connected via the current continuity equation, the scalar and vector potentials must also be connected in some way. In fact, they can both be derived from single *superpotential* functions in various ways. We illustrate for the Lorentz gauge in the following examples.

Example 7-8

Let a function $\mathbf{P}$ be defined as the time integral of $\mathbf{J}$. Express the source densities in terms of it.

Solution   From the definition,

$$\mathbf{J} = \dot{\mathbf{P}} \tag{7-21}$$

Then from the current continuity equation, Eq. (2-6),

$$\dot{\rho} = -\text{div}\,\mathbf{J} = -\text{div}\,\dot{\mathbf{P}}$$

or

$$\rho = -\text{div}\,\mathbf{P} \tag{7-22}$$

[Note the similarity to Eq. (4-5); because of this, the function $\mathbf{P}$ defined here is also called the *polarization*. However, we are here dealing with *any* charge displacement, not just that in a dielectric.]

Example 7-9

Let a function $\mathbf{Z}$ be defined as a solution of the equation

$$\nabla^2 \mathbf{Z} - \mu\epsilon\ddot{\mathbf{Z}} = -\frac{\mathbf{P}}{\epsilon} \tag{7-23}$$

Show that both $\mathbf{A}$ and U may be obtained by taking suitable derivatives of $\mathbf{Z}$.

Solution   (a) Take the time derivative of Eq. (7-23) and multiply by $\mu\epsilon$ to obtain

$$\left(\nabla^2 - \mu\epsilon\frac{\partial^2}{\partial t^2}\right)(\mu\epsilon\dot{\mathbf{Z}}) = -\mu\dot{\mathbf{P}} = -\mu\mathbf{J}$$

This is the same as Eq. (7-19), hence

$$\mathbf{A} = \mu\epsilon\dot{\mathbf{Z}} \tag{7-24}$$

(b) Take the divergence of Eq. (7-23) and multiply by -1 to obtain

$$\left(\nabla^2 - \mu\epsilon \frac{\partial^2}{\partial t^2}\right)(-\text{div } \mathbf{Z}) = \frac{\text{div } \mathbf{P}}{\epsilon} = -\frac{\rho}{\epsilon}$$

This is the same as Eq. (7-20), hence

$$U = -\text{div } \mathbf{Z} \tag{7-25}$$

Example 7-10

Express the fields in terms of the superpotential function $\mathbf{Z}$ defined in Example 7-9.

Solution   Substitute Eqs. (7-24) and (7-25) in (7-9) and (7-11).

$$\mathbf{B} = \text{curl } \mathbf{A} = \mu\epsilon \text{ curl } \dot{\mathbf{Z}} \tag{7-26}$$

$$\mathbf{E} = -\dot{\mathbf{A}} - \text{grad } U = -\mu\epsilon \ddot{\mathbf{Z}} + \text{grad div } \mathbf{Z}$$

$$= -\mu\epsilon \ddot{\mathbf{Z}} + \text{curl curl } \mathbf{Z} + \nabla^2 \mathbf{Z}$$

$$= \text{curl curl } \mathbf{Z} - \frac{\mathbf{P}}{\epsilon} \tag{7-27}$$

Note that $\mathbf{P}$ has a nonzero value only within the source distribution.

The function $\mathbf{Z}$ is called the *Hertz potential*. It is particularly useful in problems of radiation from sources of small extent.

7-4 DIMENSIONS AND UNITS

Physical quantities are measured in terms of arbitrarily chosen *units*, or fixed amounts of the same "kind" of quantity. It would be perfectly feasible to choose a separate arbitrary unit for each different kind of quantity. However, the definitions and laws of physics make it unnecessary to do this. Once certain units are chosen, others can be defined in terms of them. For example, once units of length and time are chosen, it is clearly unnecessary to choose arbitrary units of velocity or acceleration. Units of the latter quantities are *derived* from the "fundamental" ones.

A point that is often not appreciated is that the number of fundamental units is also *purely arbitrary*. Any number from one up may be used. For example, if the unit of time were chosen, the unit of length could be defined in terms of the speed of light (a fundamental physical constant), then the unit of mass in terms of the acceleration imparted by the gravitational attraction of an identical body at unit distance, and so on. The expression of the derived unit of a quantity in terms of the fundamental units is called the *dimensionality* of the quantity. This is an unfortunate choice of terminology. The word has nothing whatever to do with the three-dimensional nature of physical space, but its use has fostered the erroneous impression that three fundamental units are somehow required. From the foregoing discussion it is seen that the dimensionality of any quantity *is not inherent in the quantity itself* but depends on the system of fundamental units chosen. Of course, in any system of units, all terms in any equation must be of the same dimensionality.

In electromagnetic theory, the internationally accepted MKSA system is based on *four* fundamental units: the meter of length, kilogram of mass, second of time, and ampere of current. We are using this system throughout the book. However, there have been a variety of other systems used in the past, and one of them (the Gaussian system) still retains many devotees and, indeed, presents several advantages. All the other systems have been based on *three* fundamental units: length, mass, and time. Usually these are the centimeter, gram, and second, respectively. In this section we give a brief discussion of the dimensionalities and magnitudes of the derived units of electromagnetic quantities in some of these systems. We use braces { } to denote dimensionalities.

Electrostatic Units

The *esu* (electrostatic unit) system has three fundamental units: length in centimeters, mass in grams, and time in seconds (cgs system). The starting point is Coulomb's law [Eq. (1-3)]. The experimental law is that the force is *proportional* to the product of the charges (measured in terms of an arbitrary unit charge) divided by the square of the distance, that is,

$$F = \frac{K_1 qq'}{r^2} \tag{7-28}$$

K_1 is now *chosen* to be dimensionless and equal to 1. Then dimensionally the equation reads

$$\{MLT^{-2}\} = \left\{\frac{Q_{esu}^2}{L^2}\right\} \tag{7-29}$$

$$\{Q_{esu}\} = \{ML^3T^{-2}\}^{1/2} \tag{7-30}$$

If the reader is puzzled by the fractional exponents, he or she should bear in mind that they are a purely formal consequence of the choice of K_1 and have no independent physical meaning.

The electrostatic unit of charge is called the *statcoulomb*; it is the charge that exerts a force of 1 dyne on an equal charge 1 centimeter away.

Example 7-11

Evaluate the statcoulomb in terms of the coulomb.

Solution   From the preceding definition, charges of 1 statcoul at a distance of 1 cm (10^{-2} m) would exert a force of 1 dyne (10^{-5} N) on each other. Thus*

$$10^{-5}\,\text{N} = \frac{9 \times 10^9 q^2}{(10^{-2})^2}$$

$$q = 1\,\text{statcoul} = (\tfrac{1}{3} \times 10^{-9})\,\text{coulomb}$$

---

*3 is written in place of 2.9979 . . . , an experimental constant related to the speed of light.

Field strength and potential are defined as force per unit charge and potential energy per unit charge, respectively. The unit of potential is called the *statvolt*.

Example 7-12

Find the value and dimensions of the statvolt.

Solution The cgs unit of energy is the *erg* (10^{-7} J). Thus the esu of potential is 1 erg per statcoul, that is, 1 statvolt = 1 erg per statcoul = 10^{-7} J per $\frac{1}{3} \times 10^{-9}$ C = 300 V. Dimensionally,

$$\{U_{esu}\} = \left\{ \frac{ML^2T^{-2}}{(ML^3T^{-2})^{1/2}} \right\} = \{M^{1/2}L^{1/2}T^{-1}\} \tag{7-31}$$

Example 7-13

Evaluate the esu of electric field strength.

Solution 1 statvolt per centimeter = 300 V per 10^{-2} m = 3×10^4 V/m.

Example 7-14

Write Gauss's law, Eq. (1-10), in esu.

Solution In Eq. (1-7), K_f is replaced by unity. Thus

$$\text{div } \mathbf{E}_{esu} = 4\pi \rho_{esu} \tag{7-32}$$

Note that, in effect, ϵ_0 is replaced by $1/4\pi$.

Electromagnetic Units

The starting point here is the magnetic force between chargeamentum elements [Eq. (2-8)]. Again, experiment only establishes proportionality, so the law is written

$$F = \frac{K_2 \Xi \Xi'}{r^2} (\times \text{ angle factors}) \tag{7-33}$$

where Ξ and Ξ' are the chargeamenta of the interacting current elements. If K_2 is chosen to be dimensionless, this gives (since chargeamentum is current times distance)

$$\{MLT^{-2}\} = \{I^2_{emu}\}$$
$$\{I_{emu}\} = \{MLT^{-2}\}^{1/2} \tag{7-34}$$

The electromagnetic unit of current is called the *abampere*. (The prefix stands for *absolute*; the system was the first to be based on the fundamental units of length, mass, and time, and the name was used to distinguish it from the *practical* everyday units.) The abampere is defined operationally in terms of the *force per unit length* between two long straight parallel wires carrying equal currents; it is the current value such that this force is 2 dyn/cm when the wires are 1 cm apart.

Example 7-15

Evaluate the abampere in amperes.

Solution   From Eqs. (2-9) and (2-13),

$$dF = (I \, dl) \frac{\mu_0 I}{2\pi s}$$

For $dl = s = 1$ cm, $dF = 2$ dyn $= 2 \times 10^{-5}$ N, so

$$2 \times 10^{-5} = \frac{I^2(4\pi \times 10^{-7})}{2\pi}$$

$$I = 1 \text{ abampere} = 10 \text{ A}$$

Magnetic field strength is defined as force per unit chargeamentum. The unit is called the *gauss*.

Example 7-16

Find the value and dimensions of the gauss.

Solution   1 gauss = 1 dyne per abampere-cm $= 10^{-5}$ newton per (10 amperes $\times 10^{-2}$ meter) $= 10^{-4}$ T. Dimensionally,

$$\{B_{emu}\} = \left\{ \frac{MLT^{-2}}{(MLT^{-2})^{1/2}L} \right\} = \{M^{1/2}L^{-1/2}T^{-1}\} \tag{7-35}$$

On comparing with Eq. (7-31), it is seen that $\{B_{emu}\} = \{E_{esu}\}$.

Example 7-17

Write Ampère's circuital law in vacuum [Eq. (2-23)] in emu.

Solution   The constant $\mu_0/4\pi$ in Eq. (2-8) is replaced by unity in Eq. (7-34), or μ_0 is replaced by 4π. Thus

$$\text{curl } \mathbf{B}_{emu} = 4\pi \mathbf{J}_{emu} \tag{7-36}$$

Relation Between esu and emu

We have just defined units for two sets of quantities: charges and their fields on the one hand, and currents and their fields on the other. However, there is a further experimental law that charge and current are related. Since experiment can only establish *proportionality* between otherwise independent quantities, this must be written

$$\frac{dq_{esu}}{dt} = K_3 I_{emu} \tag{7-37}$$

The constant K_3 is *not* arbitrary, since all other units in the equation have been fixed.

Example 7-18

Determine K_3.

Solution 1   Dimensionally,

$$\frac{\{M^{1/2}L^{3/2}T^{-1}\}}{\{T\}} = \{K_3\}\{M^{1/2}L^{1/2}T^{-1}\}$$

$$\{K_3\} = \left\{ \frac{L}{T} \right\}$$

Solution 2   Numerically,

$$\frac{d}{dt} \frac{q_{coul}}{\frac{1}{3} \times 10^{-9}} = K_3 \left(\frac{I_{amp}}{10}\right)$$

Thus

$$K_3 = 3 \times 10^{10} \text{ cm/s} = c \qquad (7\text{-}38)$$

The constant c is the speed of light in vacuum. (Remember that the 3 in all these equations is only approximate; the true value is 2.9979 . . . .) Equation (7-38) shows that this universal constant plays a fundamental role in all electromagnetic phenomena.

From Eqs. (7-37) and (7-38), the current continuity relation (Section 2-2) in the present units would read

$$\frac{1}{c} \frac{\partial}{\partial t} \rho_{esu} + \text{div } \mathbf{J}_{emu} = 0 \qquad (7\text{-}39)$$

Although there is nothing wrong with this, it is considered more desirable to change the units of either ρ or J so that the factor $1/c$ disappears. For example, we could define an electrostatic unit of current as

$$I_{esu} = c I_{emu} \qquad (7\text{-}40)$$

Dimensionally, this reads

$$\{I_{esu}\} = \left\{\frac{L}{T}\right\} \{M^{1/2}L^{1/2}/T\} = \left\{\frac{Q_{esu}}{T}\right\}$$

so that the familiar charge-to-current relationship holds. In other words, the esu of current (called the *statampere*) is simply 1 statcoul per second.

Equation (7-40) says that a current measured in esu (statamperes) is equal to c times the same current measured in emu (abamperes). In other words,

$$1 \text{ abampere} = c \text{ statamperes} \qquad (7\text{-}41)$$

This strange-looking relation, with a dimensional constant connecting quantities of the same kind (currents), emphasizes the fact discussed at the beginning of this section that dimensionality is not an intrinsic property of any physical quantity. The dimensional factor here can be traced back to the arbitrary choice of *dimensionless* constants in the force laws. Since cgs units were used in both definitions, Eq. (7-31) means

$$I_{esu} = c_{cgs} I_{emu}$$

or

$$1 \text{ abampere} = 3 \times 10^{10} \text{ statamperes}$$

However, if both sets of quantities appear in any equation, the dimensional factor c must appear.

Example 7-19

Write the magnetic force law in terms of current in esu.

Solution   In view of Eq. (7-40) the force on length dl of a current filament may be written

$$d\mathbf{F} = \frac{1}{c} I_{esu} \, \mathbf{dl} \times \mathbf{B}_{emu} \qquad (7\text{-}42)$$

Example 7-20

Define a suitable esu of magnetic field strength.

Solution   We want a unit such that Eq. (7-42) will hold without the factor c. Thus

$$\mathbf{B}_{esu} = \frac{\mathbf{B}_{emu}}{c} \qquad (7\text{-}43)$$

Gaussian Units

This is a *mixed* system of units. Charge and electric field strength are measured in esu, while magnetic field strength is measured in cmu. As for current, some authors keep this in emu, but it is more common to convert it to esu. We shall follow the latter practice.

The vacuum Maxwell equations (see Section 3-6) in Gaussian units are then obtained as follows:

1. The divergence equations contain quantities from only one of the two basic systems. Equation (7-32) is written simply

$$\text{div } \mathbf{E} = 4\pi\rho \qquad (7\text{-}44)$$

and obviously

$$\text{div } \mathbf{B} = 0 \qquad (7\text{-}45)$$

2. The curl equations involve quantities from both basic systems. The Gaussian forms are obtained by writing them in one of the systems and then converting the appropriate terms. Faraday's law of induction is

$$\text{curl } \mathbf{E}_{esu} = -\dot{\mathbf{B}}_{esu}$$

which becomes on converting the right side,

$$\text{curl } \mathbf{E} = -\frac{1}{c}\dot{\mathbf{B}} \qquad (7\text{-}46)$$

For the curl $\mathbf{B}$ equation, we start with the static case, Ampère's circuital law. Equation (7-36) is converted by means of Eq. (7-40) to read

$$\text{curl } \mathbf{B}_{emu} = \frac{4\pi}{c}\mathbf{J}_{esu}$$

The displacement current term is written so that on taking the divergence, the correct continuity relation for div $\mathbf{J}$ is obtained. In view of Eq. (7-33) for ρ, the final result is

$$\text{curl } \mathbf{B} = \frac{4\pi}{c}\mathbf{J} + \frac{1}{c}\dot{\mathbf{E}} \qquad (7\text{-}47)$$

Fields in Matter

For the fields in dielectric and magnetic matter we required the dipole moment densities $\mathbf{P}(\mathbf{r})$ and $\mathbf{M}(\mathbf{r})$. Their dimensionalities in Gaussian units are

$$\{P_{esu}\} = \left\{\frac{Q_{esu}\,L}{L^3}\right\} = \{M^{1/2}L^{-1/2}T^{-1}\}$$

$$\{M_{emu}\} = \left\{\frac{I_{emu}\,L^2}{L^3}\right\} = \{M^{1/2}L^{-1/2}T^{-1}\}$$

On comparing with Eqs. (7-31) and (7-35), it is seen that these moment densities have the same dimensionality as the Gaussian units of the fields, $\{E_{esu}\}$ and $\{B_{emu}\}$. Thus it is natural to define D and H also in these same units. The quantitative relations follow, as in Chapters 4 and 6, from the effective source densities. In the dielectric case, $\rho_P = -\text{div }\mathbf{P}$, so

$$\text{div }\mathbf{E} = 4\pi(\rho + \rho_P) = 4\pi(\rho - \text{div }\mathbf{P})$$

or

$$\text{div }\mathbf{D}(\mathbf{r}) = 4\pi\rho(\mathbf{r}) \tag{7-48}$$

where

$$\mathbf{D}(\mathbf{r}) = \mathbf{E}(\mathbf{r}) + 4\pi\mathbf{P}(\mathbf{r}) \tag{7-49}$$

In the magnetic case, a bit more care is required. The relation $\mathbf{J}_M = \text{curl }\mathbf{M}$ holds in emu. However, we want $\mathbf{J}$ in esu for Gaussian units. Hence, for the static relation,

$$\mathbf{J}_{M,esu} = c\,\text{curl }\mathbf{M}$$

$$\text{curl }\mathbf{B} = \frac{4\pi}{c}(\mathbf{J} + \mathbf{J}_M)$$

$$= 4\pi\left(\frac{1}{c}\mathbf{J} + \text{curl }\mathbf{M}\right)$$

Thus

$$\text{curl }\mathbf{H}(\mathbf{r}) = \frac{4\pi}{c}\mathbf{J}(\mathbf{r}) \tag{7-50}$$

where

$$\mathbf{H}(\mathbf{r}) = \mathbf{B}(\mathbf{r}) - 4\pi\mathbf{M}(\mathbf{r}) \tag{7-51}$$

The general relation including displacement currents is

$$\text{curl }\mathbf{H} = \frac{4\pi}{c}\mathbf{J} + \frac{1}{c}\dot{\mathbf{D}} \tag{7-52}$$

MKSA System

In Chapters 1 and 2 we developed the force laws in terms of a *four-dimensional* unit system: the *meter* of length, the *kilogram* of mass, the *second* of time, and the *coulomb* of charge. Current is measured in coulombs per second (amperes). In practice, the ampere is easier to measure precisely, and it is now taken as the fundamental unit, with the coulomb defined as the ampere-second. This gives the MKSA system, which is the presently adopted SI (*Système International*).

We noted in connection with Eq. (7-38) that the speed of light plays a fundamental role. Obviously, this has to enter via the constants (no longer arbitrary!)

that appear in the force laws. If we compare the force equations (7-28) and (7-33) dimensionally and insist on the usual current continuity relation, it is evident that the constants must be related by

$$\frac{K_1}{K_2} = c^2$$

Since we have written $K_1 = 1/4\pi\epsilon_0$ and $K_2 = \mu_0/4\pi$, this reads

$$\mu_0\epsilon_0 = \frac{1}{c^2} = \frac{1}{(3 \times 10^8 \text{ m/s})^2} \tag{7-53}$$

If the ampere (or coulomb) were chosen arbitrarily, both μ_0 and ϵ_0 would be experimental constants, subject to Eq. (7-53). In present practice the ampere is *defined* so that μ_0 is *exactly* $4\pi \times 10^{-7}$ henry per meter. In other words, the *absolute ampere* is defined such that two infinitely long parallel 1-A currents spaced 1 m apart in vacuum would exert forces of 2×10^{-7} N/m on each other. In practice, of course, instead of infinitely long parallel wires two coils of accurately known dimensions are used in an arrangement called a *current balance*. Another current unit, the *international ampere*, based on the rate of electroplating of silver, was formerly used, but was discarded in 1948. The discrepancy was less than 2×10^{-4}.

Comparison of the Systems

The primary advantage of the Gaussian system is that all the fields (**E, D, P, B, H, M**) have the same unit, the gauss or statvolt per centimeter. This facilitates the comparison of the strengths of electric and magnetic effects when both are present. The disadvantages are the awkward dimensionalities and the impractical sizes of some of the units.

The primary advantage of the MKSA system is that familiar units are used throughout. This facilitates practical calculations and analysis of measurements. One disadvantage is that two constants appear in the equations, although only one is independent. This is not much of a drawback in practice. More serious to many people is the fact that quantities that intuitively seem to be of the same "kind" are assigned different dimensionalities. For example, **E** and **D** are both fields due to *charges*, the source densities being true plus polarization charges in the one case and true charges only in the other, but nevertheless charges in both cases. Despite this, the two quantities are related by the dimensional constant ϵ_0. Similar remarks apply to **B** and **H**, both being fields due to *currents*. Whether this assignment of different dimensionalities is a serious drawback or not is a matter of taste. In practice, the difference is often a help in keeping things straight.

Tables 7-1 and 7-2 give conversions of symbols in equations and of unit amounts of various physical quantities from MKSA to Gaussian units. The tables work equally well in reverse, of course.

Example 7-21

Use Table 7-1 to convert the Maxwell equation for curl H to Gaussian units.

TABLE 7-1 CONVERSION OF EQUATIONS

	MKSA	$\longrightarrow$	Gaussian
Speed of light	$\mu_0 \epsilon_0$	$\longrightarrow$	$\dfrac{1}{c^2}$
Electric field	E	$\longrightarrow$	$\dfrac{1}{\sqrt{4\pi\epsilon_0}} E$
Displacement	D	$\longrightarrow$	$\sqrt{\dfrac{\epsilon_0}{4\pi}} D$
Charge	q	$\longrightarrow$	$\sqrt{4\pi\epsilon_0} q$
Magnetic field	B	$\longrightarrow$	$\sqrt{\dfrac{\mu_0}{4\pi}} B$
Magnetic intensity	H	$\longrightarrow$	$\dfrac{1}{\sqrt{4\pi\mu_0}} H$
Magnetization	M	$\longrightarrow$	$\sqrt{\dfrac{4\pi}{\mu_0}} M$
Conductivity	g	$\longrightarrow$	$4\pi\epsilon_0 g$
Permittivity	$\dfrac{\epsilon}{\epsilon_0}$	$\longrightarrow$	ϵ
Permeability	$\dfrac{\mu}{\mu_0}$	$\longrightarrow$	μ
Resistance	R	$\longrightarrow$	$\dfrac{1}{4\pi\epsilon_0} R$
Inductance	L	$\longrightarrow$	$\dfrac{1}{4\pi\epsilon_0} L$
Capacitance	C	$\longrightarrow$	$4\pi\epsilon_0 C$

SOURCE: Adapted from J. D. Jackson, *Classical Electrodynamics*, 2nd ed., John Wiley & Sons, Inc., New York, 1975, with permission. Copyright 1975 John Wiley & Sons, Inc.

TABLE 7-2 CONVERSION OF GIVEN AMOUNTS OF QUANTITIES

MKSA		Gaussian
1 meter	$=$	10^2 cm
1 kilogram	$=$	10^3 g
1 newton	$=$	10^5 dyn
1 joule	$=$	10^7 ergs
1 coulomb	$=$	3×10^9 statcoulomb
1 ampere	$=$	3×10^9 statampere
1 volt	$=$	$\frac{1}{300}$ statvolt
1 coulomb/meter^2	$=$	$12\pi \times 10^5$ statcoulomb/cm^2
1 ohm	$=$	$\frac{1}{9} \times 10^{-11}$ s/cm
1 farad	$=$	9×10^{11} cm
1 tesla	$=$	10^4 gauss
1 ampere/meter	$=$	$4\pi \times 10^{-3}$ oersted
1 henry	$=$	$\frac{1}{9} \times 10^{-11}$ esu $= 10^9$ emu

SOURCE: Adapted from J. D. Jackson, *Classical Electrodynamics*, 2nd ed., John Wiley & Sons, Inc., New York, 1975, with permission. Copyright 1975 John Wiley & Sons, Inc.

Solution

$$\text{MKSA:} \qquad \text{curl } \mathbf{H} = \mathbf{J} + \dot{\mathbf{D}}$$

$$\text{Gaussian:} \quad \frac{1}{\sqrt{4\pi\mu_0}} \text{curl } \mathbf{H} = \sqrt{4\pi\epsilon_0}\, \mathbf{J} + \sqrt{\frac{\epsilon_0}{4\pi}}\dot{\mathbf{D}}$$

or

$$\text{curl } \mathbf{H} = \frac{4\pi}{c}\mathbf{J} + \frac{1}{c}\dot{\mathbf{D}}$$

Note that $\mathbf{J}$ converts like charge, since time is the same in both systems.

FURTHER REFERENCES

R. H. GOOD, JR., and T. J. NELSON, *Classical Theory of Electric and Magnetic Fields*, Academic Press, Inc., New York, 1971, pp. 299ff.

J. A. STRATTON, *Electromagnetic Theory*, McGraw-Hill Book Company, New York, 1941, pp. 16ff., 28ff.

W. K. H. PANOFSKY and M. PHILLIPS, *Classical Electricity and Magnetism*, Addison-Wesley Publishing Co., Inc., Reading, Mass., 1962, pp. 240ff., 459ff.

J. D. JACKSON, *Classical Electrodynamics*, 2nd ed., John Wiley & Sons, Inc., New York, 1975, pp. 811ff.

American Institute of Physics Handbook, McGraw-Hill Book Company, New York, 1957, Sec. 5c.

PROBLEMS

7-1. Show that it is a consequence of the curl $\mathbf{E}$ equation that div $\mathbf{B}$ is constant in time. *Hint:* Evaluate $(\partial/\partial t)(\text{div } \mathbf{B})$ without using the fact that div $\mathbf{B} = 0$.

7-2. Show that it is a consequence of the curl $\mathbf{H}$ equation, along with the current continuity relation, that $(\text{div } \mathbf{D} - \rho)$ is constant in time.

7-3. Show that even when the gauge (i.e., div $\mathbf{A}$) is specified, the vector potential is still subject to a certain class of gauge transformations. What are they?

7-4. In what gauge are the potentials that are derived from the Hertz superpotential in Example 7-9?

7-5. Suppose that $\mathbf{S}(\mathbf{r}, t)$ is a vector function such that curl $\mathbf{S} = \mathbf{J}$. Show that a superpotential $\mathbf{Z}'$ can be defined as the solution of an inhomogeneous wave equation

$$\left(\nabla^2 - \mu\epsilon \frac{\partial^2}{\partial t^2}\right)\mathbf{Z}' = -\mu\mathbf{S}.$$

7-6. In what gauge are the potentials found in Problem 7-5?

7-7. In a one-dimensional system of units with time as the only fundamental unit, what are the dimensionalities of acceleration and mass?

7-8. In the system of Problem 7-7, what is the dimensionality of charge?

7-9. Write, in esu, the equations for the electric field and potential due to a point charge q at the origin.

7-10. Write, in esu, the integral form of Gauss's law in free space.

7-11. What is the esu of capacitance?

7-12. Write, in emu, the equations for the magnetic field and vector potential contributions at a point **r** due to a current element I **dl** at point **r'**.

Part III

ENERGY RELATIONS

chapter 8

Electrostatic Energy

8-1 INTERACTION OF CHARGES

In this section we consider the mechanical work required to move a set of charges into some specified configuration. The charges are assumed initially to be infinitely far apart so that there is then no interaction between them. Also, the motion is assumed to be extremely slow, so that time-dependent effects are negligible. Thus the only work done is that required to overcome the electrostatic forces. This work is stored as potential energy in the final configuration, just as the work done in lifting a body is stored as gravitational potential energy in the configuration body-plus-earth.

We first calculate the work required to move a point charge q_1 from infinity into final position $\mathbf{r}_1$ in a *preexisting* field due to some set of charges other than q_1. From Eq. (1-21) and the discussion of the physical significance of the potential, we have

$$W = q_1 U'(\mathbf{r}_1) \tag{8-1}$$

where $U'(\mathbf{r})$ is the potential function due to the *other* charges. To emphasize, U' does not include the part of the potential due to q_1 itself. This must obviously be the case since q_1 does not exert any force on itself.

In the event that the moving charge is vanishingly small (call it δq_1) it makes essentially no contribution to the potential at any point. In this event,

$$U'(\mathbf{r}) = U(\mathbf{r}) \tag{8-2}$$

Thus

$$\delta W = U(\mathbf{r})\, \delta q \tag{8-3}$$

The symbol δ is a differential operator relating to a *change with time* in some quantity or point function. Its meaning is

$$\delta(\ ) = \frac{d(\ )}{dt}\, \delta t \tag{8-4}$$

The next step is to consider a *distribution* of such vanishingly small charge increments. In other words, a preexisting charge distribution $\rho(\mathbf{r})$ is altered by adding an infinitesimal charge density $\delta\rho(\mathbf{r})$ to it. Figure 8-1 gives a one-dimensional illustration of the situation.

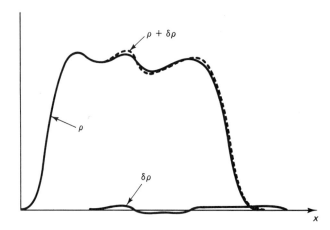

Figure 8-1 Addition of an infinitesimal charge density $\delta\rho(x)$ to a preexisting distribution $\rho(x)$.

A volume element $d^3\mathbf{r}$ at $\mathbf{r}$ receives an amount of charge $\delta\rho(\mathbf{r})\, d^3\mathbf{r}$, so the work done upon it is

$$\delta d^3 W = U(\mathbf{r})\, \delta\rho(\mathbf{r})\, d^3\mathbf{r} \tag{8-5}$$

Since $\delta\rho$ is specified to be infinitesimal, Eq. (8-2) continues to hold as we add the charge to more and more volume elements. Thus summing over all the volume elements,

$$\delta W = \iiint_\infty U(\mathbf{r})\, \delta\rho(\mathbf{r})\, d^3\mathbf{r} \tag{8-6}$$

Equation (8-6) is the basic relation on which all energy calculations are built. It holds without restrictions. However, when we attempt to integrate over the timelike differential operator δ, certain restrictions enter the picture. Suppose that we intend to build up a certain final distribution $\rho_{\text{final}}(\mathbf{r})$ by assembling differential elements.

There are infinitely many sequences by which this could be done. We could, for example, start in a single volume element and build it up to its full charge, then proceed to the neighboring volume element, and so on. This would not be a clever procedure because the *shape* of the charge distribution would keep changing. Much more appropriate for mathematical analysis is to start with a scaled-down version of the final distribution, and gradually build it up keeping the same form all the time. In other words, choose a buildup pattern such that

$$\rho(\mathbf{r}, t) = f(t)\rho_{\text{final}}(\mathbf{r}) \tag{8-7}$$

$$\delta\rho(\mathbf{r}, t) = \delta f \rho_{\text{final}}(\mathbf{r}) \tag{8-8}$$

where, obviously, $0 \leq f \leq 1$. This illustrated in Figure 8-2. Then, *assuming the medium to have linear dielectric response*, the potential at any point will simply scale up along with the charge. That is,

$$U(\mathbf{r}, t) = f(t)U_{\text{final}}(\mathbf{r}) \tag{8-9}$$

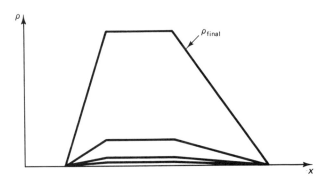

Figure 8-2 Illustration of "proportional" buildup of final charge distribution.

On inserting Eqs. (8-8) and (8-9) into (8-6), we may then carry out the timelike integration to obtain

$$W \equiv \int_0^1 \iiint (f U_{\text{final}})(\delta f \rho_{\text{final}}) \, d^3\mathbf{r}$$

$$= \iiint U_{\text{final}}\rho_{\text{final}} \, d^3\mathbf{r} \int_0^1 f \, \delta f$$

$$= \frac{1}{2} \iiint U(\mathbf{r})\rho(\mathbf{r}) \, d^3\mathbf{r} \tag{8-10}$$

where, in the last line, we have dropped the cumbersome subscript "final."

Even though we used a special build-up sequence to facilitate the calculation, the result depends only on the final charge distribution; this is another consequence of the linearity assumption. Obviously, for surface, line, or point charge distributions the appropriate densities and geometrical elements are used. Furthermore, even though the integral is formally taken over all space, the integrand is nonzero only where there is charge density.

W is the work that was done by some "outside agency" in assembling the charge distribution. It was tacitly assumed that no other work was done; for example, there was no mechanical work due to motion of material bodies, no chemical work due to changes of state, and so on. Thus, in this case, all the work done has to be stored in the final configuration. Thus, from Eq. (8-10), the *electric field energy* is

$$\mathcal{U}_{el} = \frac{1}{2} \iiint_\infty U(\mathbf{r})\rho(\mathbf{r})\, d^3\mathbf{r} \tag{8-11}$$

Example 8-1

A parallel-plate capacitor is charged to potential difference V. Calculate the stored energy.

Solution   The surface charge density on the plates is $\pm CV/A$, where A is the plate area. Thus

$$\mathcal{U}_{el} = \frac{1}{2} A(U_+\sigma_+ + U_-\sigma_-)$$

$$= \frac{1}{2} A\frac{CV}{A}[U_+ - (U_+ - V)]$$

$$= \frac{1}{2} CV^2$$

Example 8-2

Calculate the energy of a sphere of radius a with uniform charge density ρ_0.

Solution   From Problem 1-26

$$U(r) = \frac{\rho_0(3a^2 - r^2)}{6\epsilon_0} \qquad \text{for } r < a$$

$$\mathcal{U}_{el} = \frac{1}{2} \int_0^a \rho_0 \frac{\rho_0}{6\epsilon_0}(3a^2 - r^2)4\pi r^2 \, dr$$

$$= \frac{\pi\rho_0^2}{3\epsilon_0}\left(3a^2\frac{a^3}{3} - \frac{a^5}{5}\right)$$

$$= \frac{4\pi}{15}\frac{\rho_0^2 a^5}{\epsilon_0}$$

The preceding result may also be expressed in terms of the total charge q of the sphere. Since $q = \frac{4}{3}\pi a^3\rho_0$,

$$\mathcal{U}_{el} = \frac{3}{20\pi}\frac{q^2}{\epsilon_0 a}$$

Thus if we let the radius go to zero, the energy becomes infinite. This suggests that true point charges cannot exist. Nevertheless, in many respects electrons do seem to act like point charges. In current theory, the infinite energy is "canceled out" by a mathematical process called *renormalization*.

Self-Energy and Interaction Energy

If a charge distribution consists of several nonoverlapping clusters, the total energy can be written as sums of two distinct kinds of terms: (1) terms involving the charges and potentials of the individual clusters (the "self-energy" terms) and (2) terms connecting the clusters with each other (the "interaction" terms). To prove this we write the total energy as a sum of integrals over the individual charge volumes, that is,

$$\mathcal{U}_{el} = \sum_i \frac{1}{2} \iiint_{V_i} \rho(\mathbf{r})U(\mathbf{r}) \, d^3(\mathbf{r})$$

In each volume V_i, we may separate the potential into two terms,

$$U(\mathbf{r}) = U_i(\mathbf{r}) + U_i'(\mathbf{r})$$

where U_i is the potential function due to the ith cluster alone, and U_i' is that due to all the other clusters. (Clearly, this entails a *different* separation for each volume.) Thus

$$\mathcal{U}_{el} = \sum_i \left[\frac{1}{2} \iiint_{V_i} \rho(\mathbf{r})U_i(\mathbf{r}) \, d^3\mathbf{r} + \frac{1}{2} \iiint_{V_i} \rho(\mathbf{r})U_i'(\mathbf{r}) \, d^3\mathbf{r} \right] \qquad (8\text{-}12)$$

The first terms are the self-energies and the remaining sum is the interaction energy.

Example 8-3

Two concentric spherical surfaces have radii a and b ($b > a$) and are uniformly charged with charges q_a and q_b, respectively (Figure 8-3). Calculate the self-energies and the interaction energy.

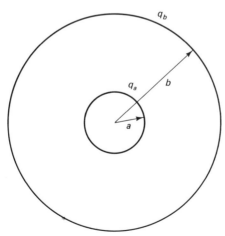

Figure 8-3 Concentric charged spherical surfaces (see Example 8-3).

Solution The potential due to a uniform spherical shell of charge is given in Example 1-12. From this we get

$$U(a) = \frac{1}{4\pi\epsilon_0}\left[\frac{q_a}{a} + \frac{q_b}{b}\right]$$

$$U(b) = \frac{1}{4\pi\epsilon_0}\left[\frac{q_a}{b} + \frac{q_b}{b}\right]$$

Thus the self-energies are

$$\mathfrak{U}_a = \frac{q_a^2}{8\pi\epsilon_0 a}$$

$$\mathfrak{U}_b = \frac{q_b^2}{8\pi\epsilon_0 b}$$

and the interaction energy is

$$\mathfrak{U}_{\text{int}} = \frac{q_a q_b}{4\pi\epsilon_0 b}$$

Example 8-4

Calculate the interaction energy among N point charges $q_1 \cdots q_N$ at positions $\mathbf{r}_1 \cdots \mathbf{r}_N$.

Solution Consider one of the charges, say the ith. The potential at any point $\mathbf{r}$ due to all the others is

$$U_i'(\mathbf{r}) = \sum_{j=1}^{N}{}' \frac{q_j}{4\pi\epsilon_0 |\mathbf{r} - \mathbf{r}_j|}$$

where the prime on the summation symbol signifies that the term $j = i$ is to be omitted. Then Eq. (8-12) gives

$$\mathfrak{U}_{\text{int}} = \frac{1}{2}\sum_{i=1}^{N}\sum_{j=1}^{N}{}' \frac{q_i q_j}{4\pi\epsilon_0 |\mathbf{r}_i - \mathbf{r}_j|}$$

Although the self-energies of the point charges are infinite, the interaction energy among them is perfectly finite. It is simply the work required to move them into position *after* they have been formed.

Charged Conductors

Consider a field set up by N separate charged conductors. Since the charges all lie on the surfaces of the conductors, Eq. (8-11) becomes

$$\mathfrak{U}_{\text{el}} = \frac{1}{2}\sum_{i=1}^{N} \oiint_{S_i} U(\mathbf{r})\sigma(\mathbf{r})\, d^2\mathbf{r}$$

But each conductor is an equipotential, so U can be taken out of each integral. The remaining integral is simply the total charge of each conductor. Thus

$$\mathcal{U}_{el} = \frac{1}{2} \sum_i U_i \oiint_{S_i} \sigma \, dS = \frac{1}{2} \sum_i U_i q_i \tag{8-13}$$

This is the *total* energy, not just the interaction part. The reason is that it is the *total potential*, not just the "external" part, that is constant throughout each conductor.

Thomson's Theorem

When charges are distributed on a set of conductors in such a way that each conductor is an equipotential, the total energy is a minimum. To show this, imagine the charges to be infinitesimally displaced along the surfaces in some arbitrary manner (Figure 8-4).

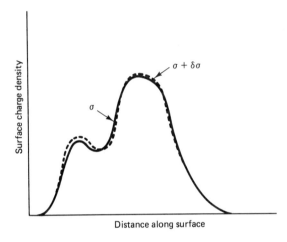

Figure 8-4 Displacement of charge on an isolated conductive surface.

Thus the surface charge densities are altered from the original $\sigma_i(\mathbf{r})$ to $\sigma_i(\mathbf{r}) + \delta\sigma_i(\mathbf{r})$. From Eq. (8-6) the change in energy is

$$\delta\mathcal{U}_{el} = \sum_i \oiint_{S_i} U_i \, \delta\sigma_i \, dS$$

Since each conductor is an equipotential, U_i may again be factored out of each integral to give

$$\delta U_{el} = \sum_i U_i \oiint_{S_i} \delta\sigma_i \, dS$$

But since charge is not added to or taken from any of the conductors, merely moved around on them, all the integrals are zero. Hence $\delta\mathcal{U}_{el} = 0$, and $\mathcal{U}_{el}$ is an *extremum* (maximum or minimum). Common sense tells us that it must be a minimum, and this can be proved rigorously though the proof is complicated. This is known as *Thomson's theorem*. It answers the often perplexing question: How do the electrons know how to distribute themselves so that the conductor will be an equipotential? Of course, they don't; they simply go where they are pushed until the energy is a minimum.

Example 8-5

Consider a field due to a set of charged conductors fixed in position (Figure 8-5a). Show that if an *uncharged* additional conductor is moved into this field (Figure 8-5b) the total energy is *lowered*.

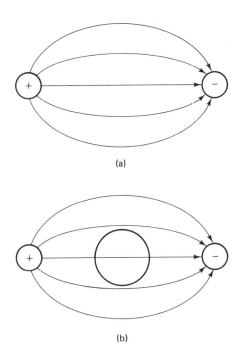

(a)

(b)

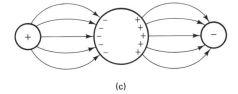

(c)

Figure 8-5 (a) Field set up by two charged conductors; (b) third conductor (uncharged) introduced, but all charges "glued" in place; (c) glue released and system equilibrated.

Solution   Imagine the charges on all the conductors (including the moving one) to be "glued" in position. Then since no charge is moved, no work is done in introducing the additional one. The conductors will then no longer be equipotentials. On releasing the glue, charges will rearrange themselves on the surfaces and, by Thomson's theorem, the energy will decrease (Figure 8-5c).

This result shows that an uncharged conductor will always be attracted to the strongest part of the field.

Energy in Terms of the Field Strengths

We now transform Eq. (8-11) into an expression in terms of field strength rather than potential and charge density. This follows from the vector identity

$$\operatorname{div} (U\mathbf{D}) = U \operatorname{div} \mathbf{D} + (\operatorname{grad} U) \cdot \mathbf{D}$$
$$= U\rho - \mathbf{E} \cdot \mathbf{D}$$

Thus

$$\mathcal{U}_{\text{el}} = \frac{1}{2} \iiint_{\infty} \mathbf{E}(\mathbf{r}) \cdot \mathbf{D}(\mathbf{r}) \, d^3\mathbf{r} + \frac{1}{2} \iiint_{\infty} \operatorname{div} (U\mathbf{D}) \, d^3\mathbf{r}$$

In all physically realistic cases the second integral vanishes, as we now prove. By the divergence theorem, it is equal to

$$\frac{1}{2} \oiint_{S(\infty)} U\mathbf{D} \cdot \hat{\mathbf{n}} \, dS$$

In a physically realistic situation, all charge lies in some *finite* region of space, while the integral is over a surface at infinite distance. Since the field must eventually fall off at least as fast as $1/r^2$ and the potential as $1/r$ (faster if there is no net charge but only higher multipole moments in the distribution), whereas the surface area increases only as r^2, the integral vanishes. The result is, therefore,

$$\mathcal{U}_{\text{el}} = \frac{1}{2} \iiint_{\infty} \mathbf{E}(\mathbf{r}) \cdot \mathbf{D}(\mathbf{r}) \, d^3\mathbf{r} \qquad (8\text{-}14)$$

It must be borne in mind that the starting expression [Eq. (8-11)] holds only a *linear* dielectric medium. Hence, more properly we should write

$$\mathcal{U}_{\text{el}} = \frac{1}{2} \iiint_{\infty} \kappa(\mathbf{r})\epsilon_0 E^2(\mathbf{r}) \, d^3\mathbf{r} \qquad (8\text{-}15)$$

Another important point is that the integral here *really* extends over all space (except inside conductors). This is in contrast to Eq. (8-11), which, although formally written as an integral over all space, really extends just over the charges. This distinction suggests that the integrand of Eq. (8-15) may be interpreted as the *energy density*, that is,

$$u_{\text{el}} = \tfrac{1}{2}\kappa\epsilon_0 E^2 \qquad (8\text{-}16)$$

There is no way of proving this interpretation, since we cannot ascertain where the energy is located. Nevertheless, it is a reasonable and useful concept. For example, we shall find later that energy can be transported through empty space by electromagnetic fields, and thus the idea that the fields "contain" the energy seems inescapable.

Example 8-6

Calculate the energy density in a charged parallel-plate capacitor, charged to potential difference V.

Solution $E = V/d$, $u_{el} = \frac{1}{2}\kappa\epsilon_0 V^2/d^2$. To check, multiply by the volume and compare with Example 8-1.

Example 8-7

Calculate the energy of a uniform spherical shell of charge, of radius a, charge q, using Eq. (8-15).

Solution   From Example 1-12, $E = 0$ inside, $E = q/4\pi\epsilon_0 r^2$ outside, $\kappa = 1$ everywhere. Thus Eq. (8-15) gives

$$\mathfrak{U}_{el} = \frac{1}{2}\int_a^\infty \frac{q^2}{16\pi^2\epsilon_0 r^4} 4\pi r^2 \, dr$$

$$= \frac{q^2}{8\pi\epsilon_0 a}$$

This is seen to be the same as was obtained in Example 8-3.

In general it is easier to calculate energies of charge configurations using Eq. (8-11) than (8-15). However, the formulation in terms of the field has great theoretical importance, and is useful in some other classes of problems.

8-2 FORCES AND TORQUES

Since the definition of the electric field is given in terms of forces it is a trivial matter to write a general expression for the force on a charged body. A point charge q at $\mathbf{r}$ experiences a force

$$\mathbf{F} = q\mathbf{E}'(\mathbf{r})$$

where $\mathbf{E}'$ is the field due to all *other* charges. Thus for a volume distribution of charge

$$\mathbf{F} = \iiint \rho(\mathbf{r})E'(\mathbf{r}) \, d^3\mathbf{r} \tag{8-17}$$

Obviously, for a surface or line distribution the appropriate dimensionality of integral would be used.

The trouble with this form is that it is sometimes tricky to separate the total field into an *external field* $\mathbf{E}'$ plus the *self-field* due to the charge element on which the force is being calculated. A more unambiguous formulation of the force expression is therefore desirable, and this can be obtained from the field energy $\mathfrak{U}_{el}$. In mechanics, force is defined in terms of the work done in a *virtual displacement*, that is, an infinitesimal shift $\delta\mathbf{r}$ in the position of the body in question. If the body is *rigid*, $\delta\mathbf{r}$ is constant over the region occupied by it, while if it is deformable, $\delta\mathbf{r}$ may be a function of the coordinates in that region. Mechanical work is the product of force by the parallel component of displacement. If $\mathbf{F}$ is the force due to the field, the agency must exert $-\mathbf{F}$ to move the body. Hence

$$\delta W_m = -\mathbf{F} \cdot \delta\mathbf{r} = -F_x \, \delta x - F_y \, \delta y - F_z \, \delta z \tag{8-18}$$

But by the *chain rule* of differentiation the change in any function $f(x, y, z)$ is given by

$$df = \frac{\partial f}{\partial x} dx + \frac{\partial f}{\partial y} dy + \frac{\partial f}{\partial z} dz$$

Hence we may identify the coefficients in Eq. (8-18) as

$$F_x = -\frac{\partial W_m}{\partial x} \text{ etc.} \tag{8-19}$$

or, on combining the three components into a vector,

$$\mathbf{F} = -\text{grad } W_m \tag{8-20}$$

Here the mechanical work W_m is understood to be expressed as a function of the position coordinates of the body or element in question.

It remains now to relate W_m to the field energy $\mathfrak{U}_{el}$ of the charge configuration. There are two distinct cases most commonly considered: The field may be set up by fixed charges plus (1) *isolated* charged conductors, or (2) conductors held at *fixed potentials*. The results are quite different in form.

Case 1 : Isolated Conductors

Because the system is isolated, no energy can enter or leave. Hence the mechanical work done adds to the stored energy. In other words,

$$\delta W_m = \delta \mathfrak{U}_{el} \tag{8-21}$$

Thus Eq. (8-20) gives

$$\mathbf{F} = -(\text{grad } \mathfrak{U}_{el})_{\text{const } q} \tag{8-22}$$

To illustrate the use of this formula, we start with an example that could be done much more easily by means of Eq. (8-17).

Example 8-8

Consider N point charges q_i at positions $\mathbf{r}_i$. Calculate the force on one of them.

Solution From Example 8-4 the interaction energy is

$$\mathfrak{U}_{\text{int}} = \frac{1}{8\pi\epsilon_0} \sum_{i=1}^{N} \sum_{j=1}^{N}{}' \frac{q_i q_j}{r_{ij}}$$

Let us designate the charge in question as charge 1 and separate out the terms pertaining to it; these will be all the terms in which either $i = 1$ or $j = 1$. (The prime on the summation sign reminds us that they cannot both be 1.)

$$\mathfrak{U}_{\text{int}} = \frac{1}{8\pi\epsilon_0} \left[2q_1 \sum_{k=2}^{N} \frac{q_k}{r_{1k}} + \sum_{i=2}^{N} \sum_{j=2}^{N}{}' \frac{q_i q_j}{r_{ij}} \right]$$

Only the terms in the first sum depend on $\mathbf{r}_1$. Equation (8-22) thus gives

$$\mathbf{F} = q_1 \sum_{k=2}^{N} -\frac{q_k}{4\pi\epsilon_0} \text{ grad } (|r_1 - r_k|)^{-1}$$

The sum is, of course, nothing but $E'(r_1)$, in agreement with Eq. (8-17). The advantage of the energy approach is that this comes out automatically without having to think about which parts of the field are to be included. The following examples bring this out in cases that are less obvious.

Example 8-9

Calculate the force per unit area on one of a pair of parallel plates carrying surface charge densities $\pm\sigma$ (Figure 8-6).

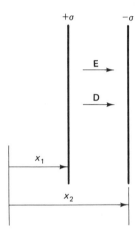

Figure 8-6 Charged parallel plates with variable spacing.

Solution   In the space between the plates, $E = \sigma/\epsilon_0$, so by Eq. (8-11), $u_{el} = \sigma^2/2\epsilon_0$. Thus for the unit area of the plates,

$$\mathcal{U}_{el} = u_{el}(x_2 - x_1) = \frac{\sigma^2}{2\epsilon_0}(x_2 - x_1)$$

The force on the negative plate is

$$\mathbf{F}_2 = -\frac{\partial U_{el}}{\partial x_2}\hat{\mathbf{i}} = -\frac{\sigma^2}{2\epsilon_0}\hat{\mathbf{i}}$$

Note that $\sigma/2\epsilon_0$ is the part of the field contributed by plate 1. Thus the result again corresponds to Eq. (8-17). In this case $E' = E/2$.

Example 8-10

Calculate the force per unit area at the surface of a charged conductor.

Solution   Since the field is zero inside, Eq. (8-15) gives

$$\mathcal{U}_{el} = \frac{\epsilon_0}{2}\iiint E^2\, d^3\mathbf{r}$$

where the integral is over all *exterior* space. If a small area A at point r is infinitesimally displace inward, say by amount δx (Figure 8-7), the exterior volume is increased by $A\,\delta x$ without significantly altering the field. Hence the

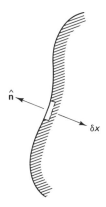

Figure 8-7   Virtual displacement of a small area of a charged conductor surface.

increase in energy is

$$\delta \mathfrak{U}_{el} = \frac{\epsilon_0}{2} A \, \delta x [\mathbf{E}(\mathbf{r})]^2$$

and

$$\frac{\mathbf{F}}{A} = -\frac{\epsilon_0}{2} E^2 \, \hat{\mathbf{i}} \tag{8-23}$$

This, too, is in agreement with Eq. (8-17). By Eq. (1-30), $\sigma = \epsilon_0 E$, so the foregoing result is $\mathbf{F} = \sigma(\mathbf{E}/2)$. That is, $\mathbf{E}' = \mathbf{E}/2$. Physically, this comes about as follows. The charge in the area under consideration produces field $\sigma/2\epsilon_0$ directed away from itself in both directions (Figure 8-8a). The inward part is canceled by the field due to all the other charges around the surface (Figure 8-8b), hence that must also amount to $\sigma/2\epsilon_0$, and the latter adds to the field outside (Figure 8-8c). Note that the surface of a conductor in a field is subjected to an *outward* stress. If the surface charge distribution is known, the net force on the body can be obtained by integrating this over the surface.

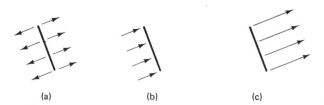

(a)   (b)   (c)

Figure 8-8   (a) Self-field of a surface charge element; (b) field of all the other charges; (c) resultant total field at the element.

For a volume charge distribution, all these complications disappear. The *force density* is simply

$$\mathbf{f}(\mathbf{r}) = \rho(\mathbf{r})\mathbf{E}(\mathbf{r}) \tag{8-24}$$

In other words, $\mathbf{E}' = \mathbf{E}$ in this case. The charge is so "dilute" that the self-field of any volume element is negligible. Only a *singularity* in the charge density gives rise to a significant self-field.

Case 2: Conductors at Fixed Potentials

The potential of each conductor is held fixed by, say, a battery (Figure 8-9). Let one of the conductors be given a vitrual displacement in which mechanical work δW_m is done by an external agency. There is now additional work done by the batteries in holding the potentials fixed. Thus the conservation of energy statement is, instead of Eq. (8-21),

$$\delta W_m = \delta \mathfrak{U}_{el} - \delta W_b \tag{8-25}$$

where δW_b is the work done by the batteries.

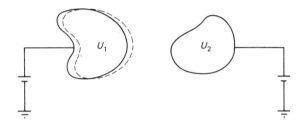

Figure 8-9   Conductors held at fixed potentials (by batteries) during a virtual displacement.

The change in field energy is

$$\delta \mathfrak{U}_{el} = \delta \left(\frac{1}{2} \sum_i U_i q_i \right)_{\text{const } U_i} \tag{8-26}$$

The batteries are required to move some charge in order to hold the potentials constant. From Eq. (8-3) the amount of work they have to do is

$$\delta W_b = \sum_i U_i \, \delta q_i \tag{8-27}$$

Thus the energy loss in the batteries is just twice the energy gain in the field, and

$$\delta W_m = \delta \mathfrak{U}_{el} - \delta W_b = -\delta \mathfrak{U}_{el} \tag{8-28}$$

Then Eq. (8-10) gives

$$\mathbf{F} = (\text{grad } \mathfrak{U}_{el})_{\text{const } U} \tag{8-29}$$

Example 8-11

Calculate the force per unit area on a plate of an air-dielectric capacitor charged to V volts.

Solution   We now want the energy expressed in terms of the potentials; from Example 8-6 this is

$$\mathfrak{U}_{el} = \frac{\epsilon_0 V^2}{2d} \text{ per unit area}$$

Then

$$\frac{F}{A} = \frac{\partial \mathfrak{U}_{el}}{\partial d} = -\frac{\epsilon_0 V^2}{2d^2}$$

The force is readily seen to be the same as in the constant-charge case (Example 8-9). This is as it should be, since once the capacitor is charged the force between plates cannot be changed simply by disconnecting the battery. In the present case, as the plates are pulled apart, the field energy decreases while the battery energy increases by just twice as much because the plate charges diminish. In the isolated-plate case, the field energy (and the potential difference) increase as the plates are pulled apart.

Torque

The *moment* of a force **F** about an origin O is defined as the product of the force and the perpendicular distance from the origin to the line of action of the force. As Figure 8-10 shows, this may be described as a vector whose direction is the axis of the rotation about O that **F** tends to produce. The *torque* on a body is defined as the *vector sum of the moments of all forces acting on it*. Thus, for a body consisting of several parts at positions $\mathbf{r}_i$ under forces $\mathbf{F}_i$,

$$\mathbf{T} = \sum_i \mathbf{r}_i \times \mathbf{F}_i \qquad (8\text{-}30)$$

Obviously, for a continuous body the sum becomes an integral.

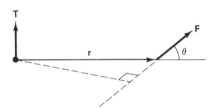

Figure 8-10   Torque **T** of a force **F** acting at a point **r**. The moment arm is $r \sin \theta$.

Example 8-12

Show that if the net force vanishes, the torque is independent of the choice of origin.

Solution   The proof follows the same lines as that of the corresponding theorem for dipole moment (Section 1-8).

Torques due to electrical forces can be calculated in terms of the field energy by considering *angular* virtual displacements. The same two major cases have to be considered.

The results for rotations about the z axis, as an example, are

$$T_z = -\left(\frac{\partial \mathfrak{U}_{el}}{\partial \phi}\right)_{\text{const } q} \qquad (8\text{-}31)$$

and

$$T_z = \left(\frac{\partial \mathfrak{U}_{el}}{\partial \phi}\right)_{\text{const } U} \qquad (8\text{-}32)$$

respectively.

8-3 DIPOLE INTERACTIONS

The interaction energy of a permanent dipole with an external field $\mathbf{E}'$ (not necessarily uniform) is readily obtained by starting with an "extended" dipole of length l centered at $\mathbf{r}_d$ (Figure 8-11). This has charges $-q$ at $\mathbf{r}_d - \mathbf{l}/2$ and q at $\mathbf{r}_d + \mathbf{l}/2$. Hence, from Eq. (8-1),

$$\mathcal{U}_{int} = (-q)U'\left(\mathbf{r}_d - \frac{\mathbf{l}}{2}\right) + qU'\left(\mathbf{r}_d + \frac{\mathbf{l}}{2}\right)$$

$$= q[\mathbf{l} \cdot \text{grad } U'(\mathbf{r}_d) + \cdots]$$

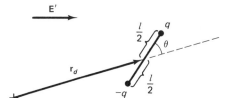

Figure 8-11 Extended (nonideal) dipole in an external field.

On letting l shrink to zero the higher terms vanish, leaving

$$\mathcal{U}_{int} = \mathbf{p} \cdot \text{grad } U'(\mathbf{r}_d) \equiv -\mathbf{p} \cdot \mathbf{E}'(\mathbf{r}_d) \tag{8-33}$$

Note that the energy depends not only on the position but also on the *orientation* of the dipole relative to the field. The energy is a minimum $(-pE')$ when the dipole points along the field.

Example 8-13

Calculate the interaction energies between two dipoles $\mathbf{p}_1$ and $\mathbf{p}_2$ at separation r_{12} (Figure 8-12). The orientations of both dipoles are arbitrary.

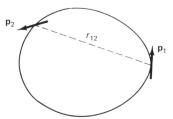

Figure 8-12 Dipole $\mathbf{p}_2$ in field produced by dipole $\mathbf{p}_1$.

Solution In Eq. (8-33) let $\mathbf{E}'$ be the field produced by $\mathbf{p}_1$. From Example 1-16, solution 1, we then get

$$\mathcal{U}_{int} = \frac{3(\mathbf{p}_1 \cdot \mathbf{r}_{12})(\mathbf{p}_2 \cdot \mathbf{r}_{12})}{r_{12}^5} - \frac{\mathbf{p}_1 \cdot \mathbf{p}_2}{r_{12}^3} \tag{8-34}$$

Note that the interaction energy is symmetrical in the two moments, as would be expected.

The force exerted on a dipole is obtained as in the preceding section. Since the charges are fixed,

$$\mathbf{F} = -\text{grad } \mathfrak{U}_{\text{int}} = \text{grad } (\mathbf{p} \cdot \mathbf{E}')$$

By using the vector identity for the gradient of a dot product, together with the facts that $\mathbf{p}$ is a constant vector and curl $\mathbf{E}' = 0$, this can be put in the form

$$\mathbf{F} = (\mathbf{p} \cdot \text{grad})\mathbf{E}' \qquad (8\text{-}35)$$

Thus there is no force if the external field is uniform in the direction of the dipole. If the external field is nonuniform, the force is the *net unbalance* of the slightly unequal forces on the two charges. In using Eq. (8-35) it must be borne in mind that since derivatives are to be taken, $\mathbf{E}'$ must be calculated *as function of the coordinates* of the dipole, not just at a fixed position. The following example illustrates this.

Example 8-14

Two dipoles are oriented as shown in Figure 8-13. Calculate the force on p_2.

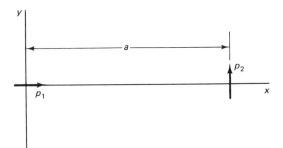

Figure 8-13   Pair of interacting dipoles of Example 8-14.

Solution   Let $\mathbf{r} = x\hat{\mathbf{i}} + y\hat{\mathbf{j}}$. (Since the problem lies entirely in the plane, we can omit z for simplicity.) From Example 1-16 the field due to $\mathbf{p}_1$ is

$$\mathbf{E}'(\mathbf{r}) = \frac{(p_1 x)(x\hat{\mathbf{i}} + y\hat{\mathbf{j}})}{(x^2 + y^2)^{5/2}} - \frac{p_1\hat{\mathbf{i}}}{(x^2 + y^2)^{3/2}}$$

Then

$$(\mathbf{p}_2 \cdot \text{grad})\mathbf{E}' = p_2\frac{\partial \mathbf{E}'}{\partial y}$$

$$= p_1 p_2 \left[\frac{x\hat{\mathbf{j}}}{(x^2 + y^2)^{5/2}} - \frac{5x(x\hat{\mathbf{i}} + y\hat{\mathbf{j}})y}{(x^2 + y^2)^{7/2}} + \frac{3\hat{\mathbf{i}}y}{(x^2 + y^2)^{5/2}} \right]$$

Finally, putting $x = a$, $y = 0$, this gives

$$\mathbf{F}_2 = \frac{p_1 p_2}{a^4}\hat{\mathbf{j}}$$

Thus there is a net upward force. This arises because the applied field has a slight upward component at the positive end and a slight downward component at the negative end of $\mathbf{p}_2$.

We next calculate the torque on a permanent dipole in a *uniform* external field. Since there is no net force in a uniform field we need not specify any particular origin about which the torque is calculated. If we consider an infinitesimal increase in the angle θ between dipole and field, then

$$T = -\frac{\partial \mathcal{U}_{\text{int}}}{\partial \theta} = -\frac{\partial}{\partial \theta}(pE' \cos \theta) = pE' \sin \theta$$

The magnitude and direction of this torque correspond to the vector relation

$$\mathbf{T} = \mathbf{p} \times \mathbf{E}'(\mathbf{r}_d) \qquad (8\text{-}36)$$

This torque is a turning couple which simply tends to rotate the dipole into the minimum-energy orientation parallel to the field.

In a nonuniform field, we must specify the origin. We could consider rotations about the origin as the most fundamental approach. However, we already know the force from Eq. (8-34), so it is simpler just to add the moment of this force about the origin to the turning couple about the dipole center, that is,

$$\mathbf{T} = \mathbf{p} \times \mathbf{E}' + \mathbf{r}_d \times (\mathbf{p} \cdot \text{grad})\mathbf{E}' \qquad (8\text{-}37)$$

Example 8-15

For the dipole arrangement of Example 8-14, calculate the torque on each dipole about the center of p_1.

Solution   On p_2, $\mathbf{p}_2 \times \mathbf{E}_1(a, 0, 0) = 0$. Thus

$$\mathbf{T}_2 = a\hat{\mathbf{i}} \times \frac{p_1 p_2}{a^4}\hat{\mathbf{j}} = \frac{p_1 p_2}{a^3}\hat{\mathbf{k}}$$

On p_1, force (if any) has zero moment arm. Hence

$$\mathbf{T}_1 = p_1\hat{\mathbf{i}} \times \mathbf{E}_2(0, 0, 0)$$

$$= p_1\hat{\mathbf{i}} \times \frac{-p_2\hat{\mathbf{j}}}{a_3} = -\frac{p_1 p_2}{a^3}\hat{\mathbf{k}}$$

Thus the torques about a specified center are equal and opposite, in accordance with Newton's third law. The turning couples (being torques about *different* centers) are not necessarily equal and opposite.

8-4 ENERGY AND FORCES IN DIELECTRICS

Ideal Dielectrics

We first consider an idealized hypothetical class of dielectric material assumed to be completely characterized by a single real dielectric constant. For such a material, the energy and force expressions of the preceding sections apply directly. Some of the complications that ensue in real dielectrics will be briefly discussed later.

Example 8-16

A dielectric slab can slide freely between a pair of capacitor plates held under potential difference V (Figure 8-14). Find the force on the slab when it is inserted partway. Neglect edge effects.

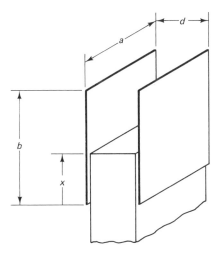

Figure 8-14 Parallel-plate capacitor with movable dielectric slab filling the gap.

Solution   From Example 8-6 the field energy is

$$\mathfrak{U}_{el} = \frac{1}{2}ad\left[(x)\kappa\epsilon_0\left(\frac{V}{d}\right)^2 + (b - x)\epsilon_0\left(\frac{V}{d}\right)^2\right]$$

$$= \frac{1}{2}\epsilon_0 a\frac{V^2}{d}[(\kappa - 1)x + b]$$

Since this is the constant-potential case, the force is $(+)$ the derivative of $\mathfrak{U}_{el}$ with respect to the displacement coordinate x, that is,

$$\mathbf{F}_x = \tfrac{1}{2}ad(\kappa - 1)\epsilon_0\mathbf{E}^2$$

Note that the force is not in the direction of the field (it is not a force on a charge) but in the direction of *decrease of dielectric constant* of the medium. This is a general result, and is always such as to draw a dielectric into the strongest part of the field.

Example 8-17

Repeat the preceding calculation for the case that the plates are charged to $\pm q$, then isolated.

Solution   First calculate the charge distribution on the plates; since the latter are equipotentials, the field strength is the same in both regions. Then use Eq. (8-22) to get the force. Naturally, it must come out the same.

Example 8-18

A pair of charged isolated capacitor plates are held apart by a (nonconductive) spring. The assembly is then immersed in a liquid dielectric. Which way do the plates move?

Solution   The force is calculated just as in Example 8-9 except that $\kappa\epsilon_0$ replaces ϵ_0 after immersion. Thus the electrostatic force is weakened and the plates move apart.

The preceding few problems have been examples in which the calculation of the energy as a function of a position coordinate of some body was particularly simple. The reason for the simplicity is that the geometry was such that the field was *uniform* (and easily calculated) throughout the material. The calculation of the fields in somewhat more complicated geometries is taken up in Chapter 11.

In the next few paragraphs we develop a general formulation of the force on an element of a dielectric material. The subject is somewhat complicated, and most of it may be skipped without loss of continuity. However, the reader should pay careful attention to the *power theorem*, as this will be used very extensively in later chapters.

The change in energy accompanying a general virtual displacement of a dielectric is obtained from Eq. (8-14):

$$\delta\mathfrak{U}_{\text{el}} = \frac{1}{2}\delta\left[\iiint \mathbf{E}\cdot\mathbf{D}\,d^3\mathbf{r}\right] = \frac{1}{2}\delta\iiint \frac{D^2}{\epsilon}\,d^3\mathbf{r}$$

$$= \frac{1}{2}\iiint\left(\frac{2D}{\epsilon}\,\delta D - \frac{D^2}{\epsilon^2}\,\delta\epsilon\right)d^3\mathbf{r} \tag{8-38}$$

A number of points about this equation require some discussion.

1. The integral is over *all space*. Of course, the integrand is zero wherever there is no field, such as in the interior of conductors.

2. ϵ is to be regarded as a function of the position coordinates (variables of integration). It has the value ϵ_0 in the empty space and $\kappa\epsilon_0$ in the dielectric. The latter is *not necessarily homogeneous*, that is, κ may also be a function of position (see Figure 8-15).

3. The quantity $\delta\epsilon$ in the second term is also a function of position; it is the amount of change that occurs at any point as a result of the motion of the body.

The integral of the first term in Eq. (8-38) vanishes if the dielectric contains no true charge, as we now show. The integral in question is

$$\iiint_\infty \mathbf{E}\cdot\delta\mathbf{D}\,d^3\mathbf{r}$$

This can be put in terms of the change in *charge density* by a procedure similar to that

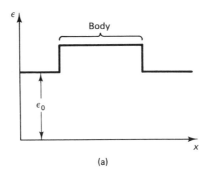

(a)

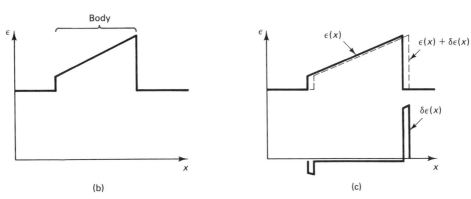

(b)

(c)

Figure 8-15 Schematic representation of $\epsilon(\mathbf{r})$ for (a) a homogeneous dielectric body; (b) an inhomogeneous dielectric body; (c) the same body displaced.

used in deriving Eq. (8-14). We use the identity

$$\text{div } (U \,\delta\mathbf{D}) \equiv U \,\text{div}(\delta\mathbf{D}) + \text{grad } U \cdot \delta\mathbf{D}$$
$$= U \,\delta\rho - \mathbf{E} \cdot \delta\mathbf{D}$$

Thus the integral is

$$\iiint_{\infty} U \,\delta\rho \, d^3\mathbf{r} - \iiint_{\infty} \text{div } (U \,\delta\mathbf{D}) \, d^3\mathbf{r} = \iiint_{\infty} U \,\delta\rho \, d^3\mathbf{r} - \oiint_{S(\infty)} U \,\delta\mathbf{D} \cdot \hat{\mathbf{n}} \, dS$$

The second term vanishes for a surface at infinity when all the bodies are located in a finite region of space, as in the derivation cited. Concerning the first term, we note that $\delta\rho$ can have a nonzero value in only two kinds of regions: the moving dielectric or the surfaces of conductors. The conditions of the problem (i.e., uncharged dielectrics) rule out the former, and the integral resulting from the latter vanishes by Thomson's theorem.

For the second term in Eq. (8-38) we need an expression for $\delta\epsilon$ in terms of the infinitesimal displacement $\delta\mathbf{r}$ imparted to the dielectric. The material that ends up at point $\mathbf{r}$ is what was originally at $\mathbf{r} - \delta\mathbf{r}$. Thus the change in ϵ at point $\mathbf{r}$ is

$$\delta\epsilon(\mathbf{r}) = \epsilon(\mathbf{r} - \delta\mathbf{r}) - \epsilon(\mathbf{r})$$
$$= -\delta\mathbf{r} \cdot \text{grad } \epsilon \tag{8-39}$$

Thus the second term in the integrand of Eq. (8-38) has nonzero value only where the dielectric constant is *varying* as a function of position. Usually this will occur only in the vicinity of the surfaces of the dielectric bodies. It is mathematically convenient, as well as physically realistic, to treat the surfaces as being somewhat "smeared out," so that ϵ is a *continuous* function. So Eq. (8-38) finally becomes

$$\delta\mathfrak{U}_{el} = \frac{1}{2} \iiint (E^2 \text{ grad } \epsilon) \cdot \delta\mathbf{r} \, d^3\mathbf{r}$$

By examining the integrand, it is evident that the force density is

$$\mathbf{f} = -\frac{1}{2} E^2 \text{ grad } \epsilon \qquad (8\text{-}40)$$

As in the examples above, it is seen that this is always perpendicular to dielectric surfaces if the dielectric bodies are homogeneous.

Real Dielectrics

The nature of the dielectric response in various types of materials was discussed in Chapter 4. It was seen there that the "dielectric constant" is not constant at all but is a function of such variables as field strength, temperature, and frequency. This leads to a variety of complications in dealing with energy and forces. We shall not be able to go into these in detail, but will mention some of them quite briefly.

One class of complications has to do with *thermal effects*, of which there are two basic types. The first arises whenever the dielectric constant is temperature dependent, for then polarization will lead to a flow of heat. To see this, consider the case of polar dielectrics (Section 4-4), which have a negative temperature coefficient. An increase in field strength decreases the randomness of the dipole orientations, and since this is just what a decrease of temperature would do, heat tends to flow out of the material. If the heat flow is impeded the temperature will rise, and this in turn will change the dielectric constant.

In order for the integrand in Eq. (8-6) to represent an amount of work, the potential at every point must remain unchanging as the increment of charge is added. Thus the dielectric properties of the medium cannot be allowed to change. From the foregoing discussion, this is seen to require that the temperature be held constant. In other words, Eq. (8-6) should strictly be written

$$(\delta W)_{\text{const } T} = \iiint_{\infty} U(\mathbf{r}) \, \delta\rho(\mathbf{r}) \, d^3\mathbf{r} \qquad (8\text{-}41)$$

From the derivation earlier in this section it is easily seen that this can also be written as

$$(\delta W)_{\text{const } T} = \iiint_{\infty} \mathbf{E}(\mathbf{r}) \cdot \delta\mathbf{D}(\mathbf{r}) \, d^3\mathbf{r} \qquad (8\text{-}42)$$

In obtaining Eq. (8-11), we used the supposition that all the assembly work was stored as field energy. But in order to hold the temperature constant, heat must be allowed to flow into or out of the material (from a "heat reservoir" assumed to be in contact with it). Any outflow of heat carries away some energy, which comes from the assembly work at the expense of the stored energy. In other words,

$$(\delta W)_{\text{const } T} = \delta \mathfrak{U}_{\text{el}} + \delta W_{\text{thermal}} \tag{8-43}$$

Thermodynamically, the combination on the right is called the *free energy*. It plays exactly the same role in respect to forces and torques as the total field energy did in the temperature-independent cases previously discussed.

Another type of thermal effect occurs when the polarization cannot instantaneously follow the field. In Section 4-4 it was seen that this situation can be described by means of a *complex dielectric response function*. We want to calculate the work done on a unit volume of such a material in a complete cycle of an alternating electric field $E_0 \cos \omega t$. From Eqs. (4-29) and (4-30) it can be seen that

$$\mathbf{D}(t) = \epsilon_0 E_0 (\kappa_{\text{re}} \cos \omega t + \kappa_{\text{im}} \sin \omega t)$$

Thus the change in a short time interval δt is

$$\delta \mathbf{D} = \omega \epsilon_0 E_0 (-\kappa_{\text{re}} \sin \omega t + \kappa_{\text{im}} \cos \omega t)\, \delta t$$

and Eq. (8-42), after integration over one cycle, gives

$$\int_0^{t=2\pi/\omega} \delta W = \omega \epsilon_0 E_0^2 \int_0^{2\pi/\omega} (-\kappa_{\text{re}} \sin \omega t \cos \omega t + \kappa_{\text{im}} \cos^2 \omega t)\, dt$$

$$= \pi \epsilon_0 E_0^2 \kappa_{\text{im}} \tag{8-44}$$

Thus the energy loss depends on the *imaginary part* of the dielectric constant. It is customary to quote the average loss rate rather than the loss per cycle. The preceding result is simply divided by the time per cycle to obtain

$$\text{average loss rate per unit volume} = \tfrac{1}{2}\omega\epsilon_0\kappa_{\text{im}}E_0^2 \tag{8-45}$$

Power Theorem

The calculation just given, namely the time average of the product of two quantities that vary sinusoidally at the same frequency, will recur in many different contexts as we proceed. Accordingly, we now digress briefly to show how it can be done much more easily in the complex exponential notation.

If $f = f_0 e^{-i\omega t}$ and $g = g_0 e^{-i\omega t}$, then

$$\overline{\text{Re}(f)\,\text{Re}(g)} = \tfrac{1}{2}\text{Re}(fg^*) \tag{8-46}$$

where the bar denotes *time average over any integral number of cycles* and the asterisk denotes *complex conjugate*. One proof is to write out the real parts, multiply, and integrate, much as was done to get Eq. (8-44). A more elegant proof is as follows. Since the real part of any complex number is one-half the sum of the number plus its complex conjugate,

$$\text{Re}(f)\,\text{Re}(g) = \tfrac{1}{4}(f + f^*)(g + g^*)$$

$$= \tfrac{1}{4}(fg + f^*g^* + fg^* + f^*g)$$

The first two terms each vary at frequency 2ω, hence average to zero. In the other two, the time factors *cancel out*, so the time averages are equal to the terms themselves. Since these terms are *complex conjugates of each other*, Eq. (8-46) follows.

Example 8-19

Derive Eq. (8-45) by means of the power theorem.

Solution $\mathbf{E} = \mathbf{E}_0 e^{-i\omega t}$, $\mathbf{D} = \kappa\epsilon_0\mathbf{E}$, so

$$\dot{\mathbf{D}} = -i\omega\kappa\epsilon_0\mathbf{E}$$

$$\overline{\mathbf{E}\cdot\dot{\mathbf{D}}} = \tfrac{1}{2}\mathrm{Re}[(\mathbf{E}_0 e^{-i\omega t})(i\omega\epsilon_0\kappa^*\mathbf{E}_0 e^{i\omega t}]$$

$$= \tfrac{1}{2}\omega\epsilon_0 E_0^2\,\mathrm{Re}\,(i\kappa^*)$$

Since $i\kappa^* = i\kappa_{\mathrm{re}} + \kappa_{\mathrm{im}}$, Eq. (8-45) follows.

In addition to energy losses and other thermal effects, a second class of difficulties in dealing with fields in matter is that the forces will generally *distort* the bodies, again leading to alteration of their dielectric properties. For example, the polarization mechanisms discussed in Section 4-4 show that compression will increase the dielectric constant. Such distortions are called *electrostriction* effects. They are often bypassed by restricting the discussion to *rigid solids* and *incompressible liquids*. Even in such idealized materials there can be internal stresses (e.g., pressure in a liquid). However, the latter give rise to *no net forces* on the bodies. Further discusson of these effects is beyond our level.

Nonlinear Dielectrics

To conclude this chapter it is worthwhile to remind the reader once again that nearly everything in it is valid only for *linear* dielectric media. In case nonlinear dielectrics must be dealt with, we have to go all the way back to Eq. (8-6) for the *incremental* work done in adding an increment of charge. This can be made somewhat more useful by putting it in terms of the field strengths. The procedure is just like that leading to Eq. (8-14) except that we deal with increments of displacement $\delta\mathbf{D}$ instead of $\mathbf{D}$ itself. The starting identity is, therefore,

$$\mathrm{div}\,(U\,\delta\mathbf{D}) = U\,\mathrm{div}\,\delta\mathbf{D} + (\mathrm{grad}\,U)\cdot\delta\mathbf{D}$$

and the final result is

$$\delta\mathfrak{U}_{\mathrm{el}} = \iiint \mathbf{E}(\mathbf{r})\cdot\delta\mathbf{D}(\mathbf{r})\,d^3\mathbf{r} \tag{8-47}$$

For linear dielectrics this obviously can be time-integrated directly to give Eq. (8-14) again. For nonlinear media, the explicit relation between $\mathbf{D}$ and $\mathbf{E}$ would have to be known.

FURTHER REFERENCES

L. D. LANDAU and E. M. LIFSCHITZ, *Electrodynamics of Continuous Media*, Pergamon Press Ltd., Oxford, 1960.

R. BECKER and F. SAUTER, *Electromagnetic Fields and Interactions*, Vol. I: *Electromagnetic Theory and Relativity*, Blaisdell Publishing Co., New York, 1964, Chap. BIII.

W. K. H. PANOFSKY and M. PHILLIPS, *Classical Electricity and Magnetism*, Addison-Wesley Publishing Co., Inc., Reading, Mass., 1960, Chap. 6.

PROBLEMS

8-1. A spherical shell of 10 cm radius has a charge of 0.001 C uniformly distributed over its surface. How much work is required to move a 1-μC point charge from infinity to the center?

8-2. How much work is required to fill the shell in Problem 8-1 with a uniform charge density of 10^{-15} C/cm^3?

8-3. For the same spherical shell, how much work was required to assemble it?

8-4. Four equal point charges are located at the corners of a square of side a. Calculate the electrostatic interaction energy.

8-5. Calculate the energy density at distance r from the center of a uniform spherical shell of radius a and total charge q.

8-6. A point charge q, initially isolated, is surrounded by a grounded hollow metallic sphere of inner radius a. What is the change in electrostatic energy?

8-7. For the uniformly charged sphere of Example 8-2, calculate the energy by means of Eq. (8-15).

8-8. A capacitor of capacitance C has charges $\pm Q$ on its plates. How much energy is stored?

8-9. A charge of 1 μC is spread uniformly on a thin flexible spherical membrane of 10 cm radius. How much external pressure is required to start collapsing the membrane?

8-10. Derive Eq. (8-35) from the equation just preceding it in the text.

8-11. For the two dipoles of Example 8-14, calculate the force on p_1.

8-12. A metal sphere of radius a is charged to potential U_0. What is the force on an atom whose polarizability is α, at a distance r from the center ($r > a$)?

8-13. Estimate the acceleration of a hydrogen atom when it is 0.2 mm from the center of a metal sphere of radius 0.1 mm charged to 10,000 V.

8-14. Two dipoles are separated by distance R and are free to pivot in all directions about their centers. What are the maximum and minimum interaction energies, and the orientations in which they occur?

8-15. One of the dipoles in Problem 8-14 is fixed perpendicular to the line of centers. What are the maximum and minimum interaction energies under this constraint?

8-16. If the two dipoles are parallel to each other, write the interaction energy as function of the angle θ that each makes with the line of centers.

8-17. Calculate the turning couple (torque about its own center) on each of the dipoles of Problem 8-16. Assume that they lie in the xy plane with their centers on the x axis.

8-18. Why is there no turning couple on either of the dipoles in Problem 8-17 when $\theta = 90°$, even though it is not the minimum-energy orientation?

8-19. Calculate the force on a polarizable molecule (see Section 4-4) in an external field.

8-20. Two time-dependent quantities are described in complex exponential notation as $(A_1 + iA_2)e^{-i\omega t}$ and $(B_1 + iB_2)e^{-i\omega t}$. Calculate the time-average value of their product.

8-21. Two time-dependent quantities are described in complex exponential notation as $R_1 e^{i(\phi_1 - \omega t)}$ and $R_2 e^{i(\phi_2 - \omega t)}$, where R_1 and R_2 are real. Calculate the time-average value of their product.

chapter 9

Magnetic Energy

9-1 INTERACTION OF CURRENTS

To obtain the energy of an electrostatic field we calculated the work necessary to build up the source charge distribution. Similarly, for the energy of a magnetostatic field we calculate the work necessary to build up the source *current* distribution. The difference is that current must be built up in finite loops rather than infinitesimal volume elements.

For a single current loop, the work done by the battery (or other source of emf) in driving a current for a short time interval δt is

$$\delta W_b = \mathcal{E}_b I \, \delta t$$

The current is governed by the *total* emf in the circuit, including the battery emf and the induced emf, and by the resistance of the circuit. By Ohm's law, Eq. (5-7),

$$I = \frac{\mathcal{E}_b + \mathcal{E}_{\text{ind}}}{R} = \frac{\mathcal{E}_b - d\Phi/dt}{R}$$

Thus, on solving for $\mathcal{E}_b$ and inserting it in the preceding equation, we obtain

$$\delta W_b = \left(I^2 R + I \frac{d\Phi}{dt} \right) \delta t \qquad (9\text{-}1)$$

The first term on the right is the energy dissipated in the resistance of the circuit. The remainder of the energy lost by the battery must be *stored in the magnetic field*. The result is easily generalized to any number of current loops, that is,

$$\delta \mathcal{U}_{\text{mag}} = \sum_i I_i \, \delta \Phi_i \qquad (9\text{-}2)$$

Example 9-1

Consider a set of fixed rigid circuits. Suppose that currents are built up "in proportion," that is, in such a manner that

$$I_i(t) = f(t) I_{i,\text{ final}} \qquad (0 \leq f \leq 1)$$

Carry out the time integration of Eq. (9-2), assuming that the medium has linear magnetic properties.

Solution   Because of the assumed linearity, the magnetic field strength at any point will build up in proportion to the currents. Hence the same is true of the flux through any of the current loops, that is,

$$\Phi_i(t) = \sum_j M_{ij} I_j(t)$$

where the M_{ij} are the mutual inductances or, when the two indexes are equal, the self-inductances (see Section 2-6) of the loops. Thus Eq. (9-2) becomes

$$\delta \mathcal{U}_{\text{mag}} = \sum_i (f I_{i,\text{ final}}) \left(\sum_j \frac{df}{dt} \delta t \, M_{ij} I_{j,\text{ final}} \right)$$

$$\mathcal{U}_{\text{mag}} = \sum_i \sum_j M_{ij} I_{i,\text{ final}} I_{j,\text{ final}} \int_0^1 f \, df$$

or, dropping the subscript "final,"

$$\mathcal{U}_{\text{mag}} = \tfrac{1}{2} \sum_i \sum_j M_{ij} I_i I_j \tag{9-3}$$

$$= \tfrac{1}{2} \sum_i I_i \Phi_i \tag{9-4}$$

Example 9-2

For two circuits, show that the magnetic energy is always positive, regardless of the magnitudes and senses of the currents or the geometry of the circuits.

Solution   Equation (9-3) is

$$\mathcal{U}_{\text{mag}} = \tfrac{1}{2}(L_1 I_1^2 + 2M I_1 I_2 + L_2 I_2^2)$$

$$= \tfrac{1}{2} I_1^2 (L_1 + 2MX + L_2 X^2)$$

where $X = I_2/I_1$ and $M = M_{12} = M_{21}$. Now the quadratic expression is certainly positive for $X = 0$. Furthermore, by Example 2-27, its discriminant is negative, so it has no real roots. Hence it is positive for all real values of X.

We can generalize Eq. (9-4) to the case where the current is distributed throughout a volume rather than in individual filaments. On using Eq. (2-31) for the flux and then replacing $I \, dl$ by $\mathbf{J} \, d^3\mathbf{r}$, we obtain

$$\mathcal{U}_{\text{mag}} = \tfrac{1}{2} \iiint \mathbf{J}(\mathbf{r}) \cdot \mathbf{A}(\mathbf{r}) \, d^3\mathbf{r} \tag{9-5}$$

This is the magnetic analog of Eq. (8-11). Again, even though the integral formally extends over all space, the integrand is nonzero only where there is current density.

Energy in Terms of Field Strengths

From Eq. (9-5) we can derive an expression for the magnetic energy in terms of the field strengths. Since $\mathbf{J} = \text{curl } \mathbf{H}$ (for static fields) and $\mathbf{B} = \text{curl } \mathbf{A}$, it seems clear that we can work the expression into the desired form. The trick is to use the identity

$$\text{div}\,(\mathbf{A} \times \mathbf{H}) = -\mathbf{A} \cdot \text{curl}\,\mathbf{H} + \mathbf{H} \cdot \text{curl}\,\mathbf{A} = -\mathbf{A} \cdot \mathbf{J} + \mathbf{H} \cdot \mathbf{B}$$

On substituting this into the equation, the term involving the divergence vanishes just as in the derivation of Eq. (8-14). This leaves

$$\mathfrak{U}_{\text{mag}} = \tfrac{1}{2} \iiint \mathbf{H}(\mathbf{r}) \cdot \mathbf{B}(\mathbf{r})\, d^3\mathbf{r} \qquad (9\text{-}6)$$

This is the magnetic analog of Eq. (8-14). Note that here the integral really does extend over all space, since $\mathbf{H}$ and $\mathbf{B}$ do not vanish at any finite distance from the sources. Again it is natural to interpret the integrand as an energy density,

$$u_{\text{mag}} = \tfrac{1}{2}\mathbf{H} \cdot \mathbf{B} \qquad (9\text{-}7)$$

Since the expression is valid only in linear magnetic media, it is more properly written

$$u_{\text{mag}} = \tfrac{1}{2}\mu\mathbf{H}^2 \qquad (9\text{-}8)$$

Example 9-3

Calculate the energy per unit length near the middle of a long solenoid.

Solution   The field of a solenoid is discussed in Examples 2-8 and 2-12. From these we obtain

$$H = \frac{NI}{L}$$

$$\mathfrak{U}_{\text{mag}} = \frac{1}{2}\,\mu_0 \left(\frac{NI}{L}\right)^2 (\pi a^2)$$

Example 9-4

A toroidal solenoid has inner radius a and square cross section of side l (Figure 9-1). Calculate the field energy when current I flows in it.

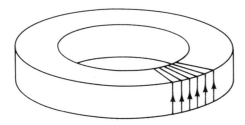

Figure 9-1   Toroidal solenoid of square cross section, discussed in Example 9-4.

Solution   From Example 2-16 the field is $H = NI/2\pi s$; thus

$$u_{\text{mag}} = \frac{1}{2}\,\mu_0 \left(\frac{NI}{2\pi s}\right)^2$$

$$\mathfrak{U}_{\text{mag}} = \frac{1}{2} \int_a^{a+l} \mu_0 \left(\frac{NI}{2\pi}\right)^2 \frac{l}{s^2}\, 2\pi s\, ds$$

$$= \frac{\mu_0}{4\pi}(NI)^2 l \ln \frac{a+l}{a}$$

The *interaction* energy obviously comprises the terms in Eq. (9-3) that involve *mutual* inductances, not the self-inductances.

Example 9-5

Calculate the interaction energy between a long straight wire carrying current I_1 and a coplanar parallel rectangular loop carrying current I_2 (Figure 9-2).

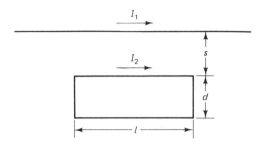

Figure 9-2   Circuit for Examples 9-5 and 9-6.

Solution   The mutual inductance is obtained from Example 2-21 as

$$M_{12} = M_{21} = \frac{\mu_0 l}{2\pi} \ln \frac{s + d}{s}$$

Thus

$$\mathfrak{U}_{\text{int}} = I_1 I_2 \frac{\mu_0 l}{2\pi} \ln \frac{s + d}{s}$$

9-2 FORCES AND TORQUES

The total magnetic force on a distribution of current is readily written from the definition of magnetic field [Eq. (2-9)].

$$\mathbf{F} = \iiint \mathbf{J}(\mathbf{r}) \times \mathbf{B}'(\mathbf{r}) \, d^3\mathbf{r} \qquad (9\text{-}9)$$

where B' is the field due to all *other* current elements. Thus, just as with electrostatic forces, we are faced with the problem of separating the *external field* $\mathbf{B}'$ acting on a given element from the *self-field* due to that element. The problem is again made easier by the use of the energy method. Thus, as in Section 8-2, we consider a virtual displacement of some part of the current distribution and express the work done in this displacement in terms of the changes in stored energy in the field and the current sources. In contrast to the electrostatic situation, however, there is only one main case to consider: that of *constant currents*. (The other case would correspond to constant fluxes; it may have some relevance to persistent current in superconductors, but we shall not discuss it.)

Thus we consider a set of current loops with the currents held constant. We wish to relate the work δW_m done in a virtual displacement of one of the loops to the changes in energy stored in the field and supplied by the current sources.

The energy balance equation is, in analogy with Eq. (8-21),

$$\delta W_m = \delta \mathcal{U}_{\mathrm{mag}} - \delta W_b' \tag{9-10}$$

where $\delta W_b'$ is the *extra* loss of energy by the batteries, over and above that required to supply the $I^2 R$ losses. From Eq. (9-2),

$$\delta W_b' = \sum_i I_i \, \delta \Phi_i \tag{9-11}$$

while from Eq. (9-4), under the conditions of constant current,

$$\delta \mathcal{U}_{\mathrm{mag}} = \tfrac{1}{2} \sum_i I_i \, \delta \Phi_i = \tfrac{1}{2} \, \delta W_b' \tag{9-12}$$

Equation (9-10) then is

$$\delta W_m = \delta \mathcal{U}_{\mathrm{mag}} \tag{9-13}$$

Hence, from Eq. (8-20), the force on the virtually displaced entity is

$$F = (\mathrm{grad} \ \mathcal{U}_{\mathrm{mag}})_{\mathrm{const} \ I} \tag{9-14}$$

In order to use this equation, $\mathcal{U}_{\mathrm{mag}}$ must, of course, be expressed as a function of the variable that is to be subjected to the displacement.

Example 9-6

Find the force on the rectangular loop in Example 9-5.

Solution

$$F = I_1 I_2 \frac{\mu_0 l}{2\pi} \, \mathrm{grad} \left[\ln \left(\frac{s+d}{s} \right) \right]$$

$$= I_1 I_2 \frac{\mu_0 l}{2\pi} \left(\frac{1}{s+d} - \frac{1}{s} \right) \hat{\mathbf{a}}_s$$

Note that the force is attractive for the current sense shown (parallel in the nearer segment). Also, it is obvious that the result is just what one would obtain directly from Eq. (9-9) using for $\mathbf{B}'$ the field due to the straight wire.

Example 9-7

Two long solenoids have n turns per meter and essentially equal radii so that they can slide freely over each other as shown in Figure 9-3. Calculate the force between them when they overlap by distance x.

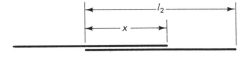

Figure 9-3 Coaxial solenoids as discussed in Example 9-7.

Solution   From Example 9-3,

$$\mathcal{U}_{\text{mag}} = \frac{1}{2}\,\mu_0\pi a^2 \left(\frac{N}{l}\right)^2 [(l_1 - x)I_1^2 + x(I_1 + I_2)^2 + (l_2 - x)I_2^2$$

$$F = \frac{1}{2}\,\mu_0\pi a^2 \left(\frac{N}{l}\right)^2 [-I_1^2 + (I_1 + I_2)^2 - I_2^2]$$

$$= \mu_0\pi a^2 I_1 I_2$$

It is seen that the force is attractive (tending to increase the overlap x) when the currents flow in the same sense.

Example 9-8

Calculate the "magnetic pressure" in a solenoid, that is, the force per unit area tending to expand the solenoid radially.

Solution   Imagine a small area A of the winding to be displaced radially outward by δs. The field strength, and thus the energy density, is essentially unchanged, but the volume is increased by $A\,\delta s$. Thus

$$\delta\mathcal{U}_{\text{mag}} = u_{\text{mag}} A\,\delta s$$

$$\text{pressure} = \frac{F}{A} = u_{\text{mag}} = \frac{1}{2}\mathbf{B}\cdot\mathbf{H} \tag{9-15}$$

Example 9-9

Evaluate the magnetic pressure corresponding to a field strength of 10 T in vacuum.

Solution $B = 10$ T, $H = (10/\mu_0)$ amperes per meter, so

$$u_{\text{mag}} = \frac{10^2}{2\times 4\pi\times 10^{-7}} \cong 3.98\times 10^7\ \text{N/m}^2$$

This is about 400 atmospheres! Thus the production of strong fields requires great mechanical strength in the windings.

Example 9-10

A cylindrical conductor of radius a carries current I uniformly distributed over its cross-sectional area. Calculate the stress (pressure or tension) at the surface.

Solution   The field is found in Example 2-13, from which

$$u_{\text{mag}} = \frac{\mu_0 I^2}{8\pi^2}\times \begin{cases} \dfrac{s^2}{a^4} & \text{for } 0\le s\le a \\[2mm] \dfrac{1}{s^2} & \text{for } s > a \end{cases}$$

The energy per unit length is thus

$$\mathcal{U}_{\text{mag}} = \frac{\mu_0 I^2}{8\pi^2}\left(\int_0^a \frac{s^2}{a^4}\,2\pi s\,ds + \int_a^\infty \frac{1}{s^2}\,2\pi s\,ds\right)$$

Now we need to know how the energy changes if we alter the radius slightly. The change in field strength is indicated in Figure 9-4. Thus

$$\delta \mathfrak{U}_{mag} = \frac{\partial \mathfrak{U}_{mag}}{\partial a} \delta a$$

$$= \frac{\mu_0 I^2}{8\pi^2} \left(\frac{2\pi}{a} + \int_0^a \frac{-8\pi s^3 \, ds}{a^5} - \frac{2\pi}{a} \right) \delta a$$

$$= -\frac{\mu_0 I^2}{4\pi a} \delta a$$

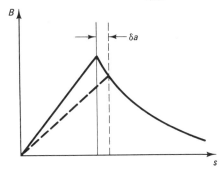

Figure 9-4 Change of magnetic field distribution on increasing radius of a bar carrying a uniform constant current.

The force per unit area is proportional to the coefficient of δa and is seen to be radially *inward*. Physically, it results from the increase of field strength from the center to the outside of the conductor. This pressure leads to the so-called *pinch effect* in fluid conductors. Current streams pinch themselves down to thin filaments rather than flowing uniformly through the material. Furthermore, these pinches are *unstable* in that localized constrictions or kinks tend to accentuate themselves rather than smooth out. These and other instabilities in gaseous plasmas have plagued efforts to develop a controlled nuclear fusion process.

Torque

Torques are treated, as in electrostatics, by considering angular virtual displacements. The analog of Eq. (8-32) is

$$T_z = \left(\frac{\partial \mathfrak{U}_{mag}}{\partial \phi} \right)_{const\ I} \tag{9-16}$$

If the moving circuit is designated as circuit 1, the partial derivative may be evaluated from Eq. (9-3) as

$$T_{z,1} = \frac{1}{2} \sum_i \sum_j I_i I_j \frac{\partial M_{ij}}{\partial \phi_1}$$

Now these derivatives have nonzero value only if either i or j (but not both) is equal to 1. Hence

$$T_{z,1} = I_1 \sum_j I_j \frac{\partial M_{ij}}{\partial \phi_1} \tag{9-17}$$

$$= I_1 \left(\frac{\partial \Phi_1}{\partial \phi_1} \right)_{const\ I} \tag{9-18}$$

where Φ_1 is the flux through circuit 1 and ϕ_1 is the rotation angle of the circuit about the z axis.

Example 9-11

A rectangular coil is pivoted about an axis in its plane. The axis is oriented perpendicular to a uniform external magnetic field $\mathbf{B}'$ as shown in Figure 9-5. Calculate the torque on the coil about the axis.

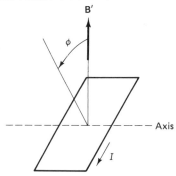

Figure 9-5   Current loop pivoted on axis in its plane, as discussed in Example 9-11.

Solution   The flux through the coil is

$$\Phi = \Phi_{\text{self}} - AB' \cos \phi$$

where A is the area of the coil. Then Eq. (9-18) gives

$$T_z = IAB' \sin \phi$$

This is just what would be obtained from a direct calculation of the forces on the segments. Note that to get the right sense of the torque (tending to increase ϕ) it is necessary to take the positive sense of the external flux to be in the same direction as the self-flux.

9-3 MAGNETIC DIPOLE INTERACTIONS

In the case of a large-scale current loop driven by a battery, Eq. (9-12) shows that any change in field energy at constant current requires *twice as large a change in the opposite sense* in the energy of the battery. The latter energy is required *in order to keep the current constant*. Now, a permanent magnetic dipole, arising from unbalanced orbital or spin motions of charged particles in atoms or molecules (see Section 6-3), is a tiny constant-current loop. Thus there must be some *internal* mechanism which plays the role of a battery in keeping the current constant. In dealing with the energy of such a dipole, we must include the energy of this internal "battery."

As the starting point, we must calculate the interaction energy of the dipole with an external field $\mathbf{B}'$ (not necessarily uniform). For simplicity, we may imagine

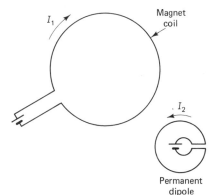

Figure 9-6   Permanent magnetic dipole interacting with an external field.

that $\mathbf{B}'$ is produced by a single large-scale magnet coil carrying current I_1. The dipole will be represented by another current loop I_2 with its internal battery as indicated in Figure 9-6. The total field energy then is, by Eq. (9-3),

$$\mathcal{U}_{\text{mag}} = \tfrac{1}{2}L_1 I_1^2 + \tfrac{1}{2}L_2 I_2^2 + M I_1 I_2$$
$$= W_{\text{self}} + I_2 \Phi'_2$$

where W_{self} is the sum of the first two terms and $\Phi'_2 = M I_1 = $ flux of $\mathbf{B}'$ through loop 2. Since I_2 is constant, the change in field energy due to any change in position or orientation of the dipole is

$$\delta \mathcal{U}_{\text{mag}} = I_2 \, \delta \Phi'_2$$

Hence the change in energy of the internal battery (negative of work done by it) is

$$-\delta W_{b2} = -2 I_2 \, \delta \Phi'_2$$

and the complete change in interaction energy is the sum of these or

$$\delta \mathcal{U}_{\text{int}} = \delta W + \delta W_{b2} = -I_2 \, \delta \Phi'_2$$

Since Φ'_2 is independent of I_2, this may be integrated to give

$$\mathcal{U}_{\text{int}} = -I_2 \Phi'_2 \qquad (9\text{-}19)$$

To express this result in terms of the dipole moment, we note that

$$\Phi'_2 = A\mathbf{B}' \cdot \hat{\mathbf{n}}$$

where A is the loop area. Thus

$$\mathcal{U}_{\text{int}} = -I_2 A\hat{\mathbf{n}} \cdot \mathbf{B}' = -\mathbf{m} \cdot \mathbf{B}' \qquad (9\text{-}20)$$

This is seen to be exactly analogous to Eq. (8-33) for electric dipoles. Hence *all force and torque relations can be taken over directly*. However, it is interesting to note the care required to get the right sign.

9-4 ENERGY AND FORCES IN MAGNETIC MATERIALS

Ideal Magnetics

As in the case of dielectrics (Section 8-4), we first consider hypothetical materials assumed to be completely characterized by a single real permeability. The theory is then exactly analogous to the dielectric case, as a simple example will illustrate.

Example 9-12

A weakly magnetic slab can slide freely between the poles of a magnet. Find the force on the slab when it is inserted partway.

Solution   This is the analog of Example 8-16. To avoid any worries about the "fringe" region near the edges of the poles, we shall simply calculate the change in energy when the slab moves a small distance δx. The energy density in the volume $al\,\delta x$, shown in dashed lines in the Figure 9-7, is

$$u_{\text{mag}} = \begin{cases} \frac{1}{2}\mu_0 H_0^2 & \text{when empty} \\ \frac{1}{2}\kappa_m\mu_0 H_0^2 & \text{when full} \end{cases}$$

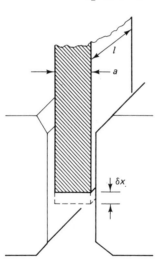

Figure 9-7 Slab of magnetic material moving between magnet poles, as discussed in Example 9-12.

Hence the net change in energy is

$$\delta\mathfrak{U}_{\text{mag}} = \tfrac{1}{2}al\mu_0 H_0^2(\kappa_m - 1)\,\delta x$$

and, by Eq. (9-14), the force is the coefficient of δx.

Note that in contrast to the dielectric case, $(\kappa_m - 1)$ may have either sign. For diamagnetic materials, $\kappa_m < 1$, so the coefficient of δx is negative. Thus, by Eq. (9-14), such materials are *repelled* by magnets. The repulsion is, however, always very weak since the diamagnetic susceptibility is always very small (Section 6-4).

Example 9-13

Estimate the force on a 1-cm² rod of a typical diamagnetic material when one end is in a field of 1 T and the other end is in a field-free region.

Solution $H_0 = B_0/\mu_0 = 10^7/4\pi$ ampere per meter. From Example 6-13 we find that $\kappa_m - 1 \cong -1 \times 10^{-6}$. Thus

$$F \cong \tfrac{1}{2} \times (0.01)^2 \times (10^7/4\pi) \times (-1 \times 10^{-6})$$

$$\cong -4 \times 10^{-5} \text{ newton}$$

The negative sign indicates that the rod tends to be rejected from the field.

The geometry of the last two examples is widely used for the measurement of susceptibility. It is seen that a fairly sensitive force measurement is needed for the weak types of magnetism.

An expression for the force density in an ideal magnetic material in a field is derived along the same lines as Eq. (8-40). We shall just briefly outline the main steps. Starting with Eq. (9-7) we have

$$\delta \mathfrak{U}_{\text{mag}} = \tfrac{1}{2}\delta \left(\iiint \mu \mathbf{H}^2 \, d^3\mathbf{r} \right)$$

$$= \tfrac{1}{2} \iiint (2\mu \mathbf{H} \cdot \delta \mathbf{H} + \mathbf{H}^2 \delta\mu) \, d^3\mathbf{r}$$

Now if all currents are held constant (which implies also that there is no true current in the moving magnetic material), the first integral can be shown to vanish as follows: The integrand is $\mathbf{B} \cdot \delta \mathbf{H}$; hence by using the identity

$$\text{div} (\mathbf{A} \times \delta \mathbf{H}) = -\mathbf{A} \cdot \text{curl } \delta \mathbf{H} + \delta \mathbf{H} \cdot \text{curl } \mathbf{A}$$

$$= -\mathbf{A} \cdot \delta \mathbf{J} + \mathbf{B} \cdot \delta \mathbf{H}$$

in a procedure much like that leading to Eq. (9-6), we obtain the integral of $\mathbf{A} \cdot \delta \mathbf{J}$, which obviously is zero since $\delta \mathbf{J}$ is postulated to be zero everywhere. The final step is to evaluate $\delta\mu$ [just like Eq. (8-39]. The final result is

$$\mathbf{f} = -\tfrac{1}{2}H^2 \text{ grad } \mu = -\tfrac{1}{2}\mu_0 H^2 \text{ grad } \kappa_m \qquad (9\text{-}21)$$

Magnetic Torques

It is a familiar fact that a nonspherical magnetizable body generally experiences a torque in a magnetic field. For example, an iron nail stands straight out from a magnet pole. Quantitative calculation of such torques is quite complicated, but a fair understanding of their origin may be gained by comparing energy densities in some simple shapes. We compare a long thin rod lying parallel to the field with a thin disk lying perpendicular to the field. On the plausible assumption that the magnetization is uniform in both cases (actually true for ellipsoids), Table 6-1 allows us to calculate the internal fields and energy densities.

For the rod:

$$\mathbf{H} = \mathbf{H}_{ext}$$

$$\mathbf{B} = \mathbf{B}_{ext} + \mu_0\mathbf{M} = \mu_0(\mathbf{H}_{ext} + \mathbf{M})$$

$$= \mu_0\mathbf{H}_{ext}(1 + \chi_m) \tag{9-22}$$

$$\mathcal{U}_{mag} = \tfrac{1}{2}\mu_0 H_{ext}^2(1 + \chi_m) \tag{9-23}$$

For the disk:

$$\mathbf{B} = \mathbf{B}_{ext} = \mu_0\mathbf{H}_{ext}$$

$$\mathbf{H} = \mathbf{H}_{ext} - \mathbf{M} = \mathbf{H}_{ext}(1 - \chi_m) \tag{9-24}$$

$$\mathcal{U}_{mag} = \tfrac{1}{2}\mu_0 H_{ext}^2(1 - \chi_m) \tag{9-25}$$

Now, assuming the external field to be held constant, we are dealing with a constant-current situation. Equation (9-16) shows that the torque is in such a direction as to *increase* the field energy. Comparison of Eqs. (9-23) and (9-25) shows that if $\chi_m > 0$ the energy is greater when the body has its long dimension along the field. In other words, paramagnetic bodies tend to align their long dimensions along the field lines, whereas diamagnetic bodies align perpendicular to the field lines. Quantitative calculations of these torques are, however, beyond our level.

Real Magnetic Materials

The nature of the magnetic response in various types of materials was discussed in Chapter 6. It was seen there that the susceptibility is not a constant but is a function of such variables as field strength, temperature, frequency, and so on. This leads to complications fairly similar to those discussed in connection with dielectrics. For example, if there is some material with a temperature-dependent susceptibility present in the field, the change of flux through any current loop will depend on whether we allow heat flow so as to maintain a constant temperature or prevent heat flow and thus cause a change of temperature. As with dielectrics, the constant-temperature assembly work goes into the free energy of the field, which is the quantity of importance in calculating forces and torques.

Nonlinear Magnetic Media

Once more, as a reminder, most of the results are valid only for *linear* media. For nonlinear media we must go back to the incremental relation (9-2). As with dielectrics, it is useful to cast this in terms of the fields. To start, we write the incremental analog of Eq. (9-5),

$$\delta\mathcal{U}_{mag} = \iiint \mathbf{J}(\mathbf{r}) \cdot \delta\mathbf{A}(\mathbf{r})\, d^3\mathbf{r}$$

$$= \iiint \text{curl } \mathbf{H}(\mathbf{r}) \cdot \delta\mathbf{A}(\mathbf{r})\, d^3\mathbf{r}$$

where we have assumed a quasi-static (i.e., very slow) variation. Now the identity preceding Eq. (9-21) leads (owing to the vanishing of the integral of the left side) to the final result,

$$\delta \mathcal{U}_{\text{mag}} = \iiint \mathbf{H}(\mathbf{r}) \cdot \delta \mathbf{B}(\mathbf{r})\, d^3\mathbf{r} \tag{9-26}$$

For linear magnetic materials this may, of course, be time-integrated to obtain Eq. (9-6).

On comparing Eq. (9-26) with Eq. (8-47), the corresponding relation for electrostatic energy, it is seen that there is an obvious similarity. However, the two relations are *not* exactly parallel, since **E** and **B** are the force fields while **D** and **H** are related to the source densities. The difference in the two energy expressions stems from the difference in the force laws between elements of the two kinds of source distributions.

Example 9-14

A soft ferromagnetic material (Section 6-4) is cycled repeatedly between equal positive and negative values of **H**. Assuming the hysteresis curve for the material to be known, show how to obtain the work done per cycle.

Solution   Since the (double-valued) relation of **M** to **H** is assumed to be known, the relation of **B** to **H** is obtained from Eq. (6-12), $\mathbf{B} = \mu_0(\mathbf{H} + \mathbf{M})$. This also takes the form of a hysteresis curve as shown in Figure 9-8. According to Eq. (9-26), the work done in moving along a small segment of this curve is the shaded horizontal bar shown in the figure. Thus, for a complete cycle, the work is the area enclosed by the **B-H** curve times the volume of the material. (*Note:* We have tacitly assumed that **H** and **B** are uniform throughout the material; actually, this will only be true for certain simple geometries.)

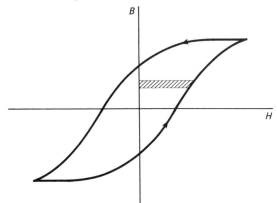

Figure 9-8 Magnetic hysteresis curve for a soft ferromagnetic material such as iron. The shaded bar indicates the increment of work done in increasing the magnetization.

Example 9-15

A magnetic material has a "generalized linear" response of the type described in Section 6-7, with a complex susceptibility. It is placed in a uniform alternating magnetic field $H_0 \cos \omega t$. Calculate the work done in a complete cycle on a unit volume of the material.

Solution   This problem is exactly analogous to Eq. (8-45), except that we are now dealing with the relative permeability, defined in Eq. (6-20), rather than the dielectric response function. In particular, it is again the *imaginary part* of the linear response coefficient that leads to the energy dissipation.

Magnetic Domains

We noted in Section 6-5 that soft ferromagnets tend to subdivide their spontaneous magnetization into small regions called *domains*. In this way, the total energy of the material plus the field is minimized. The situation is illustrated in Figure 9-9. With all the dipoles aligned there is a widespread "dipolelike" external field, and this contains a considerable amount of energy. By reversing the dipoles in one half of the bar, the field is weakened and confined, so that its energy is much less. The process continues until the added energy of the domain walls becomes greater than the reduction of external field energy. The domain wall energy is itself a balance between two opposing tendencies. The misalignment of neighboring dipoles opposes the Weiss molecular field. This energy could be reduced by spreading the rotation out over many neighbors (i.e., forming a thick wall). However, there is also an *anisotropy* energy due to the crystal structure of the material, which tends to have the dipoles aligned in certain preferred crystallographic directions. The combination of all these effects is what leads to the final minimum-energy domain configuration.

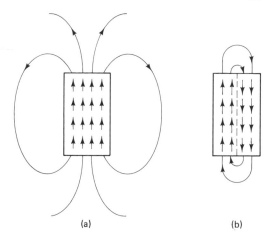

(a)   (b)

Figure 9-9   (a) Field due to a bar magnet with all dipoles aligned; (b) field when one 180-degree domain wall is introduced.

PROBLEMS

9-1. Two coils have self-inductances $L_1 = 5$ mH (millihenries), $L_2 = 3$ mH, and a mutual inductance $M_{12} = 2$ mH. If currents $I_1 = 6$ mA (milliamperes) and $I_2 = 4$ mA are flowing in the coils, how much energy is stored in the magnetic field?

9-2. If the two coils in Problem 9-1 are closely coupled, so that M_{12} has its maximum possible value, estimate the stored energy.

9-3. If two current-carrying coils are suspended freely, will they tend to move so that the coupling will increase or decrease?

9-4. A coil is embedded in a *nonlinear* magnetic medium whose properties are such that the flux through the coil is related to the current by $\Phi = KI^n$, where K and n are constants. Calculate the stored energy when current I_0 flows in the coil.

9-5. An idealized motor consists of two nearly identical circular coils, each of inductance L. One just fits inside the other and is pivoted about a common diameter. It is also connected to a commutator that reverses its current when the coils become coplanar. How much work can this motor do per revolution?

9-6. A small current loop having magnetic dipole moment m is located in a radial inverse-square magnetic field $\mathbf{B}' = Kr/r^3$. Calculate the force on the loop when it lies in a plane perpendicular to the field so that its moment is directed radially outward.

9-7. In Problem 9-6, suppose the moment makes an angle θ_0 with the direction of the field at the loop. Calculate the force.

9-8. A ferromagnetic material has a saturation magnetization (see "Ferromagnetic domains" in Section 6-4) of 1×10^6 A/m, and a "coercive field" (i.e., the value of H required to return the magnetization to zero from the saturation value) of 1000 A/m. Estimate the energy dissipated in taking a unit volume of such a material once around its hysteresis loop.

9-9. A "figure of merit" for permanent-magnet materials is often taken to be the product of the saturation magnetization and the coercive field. Why?

chapter 10

Energy Transport in Electromagnetic Fields

10-1 POYNTING'S THEOREM

In the preceding chapters we derived various expressions for energy changes in static fields. For materials that are linear and temperature independent (which are all that we shall deal with in the present chapter), the energies were expressed in terms of either the source densities and potentials [Eqs. (8-11) and (9-5)] or the *fields* [Eqs. (8-14) and (9-6)]. The expressions in terms of fields were chosen to define the energy densities [Eqs. (8-16) and (9-8)], for the reason given following Eq. (8-16).

In the present chapter we adopt the viewpoint that these expressions for energy density are valid *even under nonstatic conditions*. The idea is that in any volume element the properties can depend only on the fields that are present at any given instant. The element has no way of "knowing" whether or not those fields are going to be the same at a later instant. With this understanding, we can proceed to derive one of the important theorems of the subject.

Our aim is to calculate the rate of change of the total field energy stored within some arbitrary fixed volume V, and to express the result in terms of the fields within V and at its surface.

$$\frac{d}{dt}\mathfrak{U}(V) = \frac{d}{dt}\iiint\limits_{V} (u_{\mathrm{el}} + u_{\mathrm{mag}})\, d^3\mathbf{r}$$

$$= \iiint\limits_{V} \left(\frac{\partial u_{\mathrm{el}}}{\partial t} + \frac{\partial u_{\mathrm{mag}}}{\partial t}\right) d^3\mathbf{r}$$

Since we are restricting the discussion to linear media, this is

$$\frac{d}{dt}\mathcal{U}(V) = \iiint\limits_{V} \left(\frac{\epsilon}{2}\frac{\partial E^2}{\partial t} + \frac{\mu}{2}\frac{\partial H^2}{\partial t}\right)d^3\mathbf{r}$$

$$= \iiint\limits_{V} (\mathbf{E}\cdot\dot{\mathbf{D}} + \mathbf{H}\cdot\dot{\mathbf{B}})\,d^3\mathbf{r} \qquad (10\text{-}1)$$

This may be put in a much more perspicuous form by using the Maxwell equations to replace the time-derivative factors in the integrand.

$$\frac{d\mathcal{U}}{dt} = \iiint\limits_{V} \left[\mathbf{E}\cdot(\operatorname{curl}\mathbf{H} - \mathbf{J}) + \mathbf{H}\cdot(-\operatorname{curl}\mathbf{E})\right]d^3\mathbf{r}$$

$$= \iiint\limits_{V} \left[-\mathbf{E}\cdot\mathbf{J} - (\mathbf{H}\cdot\operatorname{curl}\mathbf{E} - \mathbf{E}\cdot\operatorname{curl}\mathbf{H})\right]d^3\mathbf{r}$$

$$= \iiint\limits_{V} \left[-\mathbf{E}\cdot\mathbf{J} - \operatorname{div}(\mathbf{E}\times\mathbf{H})\right]d^3\mathbf{r}$$

$$= -\iiint\limits_{V} \mathbf{E}\cdot\mathbf{J}\,d^3r - \oiint\limits_{S(V)} (\mathbf{E}\times\mathbf{H})\cdot\hat{\mathbf{n}}\,dS \qquad (10\text{-}2)$$

This result is known as *Poynting's theorem*. The interpretation of the terms is obvious. The first integral is, by Eq. (5-8), the rate of dissipation of energy within V. The remainder of the rate of decrease of energy must be due to *outflow* through the surface. Hence the quantity

$$\mathbf{S} \equiv \mathbf{E}\times\mathbf{H} \qquad (10\text{-}3)$$

is identified as the *energy flux*, or rate of energy flow per unit area, watts per square meter in MKS units. $\mathbf{S}$ is called the *Poynting vector*.

By writing Eq. (10-2) for an infinitesimal volume, an equivalent form is obtained:

$$\frac{\partial u}{\partial t} + \mathbf{E}\cdot\mathbf{J} + \operatorname{div}\mathbf{S} = 0 \qquad (10\text{-}4)$$

Example 10-1

A long straight cylindrical conductor (radius a, conductivity g) carries current I uniformly distributed over its cross-sectional area. Calculate the Poynting vector at its surface (Figure 10-1).

Solution The electric field strength required to drive the current through the conductor is, by Eq. (5-2), $\mathbf{E} = (I/\pi a^2 g)\hat{\mathbf{k}} = IR\hat{\mathbf{k}}$, where R is the resistance of unit length of the conductor. The magnetic field at the surface is, by Example 2-13, $\mathbf{H} = (I/2\pi a)\hat{\mathbf{a}}_\phi$. Thus

$$\mathbf{S} = \frac{1}{2\pi a}I^2 R(-\hat{\mathbf{a}}_s)$$

This shows that the energy flowing into the surface, per unit length, is just the amount that gets dissipated in that length.

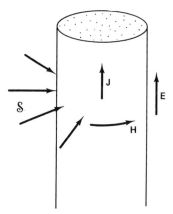

Figure 10-1   Poynting vector at the surface of a long straight cylinder.

Example 10-2

Where does the energy come from in Example 10-1?

Solution   From the battery (or other generator) where the current flow is *opposite* to the electric field (Figure 10-2).

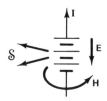

Figure 10-2   Poynting vector near a battery driving a current.

Example 10-3

For the same example, calculate div **S**.

Solution   First calculate **S** at all radial distances. **E** is constant and **H** is given in Example 2-13, so

$$\mathbf{S} = \begin{cases} \dfrac{I^2 R}{2\pi}\dfrac{1}{s}(-\hat{\mathbf{a}}_s) & \text{for } s > a \\[2mm] \dfrac{I^2 R}{2\pi}\dfrac{s}{a^2}(-\hat{\mathbf{a}}_s) & \text{for } s < a \end{cases}$$

$$\text{div } \mathbf{S} = \begin{cases} 0 & \text{for } s > a \\[2mm] -\dfrac{I}{\pi a^2}(IR) = -\mathbf{E}\cdot\mathbf{J} & \text{for } s < a \end{cases}$$

Since nothing is changing with time, this result confirms Eq. (10-4).

Example 10-4

A power line (pair of wires) runs through a room from a generator behind one wall to a motor behind the opposite wall. From observations entirely within the room, how can one find out which is which?

Solution Measure the fields in the vicinity of the line and calculate the Poynting vector. To illustrate, suppose that it is a dc power line and that the direction of current flow is as shown in cross section in Figure 10-3. Then if the generator is in front of the page, the left wire is positive and the electric field is in the sense shown. Thus the Poynting vector is directed into the page.

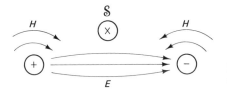

Figure 10-3 Fields and Poynting vector near a two-wire power line.

Example 10-5

In Example 10-4, suppose that it is an ac power line (of negligible resistance) and that the electric and magnetic fields in the plane perpendicular to the wires (i.e., the voltage and current) vary as $\mathbf{E} = E_0 \hat{\mathbf{i}} \cos \omega t$ and $\mathbf{H} = H_0 \hat{\mathbf{j}} \cos (\omega t + \alpha)$. Calculate the time-average Poynting vector.

Solution This is a vector example of the power theorem [Eq. (8-46)]. We write the fields in complex exponential notation as

$$\mathbf{E} = E_0 \hat{\mathbf{i}} e^{-i\omega t} \qquad \mathbf{H} = H_0 \hat{\mathbf{j}} e^{-i(\omega t + \alpha)}$$

Then

$$\mathbf{S} = \tfrac{1}{2} \operatorname{Re} (\mathbf{E} \times \mathbf{H}^*) = \tfrac{1}{2} \operatorname{Re} (E_0 H_0 \hat{\mathbf{k}} e^{i\alpha})$$
$$= \tfrac{1}{2} E_0 H_0 (\cos \alpha) \hat{\mathbf{k}}$$

10-2 ELECTROMAGNETIC MOMENTUM

The force density in a region containing some charges and currents is

$$\mathbf{f} = \rho \mathbf{E} + \mathbf{J} \times \mathbf{B}$$
$$= (\operatorname{div} \mathbf{D})\mathbf{E} + (\operatorname{curl} \mathbf{H} - \dot{\mathbf{D}}) \times \mathbf{B}$$
$$= \mathbf{E} \operatorname{div} \mathbf{D} - \mathbf{B} \times \operatorname{curl} \mathbf{H} - \dot{\mathbf{D}} \times \mathbf{B} \qquad (10\text{-}5)$$

where we have substituted for the source densities from two of the Maxwell equations. From the other two Maxwell equations, by appropriate multiplication, we obtain the identities

$$0 = -\mathbf{D} \times \operatorname{curl} \mathbf{E} - \mathbf{D} \times \dot{\mathbf{B}} \qquad (10\text{-}6)$$
$$0 = \mathbf{H} \operatorname{div} \mathbf{B} \qquad (10\text{-}7)$$

On adding these all together and assuming a homogeneous linear medium, the result is

$$\mathbf{f} = \epsilon(\mathbf{E} \operatorname{div} \mathbf{E} - \mathbf{E} \times \operatorname{curl} \mathbf{E}) + \mu(\mathbf{H} \operatorname{div} \mathbf{H} - \mathbf{H} \times \operatorname{curl} \mathbf{H}) - \mu\epsilon \frac{\partial}{\partial t}(\mathbf{E} \times \mathbf{H})$$
$$(10\text{-}8)$$

The next step is to integrate this over the volume of a region whose boundary lies well outside the entire charge and current distribution. The left side obviously gives the total force on the entire cluster of charged particles comprising the sources, so

$$\mathbf{F}_{\text{particles}} = \epsilon \iiint\limits_{V} (\mathbf{E} \operatorname{div} \mathbf{E} - \mathbf{E} \times \operatorname{curl} \mathbf{E}) \, d^3\mathbf{r}$$

$$+ \mu \iiint\limits_{V} (\mathbf{H} \operatorname{div} \mathbf{H} - \mathbf{H} \times \operatorname{curl} \mathbf{H}) \, d^3\mathbf{r} - \frac{d}{dt} \iiint \mu\epsilon(\mathbf{E} \times \mathbf{H}) \, d^3\mathbf{r} \tag{10-9}$$

The first two terms may be transformed into surface integrals by the (admittedly somewhat unfamiliar) identity

$$\iiint\limits_{V} (\mathbf{G} \operatorname{div} \mathbf{G} - \mathbf{G} \times \operatorname{curl} \mathbf{G}) \, d^3\mathbf{r} = \oiint\limits_{S(V)} [(\mathbf{G} \cdot \hat{\mathbf{n}})\mathbf{G} - \tfrac{1}{2}G^2\hat{\mathbf{n}}] \, dS \tag{10-10}$$

At the moment we are not concerned with the exact form of the surface integrals, just with the fact that they exist. So let us call them simply $\mathscr{G}_1$ and $\mathscr{G}_2$ and rewrite Eq. (10-9) as

$$\mathbf{F}_{\text{particles}} + \frac{d}{dt} \iiint \mu\epsilon(\mathbf{E} \times \mathbf{H}) \, d^3\mathbf{r} = \mathscr{G}_1 + \mathscr{G}_2 \tag{10-11}$$

Now first consider static fields; then the time-derivative term is zero, and clearly the entire force is given by the two surface integrals. In other words, the integrands of these integrals describe how the fields *transmit the force* across the bounding surface. But if this is true in the static case, then according to the viewpoint expressed in the second paragraph of this chapter, it must be equally true in the time-varying case. In other words,

$$\mathbf{F}_{\text{total}} = \mathscr{G}_1 + \mathscr{G}_2 \tag{10-12}$$

Thus Eq. (10-11) says that

$$\mathbf{F}_{\text{total}} = \mathbf{F}_{\text{particles}} + \frac{d}{dt} \iiint\limits_{V} \mu\epsilon(\mathbf{E} \times \mathbf{H}) \, d^3\mathbf{r}$$

and by Newton's laws of motion this is

$$\mathbf{F}_{\text{total}} = \frac{d}{dt} \left[\text{momentum of particles} + \iiint\limits_{V} \mu\epsilon(\mathbf{E} \times \mathbf{H}) \, d^3\mathbf{r} \right] \tag{10-13}$$

We have finally arrived at our goal. The volume integral obviously must be the momentum of something other than the sources (charges and currents) contained in the volume V. But the only other entity contained in the volume is the *electromagnetic field*. Hence the integrand may be identified as

$$\text{momentum density of field} = \mu\epsilon(\mathbf{E} \times \mathbf{H}) \tag{10-14}$$

Note that the momentum density is *proportional to the Poynting vector*. In other words, the Poynting vector describes not only energy flow but a proportional amount of momentum density.

Example 10-6

Solar energy (which is electromagnetic in nature) impinges on the top of the earth's atmosphere at the rate of about 1 kW/m². How much pressure would be exerted on (a) a perfectly absorbing surface, and (b) a perfectly reflecting surface?

Solution   Since, according to Eq. (7-53), $\mu_0\epsilon_0 = 1/c^2$, Eq. (10-14) gives the momentum density as

$$\frac{10^3 \text{ watt/meter}^2}{(3 \times 10^8 \text{ meters/second})^2} = \frac{1}{9 \times 10^{13}} \frac{\text{newton-second}}{\text{meter}^3}$$

(a) The amount of this field momentum that is lost into an absorber in time dt is the amount contained in a column of length $c\,dt$. Thus the rate of increase of momentum of the absorber, per unit area, is

$$\frac{dp_{abs}}{dt} = \frac{1}{9 \times 10^{13}} \frac{\text{newton-second}}{\text{meter}^3} \times 3 \times 10^8 \frac{\text{meter}}{\text{second}}$$

$$= \frac{1}{3 \times 10^5} \frac{\text{newtons}}{\text{meter}^2}$$

This is the force per unit area, or pressure.

(b) For a reflecting surface, the field momentum is not absorbed but is *reversed*. Hence the pressure is twice as great.

These effects are known as *radiation pressure*. Their origin lies in the forces exerted on the charged particles of the absorber or reflector by the incident electromagnetic field.

10-3 INTERPRETATION OF THE POYNTING VECTOR

In the preceding two chapters we derived integral expressions for total energies in static fields in terms of the fields themselves. These were Eqs. (8-14) and (9-6). We then elected to interpret the *integrands* of these integrals as energy densities. Integrands of other integral expressions such as Eqs. (8-11) and (9-5) were not considered, since it seems plausible to assume that the energy is concentrated where the fields are strong, not where charges and currents happen to be located. This is consistent with the obvious fact that energy can be carried through empty space by electromagnetic fields.

In the present chapter we extended the interpretation to include time-varying fields and used it to derive a further integral expression [Eq. (10-2)] for the rate of change of total energy within a given volume. Then, by once more interpreting an integrand as a density, we arrived at Eq. (10-3) for the energy flux. We now look more closely into the question of whether this is a valid and meaningful interpretation. We consider separately the two broad classes of fields: static and time varying.

Time-Varying Fields

These will be treated in detail in Chapter 13. For now, we need only note that the energy transport in such fields is readily measurable and in all cases the Poynting vector gives the correct distribution and amount. Thus the interpretation is perfectly valid and extremely useful for such fields.

Static Fields

As noted in Example 7-2, the electric and magnetic fields are decoupled under static conditions. Thus it is quite artificial to link them together in the Poynting vector. Indeed, rather strange results can be obtained, as illustrated in Examples 10-1 and 10-2. Here the Poynting vector suggests that energy flows out of the generator into empty space, then back into the wire, whereas intuitively we would imagine the energy to flow *along* the wire. In this case the putative energy flux is *not* an electromagnetic wave. Indeed, there is no known mechanism to account for it, nor any means of detecting or measuring it. On the other hand, the effect that *is* measurable, the heat production in the wire, is correctly accounted for, as Example 10-3 shows.

This sort of paradoxical energy flow pattern has led to much discussion over the years. A concise review is given by Kobe.* Many authors point out that since only div **S** is well defined, as is implicit in Eq. (10-2) and explicit in Eq. (10-4), it should be possible to add any solenoidal vector function to **S**. Such an addition would change the calculated flow *patterns*, but not the observable integrated effects. Indeed, an explicit form for such an additive term has been proposed (see references in Kobe's article). It is, however, subject to two major criticisms:

1. It explicitly contains the scalar potential and is thus not gauge invariant. In a particular gauge it can give a more intuitively appealing flow pattern, but the gauge choice is different in different geometries.
2. If **S** is altered, the neat relationship between energy flux and momentum density is destroyed. This is a very serious matter. As Feynman† emphasizes, there is a fundamental proportionality between energy flux and momentum density in *any* physical system. This is a requirement of relativistic invariance, which we shall discuss in Chapter 16. The upshot is that *no tampering with the Poynting vector is permitted!*

*D. H. Kobe, *Am. J. Phys.* **50** (1982), 1162.

†R. P. Feynman, *Lectures on Physics*, Vol. II, Addison-Wesley Publishing Co., Inc., Reading, Mass., 1964, Chap. 27.

Then what do we say about the strange-looking energy flow patterns? According to Feynman, this is just one of many cases where intuition is wrong. The momentum (and angular momentum) accompanying the Poynting vector is physically real. It had to be "invested" when the fields were set up, and can be recovered if the fields are allowed to collapse (see Problem 10-5). In this sense, the electromagnetic momentum of static fields bears the same relationship to momentum of moving bodies as potential energy does to kinetic energy. The term *potential momentum* seems very descriptive but is not in common use.

PROBLEMS

10-1. For the cylindrical conductor of Example 10-1, calculate the Poynting vector within the material as function of the radial distance s from the axis.

10-2. Show that Eq. (10-2) holds for a coaxial cylindrical volume of radius s within the conductor.

10-3. A parallel-plate capacitor is placed in a long solenoid with its plates parallel to the axis of the solenoid. After the capacitor is charged, what is the direction of energy flow in the region between the plates?

10-4. In Problem 10-3, where does the energy come from?

10-5. Suppose that the dielectric in the capacitor in Problem 10-3 is made slightly conductive (e.g., by heating) so that the electric field collapses from E_0 to zero. Calculate the momentum imparted to the capacitor by the magnetic force on the transient current.

10-6. Write the expression for the Poynting vector in Gaussian units.

10-7. Write the expression for the electromagnetic momentum density in Gaussian units.

10-8. In a region of space there exist simultaneously an electric field $\mathbf{E} = \hat{\mathbf{i}}E_1 \sin \alpha$ and a magnetic field $\mathbf{B} = \hat{\mathbf{j}}B_1 \sin \alpha$, where $B_1 = E_1/c$ and α is a function of z and t. Calculate the energy density and the Poynting vector and find the relation between them.

Part IV

SOLUTION OF STATIC FIELD PROBLEMS

chapter 11

Electrostatic Boundary-Value Problems

11-1 BOUNDARY CONDITIONS ON THE FIELDS

In Chapter 1 we saw that whenever the charge distribution is completely specified, the field at any point can be calculated directly. In Chapter 4 this idea was extended to include the effective charge density in dielectrics. We now turn to a more complicated class of problems in which the charge density (real or effective) is only incompletely specified. The additional information needed to determine the field exists in the form of the properties of different kinds of matter within or around the region of interest. In many cases the form and properties of the bounding surfaces play vital roles. Our first step, therefore, is to develop the relations satisfied by the fields at points immediately adjacent to the two sides of such boundaries. These relations come, of course, from the Maxwell equations, applied to suitable regions containing part of the interface.

Boundary Condition on **E**

The Maxwell equation gives curl **E**. Hence Stokes' theorem suggests that the appropriate type of region is a two-dimensional one (i.e., a surface). We choose this to be a small rectangle lying in a plane perpendicular to the interface (Figure 11-1) and, for each side of the Maxwell equation, integrate the component normal to this plane over the area of the rectangle to obtain

$$\iint \text{curl } \mathbf{E} \cdot \hat{\mathbf{N}} \, dS = -\iint \dot{\mathbf{B}} \cdot \hat{\mathbf{N}} \, dS$$

where $\hat{\mathbf{N}}$ is a unit vector normal to the rectangle, hence tangential to the interface.

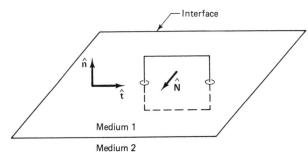

Medium 1

Medium 2

Figure 11-1 Rectangular area in a plane perpendicular to an interface between two media.

On applying Stokes' theorem to the left side, this is

$$\oint \mathbf{E} \cdot \mathbf{dl} = -\iint \dot{\mathbf{B}} \cdot \hat{\mathbf{N}} \, dS$$

Now to obtain relations for the field values at points immediately adjacent to the interface, we let the sides of the rectangle shrink to zero. Thus the area shrinks to zero, and *because* $\dot{\mathbf{B}}$ *cannot be infinite*, the right side vanishes. The integral on the left comprises the contributions of the upper and lower segments only, so

$$\mathbf{E}_1 \cdot (-\hat{\mathbf{t}}) + \mathbf{E}_2 \cdot \hat{\mathbf{t}} = 0 \qquad \text{or} \qquad (\mathbf{E}_2 - \mathbf{E}_1) \cdot \hat{\mathbf{t}} = 0 \qquad (11\text{-}1)$$

where $\hat{\mathbf{t}}$ is the tangential unit vector parallel to the sides of the rectangle. This shows that any tangential component of **E** is *continuous* (has the same value) across *any* interface.

An illustration of this boundary relation is the familiar fact that the field adjacent to an ideal conductor is perpendicular to the conductor surface (Section 1-10). All field components are zero inside the surface and the tangential ones, being continuous across the interface, have to be zero just outside. The normal component of **E** is not subject to this continuity requirement and thus may well be nonzero outside.

Boundary Condition on Potential

Since $\Delta U = \int -\mathbf{E} \cdot \mathbf{dl}$, any discontinuity in U would require an infinite field strength. Strangely enough, there is an idealized situation where this can occur. Consider a pair of charge sheets $\pm\sigma$ separated by distance d. The potential difference is $\Delta U = \sigma d/\epsilon_0$. Now if d shrinks while σ increases, keeping the product unchanged, we reach in the

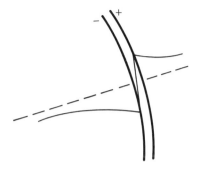

Figure 11-2   Potential jump in crossing a double layer.

limit a so-called *double layer* whose strength is the dipole moment per unit area $p/A = \sigma d$. Across such a layer, **E** is the same on both sides but U jumps by $p/\epsilon_0 A$ in crossing from the negative to the positive side (Figure 11-2). Double layers are often useful to approximate the situation at surfaces.

Boundary Condition on **D**

The Maxwell equation gives div **D**, so the appropriate type of region of application is a volume. We choose a cylinder with axis normal to the interface, and integrate both sides of the equation over the volume of the cylinder to obtain

$$\iiint \text{div } \mathbf{D} \, dV = \iiint \rho \, dV = \text{charge enclosed}$$

On applying the divergence theorem to the left side, this is

$$\oiint \mathbf{D} \cdot \hat{\mathbf{N}} \, ds = \text{charge enclosed by cylinder}$$

where $\hat{\mathbf{N}}$ is now the variable unit vector perpendicular to the surface of the cylinder (Figure 11-3). To obtain relations for points immediately adjacent to the interface, we let the height of the cylinder shrink to zero. The right side *does not necessarily vanish*, however, since there may possibly be a *surface charge* (i.e., an infinite volume

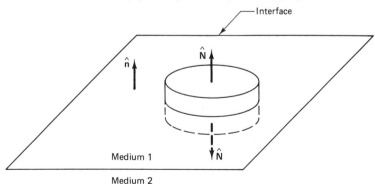

Figure 11-3   "Pillbox"-shaped volume enclosing part of an interface, for application of Gauss's law.

charge density right at the interface). Thus the limiting form, after dividing through by the area of the cylinder faces, is

$$\mathbf{D}_1 \cdot \hat{\mathbf{n}} + \mathbf{D}_2 \cdot (-\hat{\mathbf{n}}) = \sigma \qquad (11\text{-}2)$$

where $\hat{\mathbf{n}}$ is the unit normal to the interface in the direction toward medium 1. Thus the normal component of $\mathbf{D}$ is (1) continuous at an uncharged interface, and (2) discontinuous at a charged interface, the amount of discontinuity being just equal to the (true) surface charge density. This relation is illustrated in the ferroelectric slab problem, Example 4-8, where $\mathbf{D}$ was found to be the same in both media. The *effective* charge density at the dielectric surface does *not* produce a discontinuity in $\mathbf{D}$. Only *true* surface charge does so.

Example 11-1

Consider a planar interface between two linear dielectric media (Figure 11-4). The electric field in medium 2 at the interface has magnitude E_2 and makes angle θ_2 with the normal. Find the field in medium 1 at the interface.

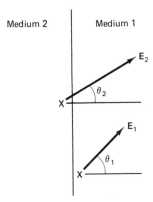

Figure 11-4 Electric fields at points adjacent to an interface (Example 11-1).

Solution   The two boundary relations give

$$E_1 \sin \theta_1 = E_2 \sin \theta_2 \qquad \text{(tangential components of } \mathbf{E}\text{)}$$

$$\epsilon_1 E_1 \cos \theta_1 = \epsilon_2 E_2 \cos \theta_2 \qquad \text{(normal components of } \mathbf{D}\text{)}$$

On dividing, we find

$$\frac{\tan \theta_1}{\tan \theta_2} = \frac{\epsilon_1}{\epsilon_2}$$

which determines θ_1.

Note that the greater angle from the normal occurs in the medium of greater permittivity. To find the magnitude of $\mathbf{E}_1$ either of the two boundary equations can then be used; from the first one

$$E_1 = \frac{E_2 \sin \theta_2}{\sin \theta_1} = E_2 \frac{\tan \theta_2 \sqrt{1 + \tan^2 \theta_1}}{\tan \theta_1 \sqrt{1 + \tan^2 \theta_2}}$$

$$= E_2 \frac{\epsilon_2}{\epsilon_1} \frac{\sqrt{1 + \left(\frac{\epsilon_1}{\epsilon_2} \tan \theta_2\right)^2}}{\sqrt{1 + \tan^2 \theta_2}}$$

11-2 GENERAL PROPERTIES OF THE POTENTIAL

We saw in Section 4-4 that the potential satisfies the Poisson equation $\nabla^2 U = -(\rho + \rho_P)/\epsilon_0$ in general, and that in regions free of true and effective charge the latter reduces to the Laplace equation $\nabla^2 U = 0$. These two important equations have been investigated very extensively by mathematicians, and not only are many explicit solutions known but also many general properties that solutions must possess. We shall now develop a few of the latter.

Absence of Extrema

The potential cannot have a maximum or minimum within a charge-free region. It is easiest to think in terms of Cartesian coordinates. In order for a maximum (or minimum) to occur at some point, (1) the gradient must vanish, and (2) all three second derivatives must be negative (or positive). But condition 2 cannot occur since the sum of the three second derivatives must be zero (Laplace equation). Note, however, that a maximum (or minimum) can occur at a *boundary point* of a charge-free region, that is, at a charge.

At a point where the electric field is zero, the gradient is zero, so the potential has a *stationary* value. However, the curvature must be positive in some directions and negative in others. This type of behavior is called a *saddle point* (Figure 11-5).

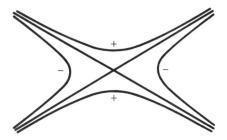

Figure 11-5 General configuration of equipotentials in vicinity of a saddle point.

Superposition

If, in a charge-free region, we have several solutions of the Laplace equation, then any linear combination of these is also a solution. Let $U_1(\mathbf{r})$, $U_2(\mathbf{r})$, ... be functions that each satisfy the Laplace equation in the given region. Let $U = C_1 U_1 + C_2 U_2 + \cdots$, where the C's are any constants. Then

$$\nabla^2 U = C_1 \nabla^2 U_1 + C_2 \nabla^2 U_2 + \cdots$$
$$= 0$$

This property is extremely useful in combining known solutions to produce new solutions that satisfy given conditions on the boundaries of the region.

The next few properties relate to the strong influence that the boundary values have on the potential function throughout a region. They are best derived with the aid of a mathematical lemma called *Green's first identity*: If $f(\mathbf{r})$ and $g(\mathbf{r})$ are any two scalar functions, then

$$\iiint_V [f\nabla^2 g + (\text{grad } f) \cdot (\text{grad } g)] \, d^3\mathbf{r} = \oiint_{S(V)} f \text{ grad } g \cdot \hat{\mathbf{n}} \, dS \qquad (11\text{-}3)$$

The integrand on the left is simply the divergence of f grad g; hence the result is simply an application of the divergence theorem to this vector function.

Constancy Within an Equipotential Surface

Consider the potential in any charge-free region surrounded by a grounded conductor. In Eq. (11-3), let $f = g = U$. Then $\nabla^2 g = 0$ throughout V and $f = 0$ at all points on the surface. Hence Eq. (11-3) becomes

$$\iiint_V (\text{grad } U)^2 \, d^3\mathbf{r} = 0 \qquad (11\text{-}4)$$

Since the integrand is positive definite, the integral can vanish only if the integrand vanishes at every point. Hence U is constant at the boundary value, zero.

If the conductor is charged and not grounded, the boundary will be at some uniform potential U_0. Now let $f = g = U - U_0$. Then the rest of the proof is exactly the same, and $U = U_0$ throughout.

Uniqueness

Consider a region containing some specified distribution of charge, with boundaries held at some specified distribution of potential (e.g., the boundaries may be conductive segments each held at some fixed potential by a battery). There is *one and only one* solution of the electrostatic problem throughout the region.

1. There certainly is one solution; the boundaries possess some distribution of surface charge and this, together with the specified charge distribution in the interior, produces a potential at every point.
2. Now assume that there are two functions $U_1(\mathbf{r})$ and $U_2(\mathbf{r})$ that coincide at the boundaries. If they are to be solutions in the interior, then $\nabla^2 U_1 = \nabla^2 U_2$ (because they must satisfy the Poisson equation for the same charge density in the interior). Thus if we let $f = g = U_1 - U_2$ in Eq. (11-3), the proof goes through the same as in the preceding two situations.

This is an important theorem, since it assures us that once *any* solution of the problem (differential equation plus boundary conditions) has been found, there is no need to look for any further generality. It should be obvious to the reader that specification of the field strength at all points of the boundary will also suffice. In fact, "mixed boundary conditions" (potential over part of the boundary, field strength over the rest) work, too. In all cases, though, complete closed boundaries are needed.

These general properties of the potential come into play in the various methods of finding solutions to be discussed in the remainder of this chapter.

11-3 METHOD OF IMAGES

The basic idea of this method is that the effects of the boundaries are *simulated* by additional charges, suitably chosen and suitably located. In other words, the original problem (charges plus boundaries) is *replaced* by an "image problem" (original charges plus "image charges"). The original problem always seeks the field in a certain "region of interest" delimited by the boundaries. The image charges must always lie *outside* this region, so that the solution of the image problem satisfies the *same* Poisson equation within the region as does the solution of the original problem. If we succeed in finding these so that the correct boundary conditions prevail, the uniqueness theorem assures us that we have the complete solution for the inside (but *not* the outside!) region.

Example 11-2

A point charge is located a distance z_1 from a grounded infinite conducting plane. Find the potential function throughout space.

Solution Choosing the origin on the plane directly under the charge, the boundary condition is $U(x, y, 0) = 0$. This can be brought about by removing the metal sheet and placing a charge $-q$ at $(0, 0, -z_1)$. Thus the image problem consists of this charge plus the original charge, with no metal. Its solution is (Figure 11-6b)

$$U_{\text{im}} = \frac{q}{4\pi\epsilon_0} \{[x^2 + y^2 + (z - z_1)^2]^{-1/2} - [x^2 + y^2 + (z + z_1)^2]^{-1/2}\}$$

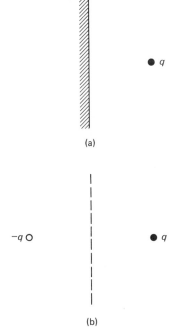

(a)

(b)

Figure 11-6 (a) Original problem and (b) image problem for Example 11-2.

This is the solution of the image problem at all points. The solution of the original problem is (Figure 11-6a)

$$U = U_{im} \text{ for } z > 0 \qquad U = 0 \text{ for } z < 0$$

Example 11-3

Find the distribution of induced surface charge density on the conductive sheet of Example 11-2.

Solution   We make use of Eq. (11-2). Since no dielectrics are present, $\mathbf{D} = \epsilon_0 \mathbf{E} = -\epsilon_0 \text{ grad } U$ in each region. The gradient in region 1 $(z > 0)$ is

$$\mathbf{E}_1(x, y, z) = \frac{q}{4\pi\epsilon_0} \left\{ \frac{x\hat{\mathbf{i}} + y\hat{\mathbf{j}} + (z - z_1)\hat{\mathbf{k}}}{[x^2 + y^2 + (z - z_1)^2]^{3/2}} - \frac{x\hat{\mathbf{i}} + y\hat{\mathbf{j}} + (z + z_1)\hat{\mathbf{k}}}{[x^2 + y^2 + (z + z_1)^2]^{3/2}} \right\}$$

and on setting $z = 0$, we obtain (Figure 11-7)

$$\sigma(x, y) = \epsilon_0 E_{1z}(x, y, 0)$$

$$= \frac{-qz_1}{2\pi(s^2 + z_1^2)^{3/2}}$$

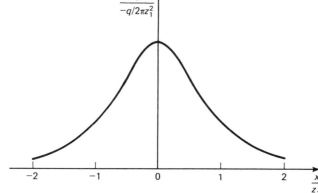

Figure 11-7   Induced charge density in Example 11-3.

Example 11-4

Find the total induced charge on the sheet.

Solution

$$q_{ind} = \iint \sigma \, ds = -\frac{qz_1}{2\pi} \int_0^\infty \frac{2\pi s \, ds}{(s^2 + z_1^2)^{3/2}} = -q$$

Example 11-5

Find the force on the point charge.

Solution   The field *acting on* this charge arises from the distribution of surface charge on the sheet. But everywhere to the right of the sheet this is the same as the field due to the image charge (since the two solutions coincide in this region). Thus we easily find

$$\mathbf{F} = q\mathbf{E}'(0, 0, z) = \frac{-q^2\hat{\mathbf{k}}}{4\pi\epsilon_0(2z_1)^2}$$

As an exercise, the reader may wish to verify this by direct calculation from the surface charge distribution calculated in Example 11-3. Divide this into rings and use Example 1-5 with a replaced by s and λ by $\sigma(s)\,ds$.

Example 11-6

An infinite conductive sheet is bent to a 90° corner, and a point charge is placed somewhere in the acute angle. Set up the appropriate image problem.

Solution A single "reflection" in either plane will make that plane an equipotential, but not the other (Figure 11-8). The further reflection of the two reflections will restore equipotentiality throughout the bounding surface.

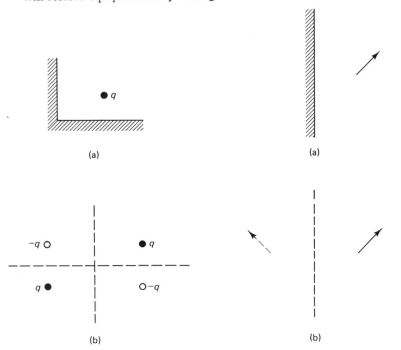

Figure 11-8 (a) Original problem and (b) image problem for 90-degree inside corner (Example 11-6).

Figure 11-9 (a) Original problem and (b) image problem for dipole near conductive plane (Example 11-7).

Example 11-7

A dipole is placed at distance z_1 from a grounded infinite conductive plane, in arbitrary orientation. Find the potential throughout the half-space.

Solution By thinking of the dipole as a pair of point charges, it is easy to see that the image dipole will have the same parallel component and the reversed perpendicular component (Figure 11-9). Thus, from Eq. (1-26),

$$U(\mathbf{r}) = \frac{1}{4\pi\epsilon_0}\left[\frac{(p_x\hat{\mathbf{i}} + p_z\hat{\mathbf{k}})\cdot(\mathbf{r} - z_1\hat{\mathbf{k}})}{|\mathbf{r} - z_1\hat{\mathbf{k}}|^3} + \frac{(p_x\hat{\mathbf{i}} - p_z\hat{\mathbf{k}})\cdot(\mathbf{r} + z_1\hat{\mathbf{k}})}{|\mathbf{r} + z_1\hat{\mathbf{k}}|^3}\right]$$

Example 11-8

A point charge is placed between a pair of parallel conductive sheets. Set up the appropriate image problem.

Solution   A reflection in either plane makes that one an equipotential, but not the other. Further successive reflections are needed, leading to an infinite series in all (Figure 11-10). Since the reflections get farther and farther away, the series for the potential is convergent. The details are left for a problem.

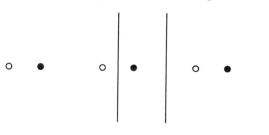

Figure 11-10   Infinite set of images needed for parallel pair of conductive planes (Example 11-8).

Example 11-9

A point charge q is located outside a grounded conductive sphere of radius a, at distance z_1 from the center. Find the potential at all exterior points.

Solution   The symmetry of the problem suggests an image charge inside the sphere somewhere on the radius toward q. Let its magnitude be q' and its distance from the center be z_2. Then the potential in the image problem has the form (Figure 11-11)

$$4\pi\epsilon_0 U(r, \theta) = q(r^2 + z_1^2 - 2rz_1 \cos \theta)^{-1/2} + q'(r^2 + z_2^2 - 2rz_2 \cos \theta)^{-1/2}$$

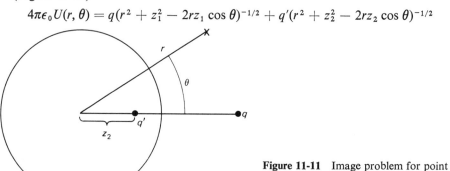

Figure 11-11   Image problem for point charge near grounded conductive sphere.

Now the boundary condition is $U(a, \theta) = 0$ for all θ. The easiest way to satisfy this is to use two specific values of θ, such as 0 and π. For these we get

$$\frac{q}{z_1 - a} + \frac{q'}{a - z_2} = 0$$

$$\frac{q}{z_1 + a} + \frac{q'}{a + z_2} = 0$$

Solving these gives $q' = -qa/z_1$, $z_2 = a^2/z_1$. Inserting these values in the expression above gives the potential (at all points in the image problem and at the exterior points in the real problem). Incidentally, the reader should verify that the boundary condition is satisfied at all values of θ, not just the two used.

Example 11-10

The sphere in Example 11-9 is held at a fixed potential U_0 instead of 0. Find the appropriate images.

Solution   An additional image charge $4\pi\epsilon_0 a U_0$ at the center obviously does the trick since it produces a uniform contribution to the potential at the sphere.

Example 11-11

A point charge is located at distance z_1 from a planar interface between two media with linear dielectric properties. Find the potential at all points.

Solution   Here we encounter a new feature: The boundary does not exclude the field. Thus we must set up *two* image problems, one for each region (Figure 11-12); in each of these, the image charges must lie outside that region. As a reason-

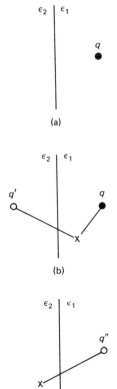

Figure 11-12   (a) Original problem, (b) image problem for region 1, and (c) image problem for region 2, for point charge near planar dielectric interface (Example 11-11).

able guess, let us try for region 1 (containing q) an image charge q' located at the reflection point, and for region 2 an image charge q'' located at the same place as q. Thus the potential is given by the two expressions

$$U_1 = \frac{1}{4\pi\epsilon_1}\left[\frac{q}{r_1} + \frac{q'}{r_2}\right]$$

$$U_2 = \frac{1}{4\pi\epsilon_2}\frac{q''}{r_3}$$

The boundary conditions are now given by Eqs. (11-1) and (11-2), so we must calculate $\mathbf{E}$ and $\mathbf{D}$. For a point on the boundary (where $r_1 = r_2 = r_3 = r$) (Figure 11-13)

$$\mathbf{E}_1 = \frac{1}{4\pi\epsilon_1 r^2}[q(\sin\theta\,\hat{\mathbf{i}} - \cos\theta\,\hat{\mathbf{k}}) + q'(\sin\theta\,\hat{\mathbf{i}} + \cos\theta\,\hat{\mathbf{k}})]$$

$$\mathbf{E}_2 = \frac{1}{4\pi\epsilon_2 r^2}q''(\sin\theta\,\hat{\mathbf{i}} - \cos\theta\,\hat{\mathbf{k}})$$

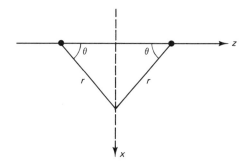

Figure 11-13 Field point on boundary of image problem for planar dielectric interface.

Also, $\mathbf{D}_1 = \epsilon_1\mathbf{E}_1$, $\mathbf{D}_2 = \epsilon_2\mathbf{E}_2$. Thus the boundary condition equations are, after canceling out common factors,

$$\frac{q+q'}{\epsilon_1} = \frac{q''}{\epsilon_2} \qquad \text{(tangential components of } E)$$

$$-q + q' = -q'' \qquad \text{(normal components of } D)$$

Solving this pair gives

$$q' = -\frac{\epsilon_2 - \epsilon_1}{\epsilon_2 + \epsilon_1}q \qquad q'' = \frac{2\epsilon_2}{\epsilon_2 + \epsilon_1}q$$

The general pattern of field lines is as shown in Figure 11-14 (for the case $\epsilon_2 > \epsilon_1$). Note that as ϵ_2 becomes infinitely large, we approach the metallic situation, $q' = -q$.

Example 11-12

Calculate the distribution of effective surface charge on the interface in Example 11-11.

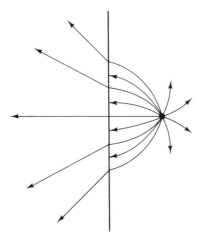

Figure 11-14   Field lines of E for positive point charge near planar dielectric interface.

Solution   Equation (4-6) gives

$$\sigma_P = (\mathbf{P}_2 - \mathbf{P}_1) \cdot \hat{\mathbf{k}} = \epsilon_0(\mathbf{E}_1 - \mathbf{E}_2) \cdot \hat{\mathbf{k}}$$

(since $\Delta \mathbf{D} \cdot \hat{\mathbf{n}} = 0$). Expressions for $\mathbf{E}$ at points on the interface are given in the solution for Example 11-11, so all we have to do is subtract the z components.

$$\sigma_P = \frac{\epsilon_0}{4\pi r^2} \cos\theta \left(\frac{q''}{\epsilon_2} + \frac{-q + q'}{\epsilon_1} \right)$$

$$= \frac{\epsilon_0 q}{4\pi(s^2 + z_1^2)} \frac{z_1}{\sqrt{s^2 + z_1^2}} \left[\frac{2\epsilon_2}{\epsilon_2(\epsilon_2 + \epsilon_1)} - \frac{2\epsilon_2}{\epsilon_1(\epsilon_2 + \epsilon_1)} \right]$$

$$= \frac{-q z_1}{2\pi \kappa_1 (s^2 + z_1^2)^{3/2}} \frac{\epsilon_2 - \epsilon_1}{\epsilon_2 + \epsilon_1}$$

The reader should compare this result with Example 11-3.

Example 11-13

Find the force on the point charge in Example 11-11.

Solution   The field acting on q is that part of $\mathbf{E}_1$ arising from q'. The details are left for a problem.

Example 11-14

An arbitrary charge distribution $\rho(r)$ lies entirely in the half-space $z > 0$. The plane $z = 0$ is a material boundary, either dielectric or conductive. Find a general expression for the potential at any point in the region containing the charge.

Solution   Here we have simply a superposition of point-charge image problems. For each source element at, say, (x', y', z') there is an image element at $(x', y',$

$-z'$). Thus

$$U(x, y, z) = \frac{1}{4\pi\epsilon_0}\left[\iiint\limits_{z'>0} \frac{\rho(x', y', z')\, dx'\, dy'\, dz'}{\sqrt{(x - x')^2 + (y - y')^2 + (z - z')^2}}\right.$$

$$\left. + \iiint\limits_{z'>0} \frac{\rho'(x', y', z')\, dx'\, dy'\, dz'}{\sqrt{(x - x')^2 + (y - y')^2 + (z + z')^2}}\right]$$

where ρ' is the appropriate image of ρ, depending on the nature of the interface.

This example shows how a simple known point-charge solution can lead to solutions of much more complicated problems. It illustrates one aspect of a very powerful method, called the *Green's function method*. Unfortunately, it is a bit too advanced to include in this book.

11-4 *SOLUTION OF THE LAPLACE EQUATION*

The Laplace equation describes the potential in a charge-free region. In typical problems, the potential is specified on various surface segments making up the boundary of the region. In some problems, the field strength instead may be specified over part or all of the boundary. In any case, the objective is to pick, out of the infinitude of possible solutions of the Laplace equation, the particular one that satisfies the boundary conditions.

The first task is to find a general solution expressed in terms of coordinates appropriate to the boundary geometry. In many cases this can be done by the method of *separation of variables*. The procedure is quite standard and comprises the following steps:

1. Assume that the potential function (a function of several variables, usually either two or three) can be written as a *product* of functions of one variable each. Then the partial derivatives of U involve only ordinary derivatives of the factor functions.

2. Manipulate the equation so that the left side involves only one of the variables and its function while the right side is completely free of this variable and its function. Then since the right side does not change with respect to this variable, neither can the left side (because the equation says that they are equal). Thus both sides are equal to the *same constant*, and the one variable has been *separated out* into an ordinary differential equation.

3. Repeat the procedure if two variables remain in the right side.

4. Solve the resulting ordinary differential equations, and multiply the corresponding solutions (i.e., those belonging to the same values of the constants of separation) together.

Solution in Cartesian Coordinates

For simplicity, we consider only problems in two dimensions. The equation then is

$$\frac{\partial^2 U}{\partial x^2} + \frac{\partial^2 U}{\partial y^2} = 0 \qquad (11\text{-}5)$$

Step 1: Assume that $U(x, y) = X(x)Y(y)$. Then the equation (using primes to denote derivatives) is

$$YX'' + XY'' = 0$$

Step 2: Dividing by XY and shifting one term to the right side, we get

$$\frac{X''}{X} = -\frac{Y''}{Y} = \text{const} \qquad (11\text{-}6)$$

Step 3: Not necessary.

Step 4: There are three cases to consider, depending on whether the constant of separation is positive, zero, or negative:

Case 1: Constant $= 0$. Then

$$X'' = 0, \quad X = Ax + B \qquad Y'' = 0, \quad Y = Cy + D$$

$$U(x, y) = (Ax + B)(Cy + D) \qquad (11\text{-}7)$$

Case 2: Constant is negative (call it $-k^2$). Then

$$\begin{aligned} X'' = -k^2 X \qquad X &= A \sin kx + B \cos kx \\ Y'' = k^2 Y \qquad Y &= C \sinh ky + D \cosh ky \end{aligned} \qquad (11\text{-}8)$$

There may well be (in fact, there usually are) numerous values of k capable of satisfying the boundary conditions. The general solution must include them all. Hence

$$U(x, y) = \sum_n (A_n \sin k_n x + B_n \cos k_n x)(C_n \sinh k_n y + D_n \cosh k_n y) \quad (11\text{-}9)$$

To simplify the notation in later work, we shall write this cumbersome expression as

$$U(x, y) = \sum_n \binom{\sin}{\cos} k_n x \binom{\sinh}{\cosh} k_n y \qquad (11\text{-}10)$$

Case 3: Constant is positive (call it k^2). Then obviously the trigonometric and hyperbolic functions are interchanged. [*Note:* If hyperbolic functions are unfamiliar, the form above is exactly equivalent to $Y = C' \exp(ky) + D' \exp(-ky)$.]

Example 11-15

An infinitely long rectangular conductive cylinder has three sides grounded, while the fourth (imagined to be insulated from the others by very tiny gaps) is held at potential U_0 (Figure 11-15). Find the potential at all interior points, $0 \le x \le a, 0 \le y \le b$.

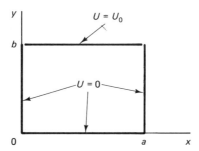

Figure 11-15 Boundary conditions for Example 11-15.

Solution The geometry obviously is appropriate for two-dimensional Cartesian coordinates, since the potential must be constant in the lengthwise (z) direction. Hence we must pick solutions out of the three cases enumerated above. However, the symmetry in the x direction clearly rules out Case 1. Also, Case 3 is ruled out since no combination of hyperbolic functions can be zero at two values of x. So we are left with Case 2 and must determine the values of k and the coefficients.

1. Since $U = 0$ at $x = 0$, $B = 0$ ($\cos 0 = 1$).
2. Since $U = 0$ at $x = a$, $k = n\pi/a$ ($\sin n\pi = 0$).
3. Since $U = 0$ at $y = 0$, $D = 0$ ($\cosh 0 = 1$).

The solution is thus reduced to the form

$$U(x, y) = \sum_n A_n \sin \frac{n\pi x}{a} \sinh \frac{n\pi y}{a} \tag{11-11}$$

The coefficients A_n are determined from the remaining boundary condition, $U = U_0$ at $y = b$. This gives

$$U_0 = \sum_n \left(A_n \sinh \frac{n\pi b}{a} \right) \sin \frac{n\pi x}{a}$$

The right side is a Fourier series and the coefficients (quantities in parentheses) are found as usual by multiplying both sides by a specific one of the sine functions [say, $\sin (m\pi x/a)$] and integrating both sides over the range 0 to a. On the right, all the resulting integrals will be zero except the one in the term $n = m$, which has the value $\frac{1}{2}a$. Thus

$$\frac{1}{2} A_m a \sinh \frac{m\pi b}{a} = \int_0^a U_0 \sin \frac{m\pi x}{a}\, dx$$

$$= \frac{aU_0}{m\pi} (1 - \cos m\pi)$$

$$= \begin{cases} \dfrac{2aU_0}{m\pi} & \text{for } m \text{ odd} \\ 0 & \text{for } m \text{ even} \end{cases}$$

Thus all the coefficients are evaluated, and the solution is complete. Unfortunately, the answer comes out in the form of an infinite series, so that numerical evaluation is a bit tedious. The final answer is (Figure 11-16)

$$A_m = \frac{4U_0}{m\pi \sinh{(m\pi b/a)}} \qquad (m \text{ odd})$$

$$U(x, y) = \frac{4U_0}{\pi} \sum_{n \text{ odd}} \frac{1}{n} \left(\sin\frac{n\pi x}{a} \right) \frac{\sinh{(n\pi y/a)}}{\sinh{(n\pi b/a)}} \qquad (11\text{-}12)$$

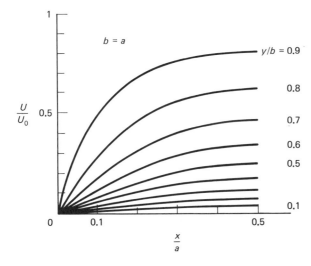

Figure 11-16   Potential in the conductive channel problem, Example 11-15, as given by Eq. (11-12).

Example 11-16

Suppose that instead of a constant U_0, the potential were some specified function $f(x)$ on the fourth side. What difference would it make?

Solution   Everything goes exactly the same up to the end, when $f(x)$ replaces U_0 in the integral.

Example 11-17

Suppose that four different constant potentials U_1, U_2, U_3, U_4 were specified on the four sides. How would that be handled?

Solution   Superimpose the solutions of the four separate problems $U_1, 0, 0, 0$; $0, U_2, 0, 0$; etc.

Example 11-18

Outline the solution of the Laplace equation in Cartesian coordinates in three dimensions.

Solution Start out just as before. After separating out $X(x)$, the yz equation is then separated. A typical solution function is exemplified by

$$A \sin k_x x \sin k_y y \sinh k_z z \qquad \text{where } k_z = (k_z^2 + k_y^2)^{1/2} \qquad (11\text{-}13)$$

Solution in Cylindrical Coordinates

For simplicity we again only consider cases where the potential is *independent of z*. The Laplace equation now takes the form

$$\frac{1}{s} \frac{\partial}{\partial s}\left(s \frac{\partial U}{\partial s}\right) + \frac{1}{s^2} \frac{\partial^2 U}{\partial \phi^2} = 0 \qquad (11\text{-}14)$$

Step 1: On substituting $U(s, \phi) = S(s)\Phi(\phi)$, this reads

$$s^{-1}\Phi(S' + sS'') + s^{-2}S\Phi'' = 0$$

Step 2: Divide by $s^{-2}S\Phi$ and shift one term, giving

$$\frac{sS' + s^2 S''}{S} = -\frac{\Phi''}{\Phi} = \text{const}$$

Step 4: There are only two cases of practical interest:

    Case 1: Constant $= 0$. Then

$$\Phi'' = 0 \qquad \Phi = A\phi + B$$
$$S' + sS'' = 0 \qquad S = C \ln s + D$$

    Case 2: Constant is positive (call it n^2). Then

$$\Phi'' = -n^2\Phi$$
$$\Phi = A \cos n\phi + B \sin n\phi$$
$$sS' + s^2 S'' = n^2 S$$
$$S = Cs^n + Ds^{-n}$$

In cases where the region of interest covers the entire angular range from 0 to 2π, the condition that U be a single-valued function forces n to be an integer.

Thus the general solution takes the form

$$U(s, \phi) = C_0 \ln s + D_0 + \sum_{n=1}^{\infty} \binom{\sin n\,\phi}{\cos n\,\phi}\left(C_n s^n + \frac{D_n}{s^n}\right) \qquad (11\text{-}15)$$

Example 11-19

An infinitely long conductive circular cylinder is split lengthwise and the two halves are held at potentials U_0 and $-U_0$ (Figure 11-17). Find the potential everywhere inside.

Solution Since the potential must remain finite at all points (because there are no charge singularities), there cannot be any log term or any negative powers of s. Thus the general form will be

$$U(s, \phi) = \sum_n s^n (A_n \cos n\phi + B_n \sin n\phi) \qquad (11\text{-}16)$$

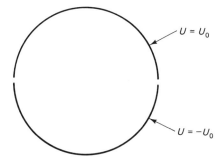

Figure 11-17 Boundary conditions for Example 11-19.

On the boundary, $s = a$,

$$U(a, \phi) = \pm U_0 = \sum_n a^n(A_n \cos n\phi + B_n \sin n\phi)$$

This is again a Fourier series. Because the left side is an odd function of ϕ (i.e., it changes sign when ϕ does) there cannot be any cosine terms, so $A_n = 0$ for all n. The B_n are determined just as in Example 11-15; the range of integration is now 0 to 2π, and the result is (Figure 11-18)

$$\pi a^n B_n = U_0 \int_0^\pi \sin n\phi \, d\phi + (-U_0) \int_\pi^{2\pi} \sin n\phi \, d\phi$$

$$= \frac{2U_0}{n} (1 - \cos n\phi)$$

Thus $B_n = 4U_0/\pi n a^n$ (n odd), $B_n = 0$ (n even), and

$$U(s, \phi) = \frac{4U_0}{\pi} \sum_{n \text{ odd}} \frac{1}{n} \left(\frac{s}{a}\right)^n \sin n\phi \tag{11-17}$$

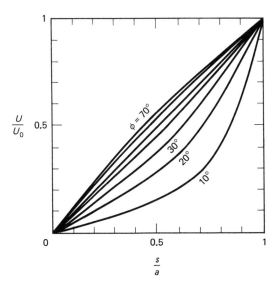

Figure 11-18 Potential values plotted along radial lines in the split-cylinder problem, Example 11-19.

Solution in Spherical Coordinates

Again for simplicity we limit the discussion. Thie time we consider only cases where the potential is *independent of ϕ.*

The Laplace equation takes the form

$$\frac{1}{r^2}\frac{\partial}{\partial r}\left(r^2\frac{\partial U}{\partial r}\right) + \frac{1}{r^2\sin\theta}\frac{\partial}{\partial\theta}\left(\sin\theta\frac{\partial U}{\partial\theta}\right) = 0 \qquad (11\text{-}18)$$

Step 1: On substituting $U(r,\theta) = R(r)P(\theta)$, this reads

$$\frac{P}{r^2}(r^2R')' + \frac{R}{r^2\sin\theta}(\sin\theta\,P')' = 0$$

Step 2: Multiply by r^2/RP and transpose, to obtain

$$\frac{(r^2R')'}{R} = -\frac{(\sin\theta\,P')'}{P\sin\theta} = \text{const}$$

Step 4: To solve these differential equations is a rather more difficult matter than in the previous cases. We start with the θ-equation. This is a very well known equation called the *Legendre equation* and its solutions have been extensively studied. For our purposes, we are interested only in the range $0 \leqq \theta \leqq \pi$, and, since we are representing a physical quantity like the potential, only interested in solutions that remain finite and well-behaved throughout that range. This turns out to be quite a stringent requirement, and it can only be satisfied for a restricted set of values of the constant, namely, positive integers of the form $n(n+1)$, where $n = 0, 1, \ldots$. The corresponding solution functions P_n are called the *Legendre polynomials*. The first few are

$$
\begin{aligned}
P_0(\cos\theta) &= 1 \\
P_1(\cos\theta) &= \cos\theta \\
P_2(\cos\theta) &= \tfrac{1}{2}(3\cos^2\theta - 1) \\
P_3(\cos\theta) &= \tfrac{1}{2}(5\cos^3\theta - 3\cos\theta)
\end{aligned}
\qquad (11\text{-}19)
$$

We see that

1. The Legendre polynomials consist of powers of the variable $(\cos\theta)$, which ranges from 1 at $\theta = 0$ to -1 at $\theta = \pi$.
2. P_n is a polynomial of the nth degree.
3. If n is even, the polynomial contains only even powers; thus its values are symmetrical about the xy plane. If n is odd, there are only odd powers, and values are anti-symmetric about the xy plane.
4. Each successive one has one more "wiggle" as θ goes from 0 to π.

Graphs of the first few Legendre polynomials are shown in Figure 11-19.

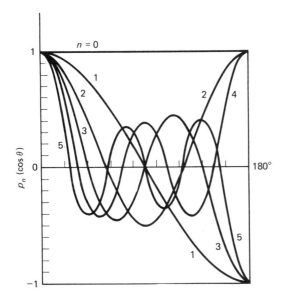

Figure 11-19 Graphs of the first six Legendre polynomials, as functions of the polar angle θ. The first four are given explicitly in Eqs. (11–19).

Returning now to the radial equation, we need consider only the same restricted set of values of the constant of separation, so this becomes

$$(r^2 R')' = n(n+1)R \qquad R(r) = Ar^n + Br^{-(n+1)}$$

Hence the general solution takes the form

$$U(r, \theta) = \sum_{n=0}^{\infty} (A_n r^n + B_n r^{-(n+1)}) P_n(\cos \theta) \qquad (11\text{-}20)$$

Potentials inside spherical boundaries can be found in much the same way as in Example 11-19 for a cylindrical boundary. However, the evaluation of the coefficients requires a detailed knowledge of the properties of the Legendre polynomials, so is beyond the level of this book. There are, though, some useful problems for fields outside spherical boundaries that can be worked quite easily.

Example 11-20

A grounded conductive sphere is placed in an initially uniform external field. Find the potential at all exterior points.

Solution In the general form of Eq. (11-20), we have to find all coefficients A_n and B_n such that (1) at large distance from the sphere the field has the uniform value $E_0 \hat{\mathbf{k}}$, and (2) on the sphere the potential is zero. The first of these is (letting r become infinite)

$$-E_0 z = \sum_{n=0}^{\infty} A_n r^n P_n(\cos \theta)$$

In order for this equation to hold at all values of θ, the coefficients of every power of $\cos \theta$ must be equal on the two sides. This is equivalent to making the coefficient of every P_n function equal on the two sides. Since $z = r \cos \theta = rP_1$, this means that $A_1 = -E_0$ and all the other A's are zero.

For the boundary condition at $r = a$, we have

$$0 = B_0 a^{-1} + (-E_0 a + B_1 a^{-2}) \cos \theta + \sum_{n=2}^{\infty} B_n a^{-(n+1)} P_n(\cos \theta)$$

Again equating coefficients of the P_n's gives $B_1 = E_0 a^3$, and all the other B's $= 0$. Thus the solution is (Figure 11-20)

$$U(r, \theta) = \left(-E_0 r + \frac{E_0 a^3}{r^2}\right) \cos \theta \qquad (11\text{-}21)$$

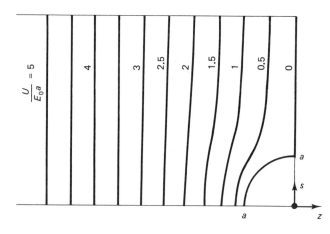

Figure 11-20 Equipotentials for grounded conductive sphere in uniform external field, Example 11-20.

Example 11-21

Calculate the surface charge density induced on the sphere.

Solution By Eq. (11-2),

$$\sigma = \epsilon_0 E_r(a, \theta) = -\epsilon_0 \left(\frac{\partial U}{\partial r}\right)_{r=a} = 3\epsilon_0 E_0 \cos \theta$$

Example 11-22

Calculate the dipole moment of the induced surface charge distribution.

Solution From the definition [Eq. (1-25)],

$$p_z = \iint \sigma z \, dS = 3\epsilon_0 E_0 \int_0^\pi (\cos \theta)(a \cos \theta)(2\pi a^2 \sin \theta \, d\theta)$$

$$= 4\pi \epsilon_0 a^3 E_0$$

Example 11-23

Identify the physical nature of the two terms in the potential (11-21).

Solution The first is the potential of the external field $U = -E_0 z$; the second is the potential due to the dipole moment induced on the sphere.

Example 11-24

Suppose that instead of being grounded the sphere is held at potential U_0.

Solution   Let $B_0 = U_0 a$, which adds a term $U_0 a/r$ to the previous solution.

Example 11-25

Concentric spheres of radii a and b are held at potentials U_a and U_b, respectively. Find the potential at all points between.

Solution   Since the symmetry prohibits any angular variation, the potential can only have the form $U(r) = A + B/r$. Hence the boundary conditions are

$$U_a = A + \frac{B}{a} \qquad U_b = A + \frac{B}{b}$$

Solving these simultaneously gives

$$A = \frac{bU_b - aU_a}{b - a} \qquad B = \frac{ab(U_a - U_b)}{b - a}$$

so

$$U(r) = \frac{1}{b - a}\left[b\left(1 - \frac{a}{r}\right)U_b - a\left(1 - \frac{b}{r}\right)U_a\right]$$

Example 11-26

A sphere of linear dielectric material is placed in an initially uniform external field. Find the potential at all points.

Solution   The potential at any point must be expressible in the form of Eq. (11-20), but the coefficients may have different values in the two regions. Hence we set

$$U_1(r, \theta) = \left(A_n r^n + \frac{B_n}{r^{n+1}}\right)P_n(\cos \theta) \qquad \text{for } r > a$$

$$U_2(r, \theta) = \left(C_n r^n + \frac{D_n}{r^{n+1}}\right)P_n(\cos \theta) \qquad \text{for } r < a$$

Now, as in Example 11-20, the condition at large r demands that $A_1 = -E_0$, and all the other A's be zero. The condition that U must be finite at $r = 0$ (since there is no point singularity there) demands that all D's be zero. Thus we are down to

$$U_1(r, \theta) = -E_0 r \cos \theta + \sum_n \frac{B_n}{r^{n+1}}P_n(\cos \theta) \tag{11-22}$$

$$U_2(r, \theta) = \sum_n C_n r^n P_n(\cos \theta) \tag{11-23}$$

Now the boundary conditions at $r = a$ may be applied. These are that the tangential component of **E** (or, alternatively, the potential) and the normal component of **D** are continuous. The tangential component of **E** comes from the

θ-derivative of U, hence

$$-E_0 a + \frac{B_1}{a^2} = C_1 a \qquad \text{for } n = 1 \tag{11-24}$$

$$\frac{B_n}{a^{n+1}} = C_n a^n \qquad \text{for } n \neq 1 \tag{11-25}$$

The normal component of D comes from the r-derivative of U multiplied by the appropriate permittivity. Hence

$$-E_0 - \frac{2B_1}{a^3} = \kappa C_1 \qquad \text{for } n = 1 \tag{11-26}$$

$$-\frac{(n+1)B_n}{a^{n+2}} = \kappa n C_n a^{n-1} \qquad \text{for } n \neq 1 \tag{11-27}$$

Equations (11-25) and (11-27) are incompatible unless $B_n = C_n = 0$ for all n except $n = 1$. This leaves only Eqs. (11-24) and (11-26) to be solved simultaneously. The solutions are

$$B_1 = \frac{\kappa - 1}{\kappa + 2} E_0 a^3 \qquad C_1 = -\frac{3}{\kappa + 2} E_0$$

$$U_1 = \left(-r + \frac{\kappa - 1}{\kappa + 2}\frac{a^3}{r^2} \right) E_0 \cos\theta \qquad (r > a)$$

$$U_2 = -\frac{3}{\kappa + 2} E_0 r \cos\theta \qquad (r < a)$$

so

$$\mathbf{E}_2 = \frac{3}{\kappa + 2} E_0 \hat{\mathbf{k}} \qquad (r < a)$$

It is interesting to note that the field is *uniform* inside the sphere, although of course with a different strength than the distant external field. This is, in fact, true for any ellipsoid even if it is not oriented parallel to the external field; however, it is beyond the level of this book to prove it. Another point of interest is that as the dielectric constant becomes infinite we again approach the metallic situation (see Example 11-11).

Example 11-27

Calculate the dipole moment of the sphere in Example 11-26.

Solution   Since the internal field is uniform and the dielectric is linear, the polarization will be uniform, given by

$$\mathbf{P} = (\kappa - 1)\epsilon_0 \mathbf{E} = 3\frac{\kappa - 1}{\kappa + 2}\epsilon_0 E_0 \hat{\mathbf{k}}$$

The dipole moment of the sphere is just the volume times this polarization. Note that the moment comes out to be just what is required to give the term $B_1 \cos\theta / r^2$ in the external potential.

11-5 APPLICATION TO CHARGE PROBLEMS

In earlier chapters we repeatedly emphasized that whenever the source distribution is completely specified the potential (and field) can be obtained by direct integration. We also noted that sometimes the form of the direct solution is somewhat inconvenient, and that it may be advantageous to express the potential in infinite-series form, the multipole expansion (Section 1-8). The methods of Section 11-4 provide us with yet another form of series solution that in some cases is even easier to obtain and more convenient to use. If the potential can be (1) readily written as a series of appropriate solutions of the Laplace equation, and (2) evaluated at some special set of points in space, the latter can be treated as "boundary" points for the evaluation of the expansion coefficients.

The following example illustrates the method in a simple problem which we solve all three ways.

Example 11-28

A uniform line charge is bent into a ring of radius a and total charge q (Figure 11-21). Find the potential at all points at distances $> a$ from the center.

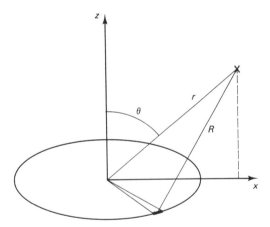

Figure 11-21 Geometry of the ring-charge problem, Example 11-28.

Solution 1 Direct integration: Since the charge distribution has axial symmetry, the potential must be independent of ϕ, so we may as well choose the xy axes so that $\phi = 0$. Then

$$U(r, \theta) = \frac{1}{4\pi\epsilon_0} \int_0^{2\pi} \frac{(q/2\pi a)a\, d\phi'}{\sqrt{(x - a\cos\phi')^2 + a^2\sin^2\phi' + z^2}}$$

$$= \frac{q}{8\pi^2\epsilon_0} \int_0^{2\pi} \frac{d\phi'}{\sqrt{r^2 + a^2 - 2ax\cos\phi'}}$$

Here our progress comes to an abrupt halt, since this simple-looking integral cannot be evaluated in closed form. It belongs to a class called *elliptic integrals*.

Thus, although a solution (this integral) was easy to write down, it takes a very inconvenient form. However, these functions are tabulated, so the form may be useful for numerical evaluation.

Solution 2 Multipole expansion: We must first evaluate the moments of the distribution:

$$\text{total charge} = q$$

dipole moment:

$$p_x = \int_0^{2\pi} \frac{q}{2\pi a}(a\,d\phi')(a\cos\phi') = 0$$

$$p_y = \int_0^{2\pi} \frac{q}{2\pi a}(a\,d\phi')(a\sin\phi') = 0$$

$$p_z = \int_0^{2\pi} \frac{q}{2\pi a}(a\,d\phi')(0) = 0$$

Thus, to the terms that we have explicitly discussed in this book,

$$U(r,\theta) = \frac{1}{4\pi\epsilon_0}\left(\frac{q}{r} + 0 + \cdots\right)$$

(There are, of course, nonzero higher-order terms, but they rapidly become cumbersome to calculate.)

Solution 3 Harmonic functions: Owing to the axial symmetry, we may obviously write, using Eq. (11-20),

$$U(r,\theta) = \sum_{n=0}^{\infty} \frac{B_n}{r^{n+1}} P_n(\cos\theta)$$

(Since the potential must be well behaved at infinity, no positive powers of r may be present.) This must hold at all values of θ, including $\theta = 0$. But at $\theta = 0$ we have $P_n(\cos\theta) = 1$ for all n, and $r = z$ ($z > 0$). Thus

$$U(z) = \sum_{n=0}^{\infty} \frac{B_n}{z^{n+1}}$$

So all we have to do is find $U(z)$ explicitly, express it in series form (inverse powers of z), and read off the coefficients.

$$U(z) = \frac{q}{4\pi\epsilon_0} \frac{1}{\sqrt{a^2 + z^2}}$$

$$= \frac{q}{4\pi\epsilon_0 z}\left(1 + \frac{a^2}{z^2}\right)^{-1/2}$$

$$= \frac{q}{4\pi\epsilon_0 z}\left[1 - \frac{1}{2}\frac{a^2}{z^2} + \left(-\frac{1}{2}\right)\left(-\frac{3}{2}\right)\frac{1}{2!}\frac{a^4}{z^4} + \cdots\right]$$

$$= \frac{q}{4\pi\epsilon_0}\left(\frac{1}{z} - \frac{a^2}{2z^3} + \frac{3a^4}{8z^5} + \cdots\right)$$

Thus $B_0 = q/4\pi\epsilon_0$, $B_1 = 0$, $B_2 = -qa^2/8\pi\epsilon_0$, and so on, and

$$U(r, \theta) = \frac{q}{4\pi\epsilon_0}\left[\frac{1}{r} - \frac{a^2}{2r^3}P_2(\cos\theta) + \frac{3a^4}{8r^5}P_4(\cos\theta) + \cdots\right]$$

For the region $r < a$, a different expansion (in positive powers of r) is needed. The details of this are left to the problems. Each expansion is convergent only in its own domain, but the two of them match at the border $r = a$. One final point: The simplicity of the method is due in very large part to the symmetry of the problem. Potentials due to less symmetrical source distributions are considerably harder to work out. Methods are discussed in the references which follow.

FURTHER REFERENCES

Solutions of the Laplace equation are treated in much more detail in all the standard advanced texts, such as those by Jackson, Good and Nelson, Panofsky and Phillips, and Konopinsky. These texts also discuss Green's functions and their applications to potential problems. For numerical solution of the Laplace equation, the "relaxation" method is described in simple terms by E. M. Purcell, *Electricity and Magnetism*, Vol. 2 of Berkeley Physics Course, McGraw-Hill, New York, 1965.

A method that we did not discuss at all, the use of "conformal transformations," is useful in a variety of two-dimensional problems. It is discussed by Lorrain and Corson, Panofsky and Phillips, and in great detail by W. R. Smythe, *Static and Dynamic Electricity*, McGraw-Hill, New York, 1939.

PROBLEMS

11-1. At one side of an interface between two materials the electric field is $\mathbf{E} = E_1\hat{\mathbf{n}} + E_2\hat{\mathbf{t}}$, where $\hat{\mathbf{n}}$ and $\hat{\mathbf{t}}$ are unit vectors normal and tangential to the interface, respectively. What is the most general form that $\mathbf{E}$ can have on the other side?

11-2. At a planar interface between two linear dielectric media, $\mathbf{D}_2$ makes angle θ_2 with the normal and has magnitude 1 C/m². Find $\mathbf{E}$ and $\mathbf{D}$ in medium 1.

11-3. Write the potential due to a point charge q at point (x_1, y_1) inside a 90-degree grounded conductive corner (Example 11-6).

11-4. Suppose that the point charge is *outside* the 90-degree corner. Can the problem now be solved by the method of images? Explain.

11-5. In Example 11-8, let the point charge be placed at the origin and let the planes be perpendicular to the x axis at positions $x = \pm a$. Write the potential at all points of the x axis.

11-6. For Example 11-9, calculate the surface charge density on the sphere.

11-7. Find the force on the point charge of Example 11-9.

11-8. A grounded conductive sheet consists of a plane with a hemispherical bulge of radius a in it. A point charge q is located directly above the center of the bulge, at distance z_1 from the plane. Find a set of images for this boundary.

11-9. Find the magnitude of the force on the point charge near the dielectric interface of Example 11-11.

11-10. In Example 11-15, let the channel have a square cross section (i.e., $b = a$) and write the explicit series for the potential at the midpoint.

11-11. How many terms of the series in Problem 11-10 are needed in order to get a numerical value accurate to three significant figures?

11-12. Show by explicit substitution that $P_2 (\cos \theta)$ is a solution of the Legendre equation.

11-13. Show by explicit substitution that $r^{-3} P_2 (\cos \theta)$ is a solution of the Laplace equation.

11-14. Show that if $f(x, y, z)$ is a solution of the Laplace equation, any partial derivative of f with respect to x, y, or z is also a solution.

11-15. A well-known solution of the Laplace equation is $1/r$ (the potential due to a point charge at the origin). What series of other solutions is generated by taking successive partial derivatives with respect to z?

11-16. Two coaxial infinitely long conductive cylinders, of radii a and b, are held at potentials U_a and U_b respectively. Find the potential at all points in between.

11-17. A grounded infinitely long conductive cylinder of radius a is placed in a uniform external field E_0 perpendicular to the cylinder axis. Find the potential at all points outside the cylinder.

11-18. Find the surface charge density induced on the cylinder of Problem 11-17.

11-19. For Example 11-28, find a general expression for the series expansion valid when $r > a$.

11-20. For the same charge distribution, find the potential for $r < a$. Verify that the two series match at $r = a$.

11-21. What do you expect happens to these series at $r = a$, $\theta = \pi/2$?

11-22. A uniformly charged circular disk has radius a, charge density σ. Find the potential at all points where $r > a$.

chapter 12

Magnetostatic Boundary-Value Problems

12-1 BOUNDARY CONDITIONS ON MAGNETIC FIELDS

In Chapter 2 we saw that whenever the current distribution is completely specified the magnetic field at any point can be calculated directly. In Chapter 6 this idea was extended to include the effective current density in magnetized matter. As in Chapter 11, we now turn to the more complicated class of problems in which the source density is only incompletely specified. Again, the first step is to develop the relations satisfied by the fields at points immediately adjacent to the two sides of boundary surfaces between different media in the fields. Again, those come from the Maxwell equations applied to suitable regions containing part of the interface.

Boundary Condition on **H**

The Maxwell equation for **H** [Eq. (7-4)] is integrated over a small rectangular area intersecting the interface as shown in Figure 12-1. This gives, after application of Stokes' theorem,

$$\oint \mathbf{H} \cdot d\mathbf{l} = \iint (\mathbf{J} + \dot{\mathbf{D}}) \cdot \hat{\mathbf{N}} \, dS$$

On letting the sides shrink to zero, the integral of $\dot{\mathbf{D}}$ vanishes. However, the integral of **J** requires a bit more care. If both media are real physical materials, **J** must remain finite and the integral vanishes in the limit. However, it is often convenient, as we have seen in electrostatic problems, to deal with fictitious *ideal* conductors which may support a sheet of *surface current*. This remains unchanged as the sides shrink, so in the limit

$$\mathbf{H}_1 \cdot (-\hat{\mathbf{t}}) + \mathbf{H}_2 \cdot \hat{\mathbf{t}} = \mathbf{j} \cdot \hat{\mathbf{N}}$$

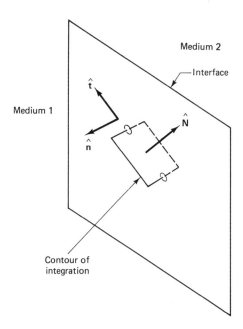

Figure 12-1 Integration contour for boundary relation of **H**. $\hat{\mathbf{N}}$ is the unit normal to the plane of the contour, and $\hat{\mathbf{n}}$ is the unit normal to the interface plane.

or

$$\mathbf{j} \cdot \hat{\mathbf{N}} = (\mathbf{H}_2 - \mathbf{H}_1) \cdot (\hat{\mathbf{n}} \times \hat{\mathbf{N}}) = (\mathbf{H}_2 - \mathbf{H}_1) \times \hat{\mathbf{n}} \cdot \hat{\mathbf{N}}$$

Since **j** has no component perpendicular to the plane, this may be solved to give

$$(\mathbf{H}_2 - \mathbf{H}_1) \times \hat{\mathbf{n}} = \mathbf{j} \qquad (12\text{-}1)$$

In using this relation, it must be borne in mind that $\hat{\mathbf{n}}$ is defined as the unit vector *from medium 2 toward medium 1*. Also, if neither medium is an ideal conductor the tangential components of **H** are *continuous*, since **j** is necessarily zero.

Boundary Condition on B

The Maxwell equation for **B**, Eq. (7-3), is integrated over the volume of a cylinder as shown in Figure 12-2. This gives, after applying the divergence theorem,

$$\oiint \mathbf{B} \cdot \hat{\mathbf{N}} \, dS = 0$$

$$\mathbf{B}_1 \cdot \hat{\mathbf{n}} + \mathbf{B}_2 \cdot (-\hat{\mathbf{n}}) = 0 \qquad (12\text{-}2)$$

Thus the normal component of **B** is *always continuous*.

Example 12-1

Show that the magnetic flux through any contour is continuous even if the contour lies in the interface between two media.

Solution   From the definition [Eq. (2-30)] the flux is seen to be an integral of the normal component of **B** over any surface bounded by the contour. Since, as we have just seen, this component is continuous at every point of every surface, so is its integral.

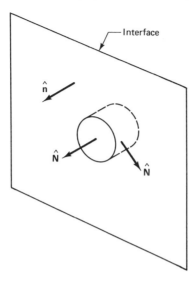

Figure 12-2 Cylindrical closed surface of integration. $\hat{\mathbf{N}}$ is the unit normal at any point of the cylinder, and $\hat{\mathbf{n}}$ is the unit normal to the interface plane.

Boundary Conditions for the Vector Potential

From Eq. (2-31), the flux is a line integral of the tangential component of **A** around the contour. Since this integral must be continuous *for any contour*, the tangential component of **A** must be continuous at every point. In other words,

$$\mathbf{A}_1 \times \hat{\mathbf{n}} = \mathbf{A}_2 \times \hat{\mathbf{n}} \qquad (12\text{-}3)$$

where $\hat{\mathbf{n}}$ is a unit vector normal to the interface. From the derivation, this is seen to be nothing more than an alternative statement of Eq. (12-2).

Example 12-2

At an interface between two *linear* magnetic media, the **B** field in medium 1 makes an angle θ_1 with the normal. Find the angle between **B** and the normal in medium 2.

Solution   From Eq. (6-20), the vector relations in the two media are $\mathbf{B}_1 = \mu_1 \mathbf{H}_1$, $\mathbf{B}_2 = \mu_2 \mathbf{H}_2$. The boundary relations are

$$B_{2n} = B_{1n}$$

$$H_{2t} = H_{1t} \qquad \frac{B_{2t}}{\mu_2} = \frac{B_{1t}}{\mu_1}$$

Thus

$$\tan \theta_2 \equiv \frac{B_{2t}}{B_{2n}} = \frac{\mu_2}{\mu_1} \frac{B_{1t}}{B_{1n}} = \frac{\mu_2}{\mu_1} \tan \theta_1$$

The fields are illustrated in Figure 12-3. Note that if μ_1 is very large, $\tan \theta_2$ is very small, so the field just outside a medium of very high permeability is nearly perpendicular to the interface. This is somewhat analogous to the electric field just outside an ideal conductor.

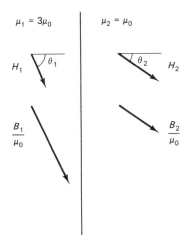

Figure 12-3 Illustration of Example 12-2 for the case $\mu_1 = 3\mu_2$.

Example 12-3

A uniform sheet of current, **j** coulombs per meter, flows in a very thin metal foil. Obtain the relation between the magnetic fields on the two sides.

Solution   Take the sheet to lie in the xy plane, the current to flow in the y direction, and region 1 to be on the $+z$ side so that $\hat{\mathbf{n}} = \hat{\mathbf{k}}$. Then Eqs. (12-1) and (12-2) give

$$(B_{2x}\hat{\mathbf{i}} + B_{2y}\hat{\mathbf{j}} + B_{2z}\hat{\mathbf{k}}) \times \hat{\mathbf{k}} = (B_{1x}\hat{\mathbf{i}} + B_{1y}\hat{\mathbf{j}} + B_{1z}\hat{\mathbf{k}}) \times \hat{\mathbf{k}} + \mu_0 j\hat{\mathbf{j}}$$

$$B_{2z} = B_{1z}$$

On equating components of the first equation, we find that

$$B_{2y} = B_{1y} \qquad B_{2x} = B_{1x} - \mu_0 j$$

Note that the field line is "refracted" clockwise as viewed along the current, as shown in Figure 12-4.

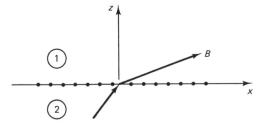

Figure 12-4 Field relations near a uniform sheet of current, Example 12-3.

Example 12-4

In Example 12-3, suppose that no other currents are present. What are the fields?

Solution

$$\mathbf{B}_1 = -\mathbf{B}_2 = \frac{\mu_0}{2} j\hat{\mathbf{i}}$$

This can be seen either from the form of the field produced by each filament of current or from the symmetry of the source distribution, along with the solenoidal character of the field.

Example 12-5

A ferromagnetic material is uniformly and permanently magnetized. Find the relations between the field on the two sides of a planar surface lying parallel to the magnetization.

Solution   Since the medium is nonlinear, **B** and **H** are *not* proportional; instead, we must use the general relation Eq. (6-12). We take the surface to be the xy plane with $\mathbf{M}_2 = -M_0\hat{\mathbf{i}}$ in medium 2. The components of this relation are

$$H_{2x} = \frac{B_{2x}}{\mu_0} - M_0 \qquad H_{2y} = \frac{B_{2y}}{\mu_0} \qquad H_{2z} = \frac{B_{2z}}{\mu_0}$$

The boundary relations give

$$B_{1z} = B_{2z} \qquad H_{1x} = H_{2x} \qquad H_{1y} = H_{2y}$$

From the first of these,

$$H_{1z} \equiv \frac{B_{1z}}{\mu_0} = \frac{B_{2z}}{\mu_0} = H_{2z}$$

so that

$$\mathbf{H}_1 = \mathbf{H}_2 = \mathbf{H}$$

Then

$$\mathbf{B}_1 = \mu_0\mathbf{H} \qquad \mathbf{B}_2 = \mu_0(\mathbf{H} + \mathbf{M}_2)$$

These relations are shown in Figure 12-5.

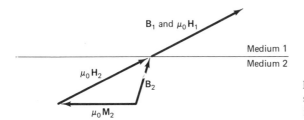

Figure 12-5   Field relations near a surface of a magnetized slab when **M** is parallel to the surface, Example 12-5.

Example 12-6

In Example 12-5, if there are no other surfaces at finite distance, nor any currents, what are the fields?

Solution   Since the effective current distribution in this example is just like the true current distribution in Example 12-4, the fields are

$$\mathbf{B}_1 = \mu_0\mathbf{H} = \mu_0\left(\frac{M_0}{2}\right)\hat{\mathbf{i}}$$

$$\mathbf{B}_2 = -\mu_0\left(\frac{M_0}{2}\right)\hat{\mathbf{i}}$$

This is shown in Figure 12-6.

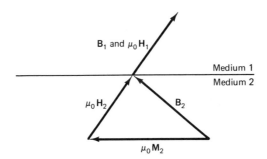

Figure 12-6   Field relations near surface of magnetized slab when no other true or effective currents are present, Example 12-6.

Example 12-7

Make a qualitative sketch of the general shapes of the **B** and **H** field lines due to an ideal bar magnet (a cylinder permanently and uniformly magnetized parallel to its axis). *Hint*: Use the results of Example 6-2 to get the **B** field on the axis, then the appropriate Maxwell equation to get an idea of the behavior off axis, then the boundary relations.

Solution   On axis, the **B** field is strongest at the middle and falls off as we go along the axis in the $\pm z$ directions. Thus on the $\pm z$ side $\partial B_z/\partial z$ is negative. Since div **B** = 0, this means that $\partial B_s/\partial s$ must be positive, that is, the radial component *increases* as we go away from the axis. Hence, the **B** lines must bow *outward*. At the end faces there is no effective current so they are not refracted. But at the cylinder surface they are refracted much as in Example 12-6. The result is as shown in Figure 12-7a. As for **H**, it is parallel to **B** outside but generally opposite inside, as indicated in Figure 12-7b. In particular, note that the tangential component of **H** is continuous everywhere, even at the end faces where the lines terminate.

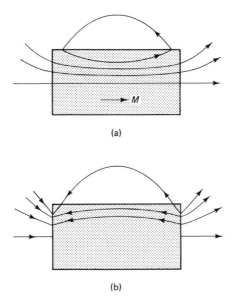

(a)

(b)

Figure 12-7   (a) Field lines of **B** and (b) field lines of **H** in and near a bar magnet.

12-2 DIFFERENTIAL EQUATIONS FOR THE VECTOR POTENTIAL

In electrostatics, the relation between field strength and source density (Gauss's law) was cast in terms of the scalar potential function to obtain the Poisson and Laplace equations. This turned out to be a very powerful approach to the solution of boundary-value problems. In magnetostatics, a similar approach is not quite as useful because (1) the equations turn out to be somewhat more complicated, and (2) as we shall see in the next section, simpler methods can often be used. Nevertheless, we shall work through the derivation of the equations and a few simple examples of their solution. The first step is to cast Ampère's law in terms of the vector potential. Recalling that $\mathbf{A}$ is a function whose curl is $\mathbf{B}$ (*not* $\mathbf{H}$), we need Ampère's law for $\mathbf{B}$. This is given in Eq. (6-10). Hence the result is

$$\text{curl curl } \mathbf{A} = \mu_0(\mathbf{J} + \mathbf{J}_M) \tag{12-4}$$

This equation is readily simplified by an appropriate choice of gauge. Since we shall certainly want to use the identity curl curl $= \text{grad div} - \nabla^2$, the obvious choice here is the Coulomb gauge

$$\text{div } \mathbf{A} = 0 \tag{12-5}$$

which reduces Eq. (12-4) to

$$\nabla^2 \mathbf{A} = -\mu_0(\mathbf{J} + \mathbf{J}_M) \tag{12-6}$$

In using this "vector Poisson equation," two things must be borne in mind:

1. It is valid *only* if Eq. (12-5) is also satisfied.
2. In evaluating the Laplacian of a vector function (expressed in some coordinate system), the differential operators will operate on the *unit vectors* as well as the components. In the special case of the *Cartesian* coordinate system, the unit vectors are constants, so the result is

$$\nabla^2 \mathbf{A} \equiv (\nabla^2 A_x)\hat{\mathbf{i}} + (\nabla^2 A_y)\hat{\mathbf{j}} + (\nabla^2 A_z)\hat{\mathbf{k}} \tag{12-7}$$

The individual terms may, of course, be calculated in any coordinate system that is convenient.

Example 12-8

Show that for a uniform linear medium Eq. (12-6) is the same as the static form of Eq. (7-17).

Solution   For a linear medium

$$\mu_0\mathbf{M} = \mathbf{B} - \mu_0\mathbf{H} = (\mu - \mu_0)\mathbf{H}$$

$$\mu_0 \text{ curl } \mathbf{M} = (\mu - \mu_0) \text{ curl } \mathbf{H} = (\mu - \mu_0)\mathbf{J}$$

Substituting this for the $\mathbf{J}_M$ term gives

$$\nabla^2 \mathbf{A} = -\mu\mathbf{J} \tag{12-8}$$

Example 12-9

Write the form of the vector Poisson equation for cases in which all currents flow in a single direction.

Solution   The source density is specified to have the form $\mathbf{J} = J\hat{\mathbf{k}}$, and the requirement that div $\mathbf{J} = 0$ [(Eq. (2-7)] means that J must be independent of z. Since, according to Eq. (2-15), any current element contributes a vector potential in its own direction, it is natural to try the form

$$\mathbf{A} = A_z(x, y)\hat{\mathbf{k}}$$

which is seen to satisfy Eq. (12-5) automatically. Then Eq. (12-8) gives

$$\left(\frac{\partial^2}{\partial x^2} + \frac{\partial^2}{\partial y^2}\right) A_z = -\mu J \qquad (12\text{-}9)$$

This is simply a two-dimensional scalar Poisson equation, so for this particularly simple pattern of current flow there is no extra difficulty due to the vector character.

Example 12-10

A steady current I is uniformly distributed over the cross section of an infinitely long cylinder made of a linear magnetic metal. Find the vector potential everywhere.

Solution   Of course, this simple problem can be done easily by Ampère's law for $\mathbf{H}$, but we want to illustrate the use of the vector Poisson equation. Naturally, we will write this in cylindrical coordinates, and since the symmetry rules out any ϕ-dependence, Eq. (12-9) becomes

$$\frac{1}{s}\frac{d}{ds}\left(s\frac{dA_z}{ds}\right) = \begin{cases} -\dfrac{\mu I}{\pi a^2} & \text{for } s < a \\ 0 & \text{for } s > a \end{cases}$$

We first consider region 1 (inside). A particular integral of the differential equation is readily found by direct integration to be $-\mu Is^2/4\pi a^2$. To this must be added a general solution of the Laplace equation. The form for cylindrical coordinates with no z-dependence was given in Eq. (11-15), and in the present case with no ϕ-dependence either it reduces to $C \ln s + D$. For region 2 (outside), there is no source density, so only the general solution pertains. Thus we have

$$A_{z1}(s) = -\frac{\mu I}{4\pi a^2} s^2 + C_1 \ln s + D_1$$

$$A_{z2}(s) = C_2 \ln s + D_2$$

Taking the curls of these two expressions gives

$$B_{\phi 1}(s) = \frac{\mu I}{2\pi a^2} s - \frac{C_1}{s}$$

$$B_{\phi 2}(s) = -\frac{C_2}{s}$$

The "boundary condition" that B remain finite at the axis gives $C_1 = 0$. Then the continuity of the tangential component of H gives $(1/\mu)(\mu I/2a\pi^2)a = -C_2/\mu_0 a$, or $C_2 = -\mu_0 I/2\pi$.

Since the additive constants D_1 and D_2 are irrelevant, the result is

$$A_z = \begin{cases} -\dfrac{\mu I}{4\pi a^2}\, s^2 & \text{for } s < a \\[2ex] -\dfrac{\mu_0 I}{2\pi}\, \ln s & \text{for } s > a \end{cases}$$

Example 12-11

Write the form of the vector Poisson equation for cases in which all currents flow in circular loops parallel to the xy plane.

Solution   The current density is specified to have the form $\mathbf{J}(r, \theta) = J_\phi(r, \theta)\,\hat{\mathbf{a}}_\phi$. Hence it seems natural to try a solution of the form $\mathbf{A} = A_\phi(r, \theta)\,\hat{\mathbf{a}}_\phi$; this also satisfies Eq. (12-5) automatically. To calculate $\nabla^2\mathbf{A}$, we can either (1) apply the Laplacian operator to this vector function, remembering that $\hat{\mathbf{a}}_\phi$ is a function of ϕ, or (2) write the function out in Cartesian components and use Eq. (12-7). The latter procedure gives

$$\mathbf{A} = A_\phi(r, \theta)(-\cos\phi\,\hat{\mathbf{i}} + \sin\phi\,\hat{\mathbf{j}})$$
$$\nabla^2\mathbf{A} = -\hat{\mathbf{i}}\,\nabla^2(A_\phi\cos\phi) + \hat{\mathbf{j}}\,\nabla^2(A_\phi\sin\phi)$$

Now

$$\nabla^2(A_\phi\cos\phi) = \left(\nabla^2 A_\phi - \frac{A_\phi}{r^2\sin^2\theta}\right)\cos\phi$$

$$\nabla^2(A_\phi\sin\phi) = \left(\nabla^2 A_\phi - \frac{A_\phi}{r^2\sin^2\phi}\right)\sin\phi$$

So

$$\nabla^2\mathbf{A} = \left(\nabla^2 A_\phi - \frac{A_\phi}{r^2\sin^2\theta}\right)\hat{\mathbf{a}}_\phi \tag{12-10}$$

Note that this is *not* simply the Laplacian of the unknown function A_ϕ. There is an additional term present, which complicates matters. The solutions of this equation are known in terms of the so-called *associated Legendre functions*. We shall not go into these in detail but will consider one simple example.

Example 12-12

Show that one solution of the vector Laplace equation $\nabla^2\mathbf{A} = 0$ is

$$\mathbf{A}(r, \theta) = \left(Cr + \frac{D}{r^2}\right)\sin\theta\,\hat{\mathbf{a}}_\phi \tag{12-11}$$

Solution   By direct substitution into Eq. (12-10), $\nabla^2[(Cr + D/r^2)\sin\theta] - (Cr + D/r^2)/r^2\sin\theta$ works out to 0.

Example 12-13

Show that the vector potential for a uniformly magnetized sphere can be written in the form of Eq. (12-11), and evaluate the coefficients and the resulting fields.

Solution   First, since there are no volume sources, the right side of Eq. (12-6) is zero. Also recall from Example 12-11 that Eq. (12-5) is satisfied. Obviously, we expect different coefficients in the two regions, hence put

$$A_{\phi 1} = \left(C_1 r + \frac{D_1}{r^2}\right) \sin \theta \qquad \text{for } r < a$$

$$A_{\phi 2} = \left(C_2 r + \frac{D_2}{r^2}\right) \sin \theta \qquad \text{for } r > a$$

It remains to see if the boundary conditions can be satisfied. For **B** to remain finite at $r = 0$ and $r \to \infty$, clearly $D_1 = 0$ and $C_2 = 0$. Then on taking the curls, we get

$$\mathbf{B}_1 = \frac{\hat{\mathbf{a}}_r}{r \sin \theta} \frac{\partial}{\partial \theta} (C_1 r \sin^2 \theta) - \frac{\hat{\mathbf{a}}_\theta}{r} \frac{\partial}{\partial r} (C_1 r^2 \sin \theta)$$

$$= 2C_1(\hat{\mathbf{a}}_r \cos \theta - \hat{\mathbf{a}}_\theta \sin \theta) = 2C_1 \hat{\mathbf{k}}$$

$$\mathbf{B}_2 = \frac{D_2}{r^3} (2 \cos \theta \, \hat{\mathbf{a}}_r + \sin \theta \, \hat{\mathbf{a}}_\theta)$$

The boundary equations (B_r and H_θ continuous at $r = a$) give

$$2C_1 = \frac{2D_2}{a^3} \qquad \text{and} \qquad \frac{-2C_1 \sin \theta}{\mu_0} - (-M_0 \sin \theta) = \frac{D_2 \sin \theta}{\mu_0 a^3}$$

and the solution of this pair is

$$C_1 = \frac{\mu_0 M_0}{3} \qquad D_2 = C_1 a^3$$

Thus the fields inside the sphere are

$$\mathbf{B}_1 = \tfrac{2}{3}\mu_0 M_0 \hat{\mathbf{k}} \qquad \mathbf{H}_1 = -\tfrac{1}{3} M_0 \hat{\mathbf{k}} \qquad (12\text{-}12)$$

and those outside are

$$\mathbf{B}_2 = \mu_0 \mathbf{H}_2 = \tfrac{1}{3}\mu_0 M_0 \left(\frac{a}{r}\right)^3 (2 \cos \theta \, \hat{\mathbf{a}}_r + \sin \theta \, \hat{\mathbf{a}}_\theta) \qquad (12\text{-}13)$$

The inside field is seen to be *uniform*, and the outside field is a *dipole field* arising from a dipole moment $(4\pi/3)M_0 a^3$, which is just the total dipole moment of the sphere.

The reader may have noted that this problem is not truly a boundary value problem at all. Since the source distribution is *completely specified* (an effective surface current of amount $\mathbf{j}_M = M_0 \sin \theta \, \hat{\mathbf{a}}_\phi$) the vector potential could in principle have been calculated directly. The resulting integral is not easy to evaluate, and the method we used here was undoubtedly quite a bit easier. However, as will be seen in the next section, even easier ones are available.

12-3 SCALAR POTENTIAL METHODS

In a region *completely free of sources*, Ampère's law reads curl **B** $= 0$. Hence **B** must be a gradient; it is conventional to write

$$\mathbf{B}(\mathbf{r}) = -\text{grad } U_B(\mathbf{r}) \qquad (12\text{-}14)$$

Since div $\mathbf{B} = 0$, U_B satisfies the Laplace equation

$$\nabla^2 U_B = 0 \tag{12-15}$$

This is obviously a very useful relation, since it enables us to take over much of electrostatic theory for certain magnetic problems.

Example 12-14

Solve the magnetized sphere problem of Example 12-13 by the use of U_B in the outside region.

Solution   Equation (11-20) gave the solution of the Laplace equation in spherical coordinates. Since there is no externally applied field in this problem, the potential will have to vanish at infinity, so we set

$$U_{B2}(r, \theta) = \sum C_n r^{-(n+1)} P_n(\cos \theta)$$

Thus

$$\mathbf{B}_2(r, \theta) = \sum (n + 1) C_n r^{-(n+2)} P_n \hat{\mathbf{a}}_r - \frac{1}{r} \sum C_n r^{-(n+1)} \frac{\partial P_n}{\partial \theta} \hat{\mathbf{a}}_\theta$$

For the inside region we cannot use the scalar potential since a source density is present (on the surface). However, our experience with electrostatic problems (e.g., Example 11-26) suggests that $\mathbf{B}$ and $\mathbf{H}$ are uniform, so we try the form

$$\mathbf{B}_1 = B_0 \hat{\mathbf{k}} \qquad \mu_0 \mathbf{H}_1 = (B_0 - \mu_0 M_0) \hat{\mathbf{k}}$$

Then the continuity of B_r at $r = a$ gives

$$B_0 \cos \theta = \sum (n + 1) C_n a^{-(n+2)} P_n(\cos \theta)$$

from which, by equating coefficients of the P_n's,

$$C_n = 0 \text{ for } n \neq 1 \qquad \frac{2C_1}{a^3} = B_0$$

The continuity of H_θ at $r = a$ gives

$$(B_0 - \mu_0 M_0)(-\sin \theta) = -\frac{C_1}{a^3} \frac{\partial P_1}{\partial \theta}$$

$$= (C_1/a^3) \sin \theta$$

Finally, solving the two equations for B_0 and C_1 gives

$$B_0 = \tfrac{2}{3} \mu_0 M_0 \qquad C_1 = \tfrac{1}{3} \mu_0 M_0 a^3$$

Thus the fields are the same as we found in Example 11-26.

In a *current-free* region, H can also be expressed in terms of a scalar potential. Since curl $\mathbf{H} = \mathbf{J}$ (under static conditions), wherever $\mathbf{J} = 0$ we have curl $\mathbf{H} = 0$. Hence $\mathbf{H}$ must be a gradient in such regions; it is conventional to write

$$\mathbf{H}(\mathbf{r}) = -\text{grad } U_H(\mathbf{r}) \tag{12-16}$$

Since div $\mathbf{B} = 0$ (always), div $\mathbf{H} = -$div $\mathbf{M}$; hence

$$\nabla^2 U_H = \text{div } \mathbf{M} \tag{12-17}$$

Thus U_H satisfies a Poisson equation with $(-\text{div } \mathbf{M})$ playing the role of the source density.

Wherever div $\mathbf{M} = 0$, U_H satisfies the Laplace equation. This can occur in two common types of situations: (1) in any *linear* magnetic medium of uniform permeability since $\mathbf{M}$ is then proportional to $\mathbf{B}$, and (2) in a *uniformly magnetized* region.

Example 12-15

Solve the uniformly magnetized sphere problem one more time, this time by the use of U_H.

Solution Write the general solution of the Laplace equation, given in Eq. (11-20), for each region. With an eye to the boundary conditions at $r = 0$ and ∞ we set

$$U_{H1} = \sum_0^\infty D_n r^n P_n (\cos \theta)$$

$$U_{H2} = \sum_0^\infty C_n r^{-(n+1)} P_n (\cos \theta)$$

The negative gradients of these potentials are

$$\mathbf{H}_1 = \sum_1^\infty \left[-n D_n r^{n-1} P_n \, \hat{\mathbf{a}}_r - D_n r^{n-1} \frac{dP_n}{d\theta} \hat{\mathbf{a}}_\theta \right] \tag{12-18}$$

$$\mathbf{H}_2 = \sum_0^\infty \left[(n+1) C_n r^{-(n+2)} P_n \, \hat{\mathbf{a}}_r - C_n r^{-(n+2)} \frac{dP_n}{d\theta} \hat{\mathbf{a}}_\theta \right] \tag{12-19}$$

[Note that the $n = 0$ terms are absent from Eq. (12-18) because $(d/dr)(r^0) = 0$ and $dP_0/d\theta = 0$.]

The expressions for $\mathbf{B}$ are

$$\mathbf{B}_1 = \mu_0 (\mathbf{H}_1 + M_0 \hat{\mathbf{k}})$$
$$= \mu_0 [\mathbf{H}_1 + M_0 (\cos \theta \, \hat{\mathbf{a}}_r - \sin \theta \, \hat{\mathbf{a}}_\theta)]$$
$$\mathbf{B}_2 = \mu_0 \mathbf{H}_2$$

We are now in position to write out the boundary relations at $r = a$ (H_θ and B_r continuous). As in Example 11-26, the coefficients of *each individual angular function* must be equal. Thus we get

$$H_\theta: \quad n = 0: \quad \text{No information}$$
$$n = 1: \quad -D_1 = -C_1 a^{-3}$$
$$n = 2: \quad -D_2 a = -C_2 a^{-4} \quad \text{etc.}$$
$$B_r: \quad n = 0: \quad 0 = C_0 a^{-2}$$
$$n = 1: \quad -D_1 + M_0 = 2C_1 a^{-3}$$
$$n = 2: \quad -2D_2 a = 3C_2 a^{-4} \quad \text{etc.}$$

Solving the pair of equations for each n gives

$$C_1 = \tfrac{1}{3} M_0 a^3 \qquad D_1 = \tfrac{1}{3} M_0$$

and all the other coefficients are zero (actually D_0 is not determined but since it gives a constant term in the potential, it does not affect the fields). These again give the same fields as were found previously.

If a current loop threads through the region of interest, U_H must be a *multiple-valued* function. Application of Ampère's circuital law to a contour linking the current loop yields

$$\oint \mathbf{H} \cdot \mathbf{dl} = I$$

$$\oint (\text{grad } U_H) \cdot \mathbf{dl} = -I$$

Thus on coming back to the starting point, U_H must have changed by amount $-I$.

To find the form of the scalar potential function for the field due to a current loop, the trick is to divide the loop into a *sheet of many small loops* by means of a mesh as shown in Figure 12-8. This does not change anything because the currents in all interior segments of the mesh cancel. Now each small loop is a magnetic dipole. Hence we can write the potential as the sum of the *dipole scalar potentials*, that is,

$$U_H = \frac{1}{4\pi} \iint \frac{(I\,dS\,\hat{\mathbf{n}}) \cdot \mathbf{R}}{R^3}$$

$$= -\frac{I}{4\pi} \iint \frac{dS\,\hat{\mathbf{n}} \cdot (-\hat{\mathbf{a}}_R)}{R^2}$$

But the integrand is just the *solid angle* subtended at the field point by the area dS. Hence the integral is the total solid angle Ω subtended by the loop, so

$$U_H = -\frac{I\Omega}{4\pi} \qquad H = \frac{I}{4\pi} \text{ grad } \Omega \tag{12-20}$$

This result is useful mainly as an aid in visualizing field patterns, since the solid angle is intuitively quite obvious. It is not especially useful for computational purposes, because the solid angle at a general field point is usually quite cumbersome to calculate. By and large, when currents are present the vector potential method is just as good. However, we shall do one simple example to illustrate this potential.

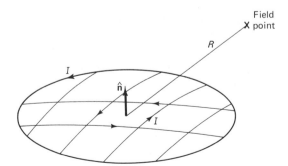

Figure **12-8**   Division of a current loop into a sheet of magnetic dipoles.

Example 12-16

Find the field at a point on the axis of a planar circular current loop by use of U_H (Figure 12-9).

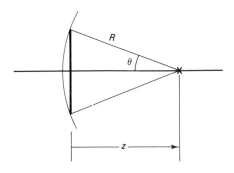

Figure 12-9 Geometry for calculation of solid angle in Example 12-16.

Solution   The solid angle subtended is

$$\Omega = \frac{\text{spherical area}}{(\text{radius})^2}$$

$$= \int_0^\theta \frac{2\pi R^2 \sin \theta' \, d\theta'}{R^2}$$

$$= 2\pi(1 - \cos \theta) = 2\pi \left(1 - \frac{z}{\sqrt{a^2 + z^2}} \right)$$

On inserting this in Eq. (12-20), we obtain

$$H_z(0, 0, z) = -\frac{dU_H}{dz} = \frac{I}{2} \frac{a^2}{(a^2 + z^2)^{3/2}}$$

which is the same as was found in Example 2-7.

12-4 MAGNETIC POLES

The results of Section 12-3 show that magnetostatic problems can be treated exactly like electrostatic ones, except for the more complicated nature of the scalar potential function when currents flow through the region of interest. Of course, there is no analog of true charge density. However, Eq. (12-17) shows that $(-\text{div } \mathbf{M})$ plays the role of an "effective" or "bound" source density. Therefore, by analogy with Eqs. (4-5) and (4-6) for dielectrics, $\mathbf{M} \cdot \hat{\mathbf{n}}$ must play the role of a surface density of effective sources. These sources are called "magnetic poles." We write

$$\rho_M(\mathbf{r}) = -\text{div } \mathbf{M}(\mathbf{r}) \tag{12-21}$$

$$\sigma_M(\mathbf{r}) = \mathbf{M}(\mathbf{r}) \cdot \hat{\mathbf{n}} \tag{12-22}$$

With these, we proceed exactly as in dielectric problems.

Example 12-17

Calculate the fields at points along the axis of an ideal bar magnet.

Solution $\sigma_M = M_0$ on one end face, $-M_0$ on the other. The corresponding dielectric problem was solved in Example 4-2. Hence if we replace P/ϵ_0 in that solution by M_0, we obtain H for the present problem. The result should be compared with Example 6-7, which treated the same situation in terms of the effective currents.

Example 12-18

A bar magnet is located perpendicular to an "infinitely permeable" flat surface. Find the field on the axis just outside the surface.

Solution Example 12-2 showed that infinite permeability acts on magnetic fields just as ideal conductivity does on electrostatic fields. Thus we have here an image problem as shown in Figure 12-10. At the surface the "image poles" contribute the same amount as the real ones, hence, from Example 12-17,

$$\mathbf{B} = \mu_0 \mathbf{M}_0 \left[\frac{-z}{\sqrt{z^2 + a^2}} + \frac{z + L}{\sqrt{(z + L)^2 + a^2}} \right]$$

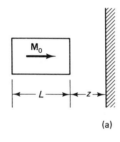

(a)

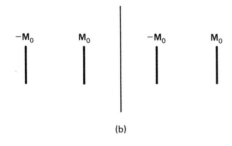

(b)

Figure 12-10 (a) Bar magnet near infinitely permeable material (Example 12-18). (b) Corresponding image problem, indicating the effective pole density distribution.

Example 12-19

Estimate the force with which a very long bar magnet adheres to an infinitely permeable sheet.

Solution In the limit as $L \longrightarrow \infty$ and $z \longrightarrow 0$, the field strength in Example 12-18 becomes $\mathbf{B} = \mu_0 \mathbf{M}_0$, $\mathbf{H} = \mathbf{M}_0$. Hence the energy density at the axis becomes, by Eq. (9-8), $\frac{1}{2}\mu_0 M_0^2$. We make the approximation that the energy density is the same over the entire area of the face (although actually it decreases toward the edge). Thus on drawing the magnet away by a small distance z, the total energy will increase by about

$$\delta \mathcal{U}_{\text{mag}} \cong \tfrac{1}{2}\mu_0 M_0^2 \pi a^2 z$$

Since this problem is the analog of the constant-charge situation in electrostatics, the force is the negative gradient of the energy, or

$$F = -\tfrac{1}{2}\mu_0 M_0^2 \pi a^2$$

Thus the *lifting power* of a magnet is proportional (roughly) to the pole area times the square of the magnetization.

Example 12-20

A bar magnet of length L is bent into a torus with a small gap of length l left between the ends as shown in Figure 12-11. Estimate the field strength in the middle of the gap. Assume that the magnetization bends around with the bar.

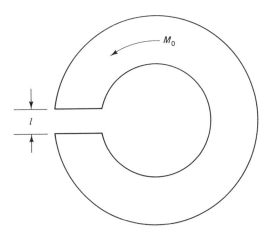

Figure 12-11 Idealized horseshoe magnet of Example 12-20.

Solution As a first approximation, take the pole faces to be parallel and infinite in extent; then the fields will be

$$H_{\text{gap}} = \sigma_M = \mathbf{M} \cdot \mathbf{\hat{n}} = M_0 \qquad B_{\text{gap}} = \mu_0 M_0$$

However, we can see that this cannot be exact if we use Eq. (12-2) to find the fields in the material, that is,

$$B_{\text{mat'l}} = \mu_0 M_0 \qquad H_{\text{mat'l}} = 0$$

For then the contour integral of H around the entire axis comes out to

$$\oint \mathbf{H} \cdot \mathbf{dl} = M_0 l$$

rather than the required value of zero. To correct this in the next approximation, we can reduce the gap field in proportion to the ratio of gap length to material length, namely

$$H_{\text{gap}} = M_0 \left(1 - \frac{l}{L} \right)$$

The reader should verify that this leads to $H_{\text{mat'l}} = -M_0 l/L$ and makes the integral come out nearly to zero.

Note that the gap *reduces* the value of B and H (i.e., it has a *demagnetizing* influence). Also note that it causes **H** in the material to be opposite in direction to **M**.

12-5 MAGNETIC CIRCUITS

It is probably evident by now that exact field calculations for practical magnet designs are well nigh impossible. However, in many cases a fair approximation may be obtained by using the concept of a *magnetic circuit*. This is based on the fact that a bar of highly permeable material will tend to *concentrate the magnetic flux within itself*. The reason is that the continuity of the tangential component of **H** makes the tangential component of **B** much larger inside than outside.

Example 12-21

A short coil of N turns of wire carrying current I is wound on a toroidal "core" of high-permeability material with minor radius a and major radius b, as shown in Figure 12-12. Estimate the magnetic flux, assuming it to be drawn completely into the material.

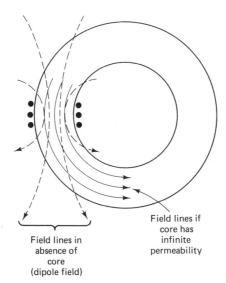

Field lines in
absence of
core
(dipole field)

Field lines if
core has
infinite
permeability

Figure 12-12   Magnetizing coil wound on a high-permeability toroidal core, as in Example 12-21.

Solution   Under the stated assumption, **B** and **H** have uniform magnitudes all around the core. Hence, Ampère's circuital law reads

$$NI = \oint \mathbf{H} \cdot d\mathbf{l} = 2\pi b \frac{B}{\mu}$$

Thus

$$\Phi \equiv \pi a^2 B = \mu \left(\frac{\pi a^2}{2\pi b}\right)(NI)$$

Of course, the basic assumption is never exact; in practice there is always some "leakage flux" outside the core.

The foregoing result may be cast in a form analogous to the current in a resistive circuit. If A is the cross-sectional area of the core and l the length around its axis, then

$$\Phi = \mu \frac{A}{l}(NI) \tag{12-23}$$

This is of the same form as Ohm's law [Eq. (5-5)],

$$I = g \frac{A}{l} V$$

Thus (NI) plays the role of the emf; it is called the *magnetomotive force* or *mmf*. Also $l/\mu A$ plays the role of the resistance; it is called the *reluctance* $\mathcal{R}$. Finally, the magnetic flux plays the role of the current.

Example 12-22

Show that in a core made up of two segments in series, the reluctances add like resistances in an electric circuit.

Solution   Ampère's circuital law now reads

$$NI = H_1 l_1 + H_2 l_2$$

On the assumption that the flux is confined within the core and is therefore the same in both segments (see Example 12-1), we have

$$H_1 = \frac{B_1}{\mu_1} = \frac{\Phi}{\mu_1 A_1} \qquad \text{etc.}$$

$$NI = \Phi \left(\frac{l_1}{\mu_1 A_1} + \frac{l_2}{\mu_2 A_2}\right) = \Phi(\mathcal{R}_1 + \mathcal{R}_2)$$

Example 12-23

A small "air gap" of length l_2 is cut in a core of length l_1. Estimate the field in the gap.

Solution   If the gap is short enough we can neglect the "fringing" of the field and take the effective area to be the same as that of the core material. Thus

$$B = \frac{\Phi}{A} = \frac{NI}{l_1/\mu + l_2/\mu_0} = \frac{\mu NI}{l_1 + \kappa_m l_2}$$

Thus the effect of the air gap in reducing the field is due to its length multiplied by the permeability of the core material.

In many cases magnetic circuits contain permanent magnets as "energizing agents" instead of current coils.

Example 12-24

A magnetic circuit comprises a permeable "yoke" (length l_1, area A_1, permeability μ), an air gap (l_2, A_2), and a permanent bar magnet (l_3, A_3) (Figure 12-13). Calculate the flux in terms of the value of H in the bar magnet.

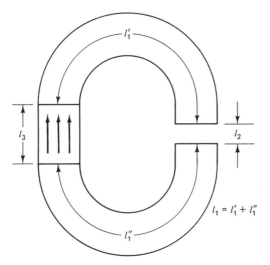

Figure 12-13 Magnetic circuit energized by a permanent magnet, Examples 12-24 and 12-25.

Solution   Ampère's circuital law now reads

$$H_1 l_1 + H_2 l_2 + H_3 l_3 = 0$$

$$H_1 l_1 + H_2 l_2 = -H_3 l_3$$

Thus ($-H_3 l_3$) is now the mmf instead of NI. From Example 12-22, we then obtain

$$\Phi = -\frac{H_3 l_3}{\mathcal{R}_1 + \mathcal{R}_2} \tag{12-24}$$

This, of course, still leaves us with the problem of determining H_3.

Example 12-25

Assuming the bar magnet in Example 12-24 to be ideal (magnetization M_0 independent of field strength), find the value of H_3 from the relation among B, H, and M, and thus determine the magnetic flux.

Solution   In the bar magnet, Eq. (12-24) gives

$$B_3 = \frac{\Phi}{A_3} = -\frac{H_3 l_3}{A_3(\mathcal{R}_1 + \mathcal{R}_2)}$$

Then

$$H_3 = \frac{B_3}{\mu_0} - M_0 = \frac{-H_3 l_3}{\mu_0 A_3} \frac{1}{\mathcal{R}_1 + \mathcal{R}_2} - M_0$$

Solving for H_3 gives

$$H_3 = -M_0 \frac{\mathcal{R}_1 + \mathcal{R}_2}{\mathcal{R}_1 + \mathcal{R}_2 + \mathcal{R}_3}$$

where $\mathcal{R}_3 = l_3/\mu_0 A_3$. On inserting this into Eq. (12-24), the result is

$$\Phi = \frac{M_0 l_3}{\mathcal{R}_1 + \mathcal{R}_2 + \mathcal{R}_3} \qquad (12\text{-}25)$$

Example 12-26

Rework Example 12-20 by the magnetic circuit method.

Solution We have $l_1 = 0$, $l_2 = l$, $l_3 = L$, $A_1 = A_2 = A_3$.

$$B = \frac{\Phi}{A} = \frac{\mu_0 M_0}{1 + l/L}$$

To first order in the small quantity l/L, this is the same as the previous result.

FURTHER REFERENCES

Most of the material in this chapter is covered in all standard texts. Magnetic circuits are often omitted, but are discussed by Lorrain and Corson, and J. R. Reitz, F. J. Milford, and R. W. Christy, *Foundations of Electromagnetic Theory*, Addison-Wesley, New York, 1979, as well as A. Shadowitz, *The Electromagnetic Field*, McGraw-Hill, New York, 1975, and, in much detail, by Smythe.

PROBLEMS

12-1. At one side of an interface between two materials the magnetic field is $\mathbf{B} = B_1 \hat{\mathbf{n}} + B_2 \hat{\mathbf{t}}$, where $\hat{\mathbf{n}}$ and $\hat{\mathbf{t}}$ are unit vectors normal and tangential to the interface, respectively. What is the most general form that $\mathbf{B}$ can have on the other side?

12-2. A ferromagnetic material has a uniform magnetization $M_0 \hat{\mathbf{k}}$. A planar surface is cut so that the normal lies in the xz plane at angle θ to the z axis. Prove that $B_{2z} = B_{1z} - \mu_0 M_0 \sin^2 \theta$, where the subscript 2 refers to the outside region (empty space).

12-3. A square loop of side a lies in the xy plane with its center at the origin. Write an integral expression for the solid angle subtended by the loop at point (x, y, z).

12-4. Show that

$$(A r^n + B r^{-(n+1)}) \sin \theta \frac{dP_n}{d(\cos \theta)} \hat{\mathbf{a}}_\phi$$

is a solution of the vector Laplace equation.

12-5. A small circular current loop is located near the planar surface of an infinitely permeable material. Describe the images if the loop is (a) parallel to the interface, and (b) perpendicular to the interface.

12-6. A soft iron ring has a mean length of 25 cm and a cross-sectional area of 1 cm². It is cut so as to leave a 1-mm gap. A 100-turn coil carrying 1 A is wrapped around part of the loop. Calculate the magnetic field in the gap, assuming the iron to have an effective permeability of $3000\mu_0$.

12-7. A magnetic circuit consists of the following parts:

1. A soft iron "yoke" of length 50 cm, cross-sectional area 1 cm², and permeability 3000.

2. An air gap of length 1 mm.

3. An alnico bar magnet of length 1 cm, cross-sectional area 1 cm². Part of the hysteresis curve for Alnico is shown in Figure 12-14.

Calculate the magnetic field in the gap.

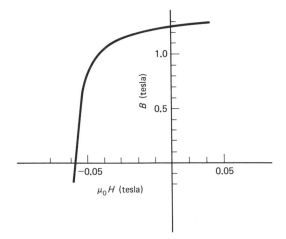

Figure 12-14   Part of the hysteresis curve for Alnico, for Problem 12-7.

Part V

TIME-VARYING FIELDS

chapter 13

Electromagnetic Waves

13-1 DERIVATION OF THE WAVE EQUATION

At this point it would be a good idea for the reader to review Sections 7-1 and 7-2, where the Maxwell equations and their formulation in terms of potentials were discussed. As a starting point we shall need the equations satisfied by the potentials in regions *free of charges and currents*. We write these in the Lorentz gauge because of the symmetry it affords; thus Eqs. (7-19) and (7-20) become

$$\nabla^2 \mathbf{A} - \mu\epsilon\ddot{\mathbf{A}} = 0 \qquad \text{(7-19) (13-1)}$$

$$\nabla^2 U - \mu\epsilon\ddot{U} = 0 \qquad \text{(7-20) (13-2)}$$

It should be borne in mind that these are valid only in conjunction with the Lorentz condition (7-18) and that the medium was assumed to have *linear* dielectric and magnetic properties.

Example 13-1

By taking suitable derivatives of the foregoing, show that $\mathbf{E}$ and $\mathbf{B}$ satisfy equations of the same form.

259

Solution Take the curl of Eq. (13-1). Since the order of differentiations may be interchanged, this gives

$$\nabla^2 \,\mathrm{curl}\, \mathbf{A} - \mu\epsilon\frac{\partial^2}{\partial t^2}\,\mathrm{curl}\, \mathbf{A} = 0$$

which, from Eq. (7-9) is

$$\nabla^2\mathbf{B} - \mu\epsilon\ddot{\mathbf{B}} = 0 \tag{13-3}$$

Next take the time derivative of Eq. (13-1) and the gradient of Eq. (13-2) and add them. This gives

$$\left(\nabla^2 - \mu\epsilon\frac{\partial^2}{\partial t^2}\right)(\dot{\mathbf{A}} + \mathrm{grad}\, U) = 0$$

which, from Eq. (7-11), is

$$\nabla^2\mathbf{E} - \mu\epsilon\ddot{\mathbf{E}} = 0 \tag{13-4}$$

Thus it is seen that the fields and potentials all satisfy the same equation (in source-free regions only).

Equations (13-3) and (13-4) may also be derived directly from the Maxwell equations. The procedure is to take the curl of one of the curl equations, then substitute from the other Maxwell equations. For example,

$$\mathrm{curl\ curl}\ \mathbf{H} = \mathrm{curl}\ \dot{\mathbf{D}} = \epsilon\,\mathrm{curl}\ \dot{\mathbf{E}} = -\epsilon\ddot{\mathbf{B}}$$

But curl curl $= \mathrm{grad\ div} - \nabla^2$ and div $\mathbf{H} = \mathrm{div}\ \mathbf{B}/\mu = 0$, so

$$\frac{-\nabla^2\mathbf{B}}{\mu} = -\epsilon\ddot{\mathbf{B}}$$

which is Eq. (13-3). A similar procedure starting with the curl $\mathbf{E}$ equation (remembering that div $\mathbf{E} = 0$ in a charge-free region) leads to Eq. (13-4). It is particularly to be noted that (1) all four Maxwell equations are used, and (2) the displacement current term (which, we recall, Maxwell invented) plays an essential role.

In an important class of problems, the fields vary in one spatial direction only (say the z direction). Then, since the partial derivatives with respect to x and y are zero, the equation becomes

$$\left(\frac{\partial^2}{\partial z^2} - \mu\epsilon\frac{\partial^2}{\partial t^2}\right)f(z, t) = 0 \tag{13-5}$$

It is easy to show that *any function* of the form $f(z, t) = f_1(z - t/\sqrt{\mu\epsilon})$ is a solution to this equation. Let $\alpha = z - t/\sqrt{\mu\epsilon}$. Then

$$\frac{\partial f}{\partial z} = \frac{df_1}{d\alpha}\frac{\partial\alpha}{\partial z} = \frac{df_1}{d\alpha} \qquad \frac{\partial^2 f}{\partial z^2} = \frac{d^2f_1}{d\alpha^2}$$

$$\frac{\partial f}{\partial t} = \frac{df_1}{d\alpha}\frac{\partial\alpha}{\partial t} = -\frac{1}{\sqrt{\mu\epsilon}}\frac{df_1}{d\alpha} \qquad \frac{\partial^2 f}{\partial t^2} = \frac{1}{\mu\epsilon}\frac{d^2f_1}{d\alpha^2}$$

On substituting these, the equation is seen to be satisfied identically. Obviously, $f_2(z + t/\sqrt{\mu\epsilon})$ is also a solution. The general solution is a sum of the two.

Example 13-2

Sketch some arbitrary curve to represent f_1 versus z at a fixed time t_1. From this obtain f_1 versus z at a later time t_2.

Solution   Since f_1 depends only on the *combined* variable α, an increase of t requires an increase of z to bring α, and thus f_1, back to the same value. In other words, the function *retains its shape* but *moves along the z axis with constant velocity* as illustrated in Figure 13-1. This behavior is characteristic of a *wave* and therefore the equation is called a *wave equation*.

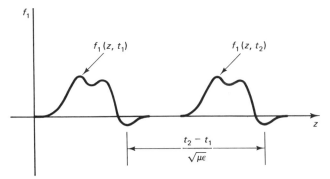

Figure 13-1   Time evolution of a solution of Eq. (13-5).

Example 13-3

Find the velocity of the wave.

Solution   The velocity is the quotient (distance moved by any "feature" of the function)/(time elapsed). Thus

$$c = \frac{z_2 - z_1}{t_2 - t_1} = \frac{1}{\sqrt{\mu\epsilon}} \tag{13-6}$$

On referring back to Eq. (7-53), we see that if the medium is empty space, c is the speed of light in vacuum. From now on we will use the same symbol to denote the speed in *whatever medium we are discussing*. Remember that the medium is assumed to have *linear* dielectric and magnetic properites, and thus to be characterized by *constant* values of ϵ and μ.

At this point, we see that Maxwell's equations *predict* the existence of waves that travel through vacuum with the speed of light. This is the first step in the identification of light waves as being electromagnetic in nature. Later we shall see that all other known properties of light waves are also correctly predicted.

13-2 HARMONIC PLANE WAVES

We now consider the case of an infinite train of harmonic (i.e., sinusoidal) waves. This is described by a function of the form

$$f(z, t) = C \cos K\alpha = C \cos (Kz - Kct)$$
$$= C \cos (Kz - \omega t) \tag{13-7}$$

where

$$\omega = Kc = \frac{Kc_0}{n} \tag{13-8}$$

where $c_0 = (\mu_0 \epsilon_0)^{-1/2} =$ speed of light in vacuum
$n = $ *index of refraction* of the medium

Since the wave function of Eq. (13-7) is constant on planes parallel to the xy plane, it is called a *plane wave*. Later we shall extend the description to include any direction of propagation, not just the z axis.

It is usually more convenient to use the complex exponential notation (see Section 4-6). In this form

$$f(z, t) = Ce^{i(Kz - \omega t)} = Ce^{iKz}e^{-i\omega t} \tag{13-9}$$

Such a wave is characterized by its *wavelength* (spatial period) and *frequency* (inverse of temporal period). The wavelength is the distance between successive maxima (or minima) *at a given time*. Thus the *phase* (argument of the cosine or exponential function) must increase by 2π when the distance increases by one wavelength λ, that is,

$$K(z + \lambda) = Kz + 2\pi$$
$$\lambda = \frac{2\pi}{K} \tag{13-10}$$

The period is the time interval between successive maxima at a given place, that is,

$$\omega(t + T) = \omega t + 2\pi \qquad T = \frac{2\pi}{\omega}$$

The *frequency* (cycles per second, or *hertz*) is thus

$$\nu = \frac{\omega}{2\pi} \tag{13-11}$$

Thus the *propagation constant K* is inversely related to the wavelength, and ω is the *angular frequency* (radians per second). Also, Eq. (13-8) is equivalent to

$$\text{speed} = \text{frequency} \times \text{wavelength}$$

which is obviously correct for any kind of wave.

We next consider a plane electromagnetic wave in which the fields are each *constant in direction*. Maxwell's equations impose quite stringent restrictions on the possible orientations of the fields.

The wave fields will have the form

$$\mathbf{E}(z, t) = \mathbf{E}_0 e^{i(Kz - \omega t)}$$
$$\mathbf{B}(z, t) = \mathbf{B}_0 e^{i(Kz - \omega t)}$$

$$\text{(13-12)}$$

where $\mathbf{E}_0$ and $\mathbf{B}_0$ are constant vectors. Since div $\mathbf{E} = 0$ (remember that we are assuming a charge-free medium),

$$\text{div } \mathbf{E} = iK(\hat{\mathbf{k}} \cdot \mathbf{E}_0)e^{i(Kz - \omega t)} = 0$$

Hence

$$\hat{\mathbf{k}} \cdot \mathbf{E}_0 = 0 \qquad \text{(13-13)}$$

Similarly, div $\mathbf{B} = 0$ leads to

$$\hat{\mathbf{k}} \cdot \mathbf{B}_0 = 0 \qquad \text{(13-14)}$$

Finally, the equation curl $\mathbf{E} = -\dot{\mathbf{B}}$ gives

$$iK(\hat{\mathbf{k}} \times \mathbf{E}_0)e^{i(Kz - \omega t)} = i\omega \mathbf{B}_0 e^{i(Kz - \omega t)}$$

or

$$\mathbf{B}_0 = \frac{K}{\omega}(\hat{\mathbf{k}} \times \mathbf{E}_0) = \frac{1}{c}(\hat{\mathbf{k}} \times \mathbf{E}_0) \qquad \text{(13-15)}$$

Thus both fields must be *perpendicular to the direction of propagation* ("transverse waves") *and perpendicular to each other*. Note also that the amplitudes are related to each other via Eq. (13-15). Waves of the type just described are called *plane-polarized*. The *plane of polarization* is defined as any one of the set of parallel planes containing $\mathbf{E}_0$ and the direction of propagation. A graph of the fields is shown in Figure 13-2. It will be noted that the Poynting vector, defined in Eq. (10-3), is at all times in the

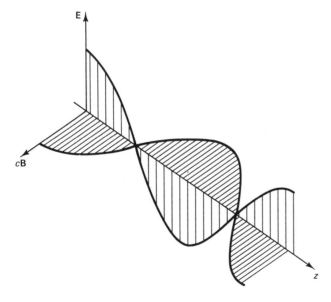

Figure 13-2 Field variation in a plane-polarized plane wave. The pattern moves along the z direction with speed c.

direction of propagation. Thus there is a flow of energy along with the wave. This will be discussed in detail in Section 13-4 and in subsequent chapters.

Various other types of waves may be obtained by superimposing two plane-polarized waves.

Example 13-4

Two waves having the same frequency, amplitude, and direction of polarization, but traveling in opposite directions, are present in the medium. Find the resultant field.

Solution

$$\mathbf{E} = \mathbf{E}_0 e^{iKz} e^{-i\omega t} + \mathbf{E}_0 e^{iK(-z)} e^{-i\omega t}$$

$$= 2\mathbf{E}_0 (\cos Kz) e^{-i\omega t}$$

This is a *standing wave*; the pattern *does not move*, nor is there any energy flow, but the field at any point oscillates with angular frequency ω.

Example 13-5

Two waves have the same frequency and direction of propagation, and are in phase with each other. However, they have different amplitudes and their planes of polarization are perpendicular to each other. Find the resultant field.

Solution   Take the directions of polarization to be the x and y directions. Then

$$\mathbf{E} = E_1 \hat{\mathbf{i}} e^{i(Kz - \omega t)} + E_2 \hat{\mathbf{j}} e^{i(Kz - \omega t)}$$

$$= (E_1 \hat{\mathbf{i}} + E_2 \hat{\mathbf{j}}) e^{i(Kz - \omega t)}$$

This is simply a single plane-polarized wave with amplitude $(E_1^2 + E_2^2)^{1/2}$ and with its plane of polarization rotated by angle $\tan^{-1}(E_2/E_1)$ from the xz plane.

Example 13-6

Let the two waves of Example 13-5 have equal amplitudes but now be 90 degrees out of phase with each other. Find the resultant field.

Solution   Recall from Eq. (4-33) that in the complex exponential notation a 90-degree phase shift is represented by a factor i. Thus, for example,

$$\mathbf{E} = E_0 \hat{\mathbf{i}} e^{i(Kz - \omega t)} - i E_0 \hat{\mathbf{j}} e^{i(Kz - \omega t)}$$

$$= (E_0 \hat{\mathbf{i}} - i E_0 \hat{\mathbf{j}}) e^{i(Kz - \omega t)}$$

To understand the nature of this field it is best to write out the real function explicitly; it is

$$\mathbf{E} = E_0 (\hat{\mathbf{i}} \cos K\alpha + \hat{\mathbf{j}} \sin K\alpha)$$

where, as before, $\alpha = z - ct$. Thus the field has a constant magnitude E_0 but its *direction* changes as the wave advances. This is called a *circularly polarized* wave. In the present case, as α increases the field vector rotates in a clockwise sense. Thus in an instantaneous diagram the tip of the vector traces out a right-handed helix, and the wave is called *right-circularly polarized*. The behavior of

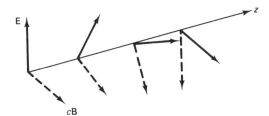

Figure 13-3 Field vectors in a right-circularly polarized wave propagating in the z direction.

the field vectors is shown in Figure 13-3. Obviously, a *left-circularly polarized* wave is obtained by reversing the sign of the phase shift.

If the two plane-polarized waves have different amplitudes, or if the phase shift is different from 90 degrees, or both, the resultant wave is *elliptically polarized*. At any point the field vector traces out an ellipse as the wave advances past the point. The reader may wish to work out the field expressions for these cases.

We now write expressions for the fields in a plane-polarized plane wave that is moving in an *arbitrary* direction given by a unit vector $\hat{\mathbf{u}}$. Heretofore we have been discussing the case of propagation in the z direction; hence the unit vector is $\hat{\mathbf{k}}$ and the distance along it is $z = \hat{\mathbf{k}} \cdot \mathbf{r}$. To generalize this we simply replace $\hat{\mathbf{k}}$ by $\hat{\mathbf{u}}$ to obtain

$$\mathbf{E} = \mathbf{E}_0 e^{i(K\hat{\mathbf{u}} \cdot \mathbf{r} - \omega t)} \qquad (\mathbf{E}_0 \cdot \hat{\mathbf{u}} = 0) \qquad (13\text{-}16)$$

$$\mathbf{B} = \mathbf{B}_0 e^{i(K\hat{\mathbf{u}} \cdot \mathbf{r} - \omega t)} \qquad \left(\mathbf{B}_0 = \hat{\mathbf{u}} \times \frac{\mathbf{E}_0}{c}\right) \qquad (13\text{-}17)$$

To check that these are correct, we see that the fields have constant values along planes perpendicular to $\hat{\mathbf{u}}$ and are perpendicular both to the direction of propagation and to each other.

13-3 PLANE WAVES IN CONDUCTIVE MEDIA

We can readily extend the theory to include the presence of a certain class of current, namely current *driven by the electric field of the wave* in a linearly conductive medium. We now have $\mathbf{J} = g\mathbf{E}$, which appears on the right side of Eq. (13-1) and in the Maxwell equation for curl $\mathbf{H}$. By following either the procedure of Example 13-1 or the alternative of taking the curls of the Maxwell curl equations, we can derive the wave equations for the fields in this case. The results are

$$\nabla^2 \mathbf{H} - \mu g \dot{\mathbf{H}} - \mu \epsilon \ddot{\mathbf{H}} = 0 \qquad (13\text{-}18\text{a})$$

and

$$\nabla^2 \mathbf{E} - \mu g \dot{\mathbf{E}} - \mu \epsilon \ddot{\mathbf{E}} = 0 \qquad (13\text{-}18\text{b})$$

These differ from the simple wave equation in that a first time-derivative term is present. These equations are *not* satisfied by general functions of form $f(z \pm ct)$, but *are* satisfied by *complex exponential* functions of this form, namely

$$f(r, t) = f_0 e^{iK\hat{u} \cdot r - i\omega t} \tag{13-19}$$

For these, $\nabla^2 f = -K^2 f$, $\dot{f} = -i\omega f$, $\ddot{f} = -\omega^2 f$. Substituting these gives

$$-K^2 + i\omega\mu g + \omega^2 \mu\epsilon = 0 \tag{13-20}$$

Thus the wave vector K is seen to be *complex*.

Example 13-7

Interpret the meaning of the real and imaginary parts of a complex wave vector.

Solution   Let

$$K = \alpha + i\beta \tag{13-21}$$

Then the wave function is

$$f(z, t) = C e^{i(\alpha + i\beta)\hat{u} \cdot r} e^{-i\omega t}$$
$$= C e^{-\beta\hat{u} \cdot r} e^{i(\alpha\hat{u} \cdot r - \omega t)} \tag{13-22}$$

The wave amplitude is thus *attenuated* (diminished) by the factor $e^{-\beta\hat{u} \cdot r}$ as it progresses. Physically, this comes about because the dissipation in the medium extracts energy from the field.

It is clear from Eq. (13-20) that the real and imaginary parts of K do not have to lie in the same direction. Thus the direction of propagation is not necessarily the same as the direction of most rapid attenuation. However, such cases can arise only in the presence of surfaces of discontinuity in the constitutive parameters of the medium. These will be dealt with in Section 14-2. For now, we confine ourselves to cases in which there is just one direction. Then

$$\mathbf{K} = K\hat{u} = (\alpha + i\beta)\hat{u} \tag{13-23}$$

The wave parameters α and β are related to the material parameters ϵ, g, and μ by the wave equation (13-20). In solving Eq. (13-20) for α and β, it is essential to keep in mind that ϵ, g, and μ are themselves complex and frequency dependent, as discussed in Sections 4-6, 5-6, and 6-7. ϵ and g both relate to displacement of the electrons (or ions) of the medium. At zero frequency these are clearly distinct, g being due to the free electrons and ϵ to the bound ones. But at high frequency the distinction is blurred, since *all* the electrons execute oscillatory motions. Thus it is best to describe these effects by a single complex response function, the *effective dielectric response function*, which is defined implicitly in Eq. (5-42). We write this

$$\epsilon_{\text{eff}}(\omega) \equiv \epsilon'(\omega) + i\epsilon''(\omega) \tag{13-24}$$

where

$$\epsilon'(\omega) = \epsilon_{\text{re}}(\omega) - \frac{g_{\text{im}}(\omega)}{\omega} \tag{13-25a}$$

$$\epsilon''(\omega) = \epsilon_{\text{im}}(\omega) + \frac{g_{\text{re}}(\omega)}{\omega} \tag{13-25b}$$

As for the permeability μ, it is usually adequate to take it as real and equal to μ_0, since the strong magnetic responses cannot follow high frequencies. There are exceptions, of course, notably in magnetic resonance effects, but we shall not deal with these.

In terms of ϵ_{eff}, the wave equation (13-20) reads

$$-K^2 + \omega^2\mu\epsilon_{eff} = 0 \tag{13-26}$$

It is customary to define a *complex index of refraction* n_c by analogy with Eq. (13-8), that is,

$$n_c = \frac{Kc_0}{\omega} = \frac{K}{\omega\sqrt{\mu_0\epsilon_0}} \equiv n + ik \tag{13-27}$$

In these terms, Eq. (13-26) reads

$$(n + ik)^2 = \frac{\mu}{\mu_0}\frac{\epsilon' + i\epsilon''}{\epsilon_0}$$

$$= \frac{\mu\epsilon'}{\mu_0\epsilon_0}(1 + i\xi) \tag{13-28}$$

where $\xi = \epsilon''/\epsilon'$. The real and imaginary parts of Eq. (13-28) are

$$n^2 - k^2 = \frac{\mu\epsilon'}{\mu_0\epsilon_0} \tag{13-29a}$$

$$2nk = \frac{\mu\epsilon'}{\mu_0\epsilon_0}\xi \tag{13-29b}$$

The latter pair is easily solved by substituting from Eq. (13-29b) into (13-29a) and solving the resulting biquadratic equation.

The results are

$$n, k = \sqrt{\frac{\mu\epsilon'}{\mu_0\epsilon_0}}\sqrt{\frac{\sqrt{1 + \xi^2} \pm 1}{2}} \tag{13-30}$$

The parameter ξ is defined in Eq. (13-28) as the ratio of imaginary to real parts of the *effective* dielectric response function. In the frequency range $\omega \ll \gamma$, where γ is the collision frequency $1/\tau$, ϵ and g are real. In this range

$$\xi = \frac{g}{\omega\epsilon} \qquad \text{for } \omega \ll \gamma \tag{13-31}$$

To illustrate the types of behavior of n and k as functions of frequency, we consider a few specific examples.

Example 13-8

Obtain approximate expressions for n and k in the case of a nearly lossless medium.

Solution   Since $\xi \ll 1$, we may make binomial expansions of the square roots in Eq. (13-30) to obtain

$$n \cong \sqrt{\frac{\mu\epsilon'}{\mu_0\epsilon_0}}\left(1 + \frac{\xi^2}{8}\right) \cong \sqrt{\frac{\mu\epsilon'}{\mu_0\epsilon_0}} \tag{13-32a}$$

$$k \cong \sqrt{\frac{\mu\epsilon'}{\mu_0\epsilon_0}}\frac{\xi}{2} \tag{13-32b}$$

In terms of the wave vector ($K = \alpha + i\beta$), these are

$$\alpha \cong \omega\sqrt{\mu\epsilon'} \tag{13-32c}$$

$$\beta \cong \frac{\xi}{2}\alpha \tag{13-32d}$$

The wave is essentially the same as in the lossless medium except for the slight attenuation described by k or β. The wave speed is

$$c = \frac{\omega}{\alpha} = \frac{c_0}{n} = \frac{1}{\sqrt{\mu\epsilon'}} \tag{13-33}$$

Example 13-9

Obtain expressions for n and k in a highly conductive medium.

Solution   In this case $\xi \gg 1$, so Eq. (13-30) reduces to

$$n \cong k \cong \sqrt{\frac{\mu\epsilon'}{\mu_0\epsilon_0}}\sqrt{\frac{\xi}{2}} = \sqrt{\frac{\mu\epsilon''}{2\mu_0\epsilon_0}} \tag{13-34a}$$

$$\beta \cong \alpha \cong \omega\sqrt{\mu\epsilon'}\sqrt{\frac{\xi}{2}} = \omega\sqrt{\frac{\mu\epsilon''}{2}} \tag{13-34b}$$

For real ϵ and g,

$$\beta \cong \alpha = \omega\sqrt{\frac{\mu g}{2\omega}} = \sqrt{\frac{\omega\mu g}{2}} \tag{13-35}$$

The wave is rapidly attenuated since k and β are large. The distance to decrease by $1/e$ in amplitude is called the *skin depth* and is given by

$$\delta \equiv \frac{1}{\beta} = \sqrt{\frac{2}{\omega\mu g}} \tag{13-36}$$

The wavelength and velocity are still defined in terms of the oscillatory factor, so that

$$\lambda = \frac{2\pi}{\alpha} = 2\pi\delta \tag{13-37}$$

$$\frac{c_0}{n} = \frac{\omega}{2\pi}\lambda = \frac{\omega}{\alpha} = \omega\delta$$

$$= \sqrt{\frac{2\omega}{\mu g}} = \sqrt{\frac{2}{\mu\epsilon''}} \tag{13-38}$$

In any of these equivalent expressions, the velocity and wavelength are seen to be extremely *small* compared to a wave of the same frequency in a nonconductive medium.

Example 13-10

Estimate the skin depth and wave velocity in copper (conductivity $= 6 \times 10^7$ mho/m) at a frequency of 1 gigahertz (10^9 Hz).

Solution Copper is essentially nonmagnetic, so $\mu \cong \mu_0$. Also, $\epsilon \cong \epsilon_0$ since only the tightly bound core electrons can contribute to the polarization (see

Section 4-4). Thus

$$\frac{g}{\omega\epsilon} = \frac{6 \times 10^7}{2\pi \times 10^9 \times 8.85 \times 10^{-12}} \cong 10^9$$

so the condition for good conductivity ($\xi \gg 1$) is very amply met, and Eqs. (13-34), (13-36), and (13-38) apply.

$$\delta = \sqrt{\frac{2}{2\pi \times 10^9 \times 4\pi \times 10^{-7} \times 6 \times 10^7}} = 2 \times 10^{-6} \text{ m}$$

$$c = \sqrt{\frac{4\pi \times 10^9}{4\pi \times 10^{-7} \times 6 \times 10^7}} = 1.3 \times 10^4 \text{ m/s}$$

Thus the penetration is only a few micrometers, and the speed is only about one-ten-thousandth that in vacuum!

In a conductor, as in any other medium, the spatial and temporal relation between the fields is obtained from the Maxwell equations. Since the fields are still solenoidal (divergenceless), they are transverse as in Eqs. (13-13) and (13-14). And from the curl equations they are still perpendicular to each other as in Eq. (13-15). However, since K is now complex, the latter equation gives

$$\frac{B_0}{E_0} = \frac{K}{\omega} = \frac{n_c}{c_0} = \frac{n + ik}{c_0} = \sqrt{\mu_0 \epsilon_0}\,(n + ik) \tag{13-39}$$

For a highly conductive medium this is, by Eq. (13-32),

$$\frac{B_0}{E_0} = \sqrt{\frac{\mu\epsilon''}{2}}(1 + i) = \sqrt{i\mu\epsilon''} \tag{13-40}$$

and if ϵ and g are real, by Eqs. (13-38),

$$\frac{B_0}{E_0} = \sqrt{\frac{i\mu g}{\omega}} = \frac{\sqrt{2i}}{c} = \frac{\sqrt{2i}\,n}{c_0} \tag{13-41}$$

The factor $\sqrt{i}$ corresponds to a phase shift of 45 degrees, since $\sqrt{i} = e^{i\pi/4}$. Also, it is seen that $c_0 B_0 \gg E_0$. A sketch of the waveform is shown in Figure 13-4.

The frequency range for normal metallic behavior (i.e., real ϵ and g) is, as stated above, $\omega \ll \gamma$, where $\gamma = \tau^{-1}$ is the collision frequency (Section 5-4). According to Problem 5-11, τ is estimated as about 10^{-13} s in a good metallic conductor at room temperature. Thus the frequency range extends to about 10^{13} s^{-1}, which corresponds to the very far infrared region of the spectrum. For poorer conductors, such as some alloys, it may extend a decade or so higher.

In the next few examples, we consider conductors in higher-frequency ranges, where g becomes complex.

Example 13-11

Calculate the effective dielectric response function in a free-electron metal at frequencies around the collision frequency and extending upward toward the plasma frequency.

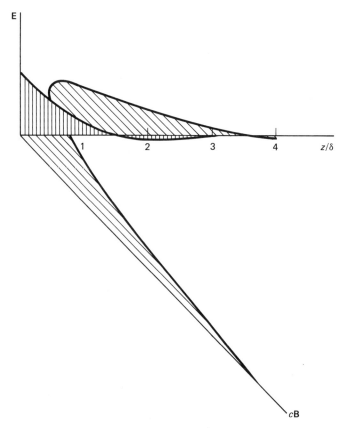

Figure 13-4 Field variation in a plane wave in a good conductor. (Compare with Figure 13-2.)

Solution From Problem 5-17 it is seen that the plasma frequency in a typical metal is around 10^{16} s^{-1}, which is considerably greater than the collision frequency τ^{-1}. From Eqs. (5-41) and (5-44), the conductivity may be written

$$g = \epsilon \tau \omega_p^2 \frac{1 + i\omega\tau}{1 + \omega^2\tau^2} \qquad (13\text{-}42a)$$

ϵ is essentially equal to ϵ_0, since the frequency is far below any resonances of the core electrons. Thus Eqs. (13-25) give

$$\epsilon' = \epsilon_0 \left(1 - \frac{\omega_p^2\tau^2}{1 + \omega^2\tau^2}\right) \qquad (13\text{-}42b)$$

$$\epsilon'' = \epsilon_0 \frac{\omega_p^2\tau^2}{\omega\tau(1 + \omega^2\tau^2)} \qquad (13\text{-}42c)$$

These functions are plotted in Figure 13-5. It is seen that ϵ' starts out large and negative, and decreases in magnitude quite rapidly with increasing ω. ϵ'' starts out positive and even larger, but decreases more rapidly.

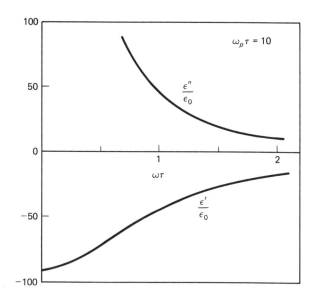

Figure 13-5 Effective dielectric response function for a plasma at frequencies near the collision frequency $\gamma = 1/\tau$.

Example 13-12

Discuss the complex index of refraction for Example 13-11.

Solution   With reference to Eqs. (13-42) and Figure 13-5:

1. At very low frequencies ($\omega\tau \ll 1$), the large value of ϵ'' means that the ± 1 in the square root is quite negligible. Thus $n \cong k$ and both are large, as was found in Example 13-9.

2. At $\omega\tau = 1$, Eqs. (13-42b) and (13-42c) show that $-\epsilon'$ and ϵ'' are essentially equal (assuming that $\omega_p\tau \gg 1$). Thus Eq. (13-30) gives

$$(n, k) = \sqrt{\frac{\mu\epsilon''}{\mu_0\epsilon_0}} \sqrt{\frac{\sqrt{2} \pm 1}{2}}$$

$$\cong (1.1, 0.45)\sqrt{\frac{\mu\epsilon''}{\mu_0\epsilon_0}}$$

$$\cong (1.1, 0.45)\sqrt{\frac{\mu}{\mu_0}}\frac{\omega_p\tau}{2}$$

3. For $\omega\tau \gg 1$ (but still $\omega \ll \omega_p$), Eqs. (13-42) along with (13-29) show that $2nk$ is positive but very small, while $n^2 - k^2$ is negative. Thus, roughly,

$$k \cong \sqrt{-\frac{\mu\epsilon'}{\mu_0\epsilon_0}} = \sqrt{\frac{\mu}{\mu_0}}\frac{\omega_p}{\omega}$$

$$n \cong 0$$

A plot of n and k versus $\omega\tau$ is shown in Figure 13-6.

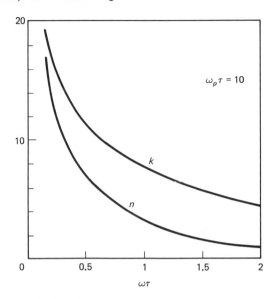

Figure 13-6 Complex index of refraction corresponding to Figure 13-5.

Example 13-13

Continue Example 13-12 to still higher frequencies, approaching and exceeding ω_p.

Solution   When $\omega\tau \gg 1$, Eqs. (13-42) become

$$\epsilon' = \epsilon_0\left(1 - \frac{\omega_p^2}{\omega^2}\right) \tag{13-43a}$$

$$\epsilon'' = \epsilon_0\frac{1}{\omega_p\tau}\left(\frac{\omega_p}{\omega}\right)^3 \tag{13-43b}$$

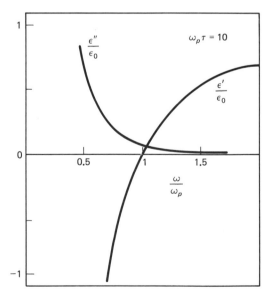

Figure 13-7 Effective dielectric response function for a plasma at frequencies near the plasma frequency.

These are plotted in Figure 13-7 and the corresponding values of n and k from Eqs. (13-30) are plotted in Figure 13-8 for the case $\omega_p\tau = 10$.

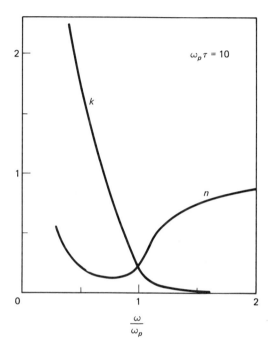

Figure 13-8   Complex index of refraction corresponding to Figure 13-7.

For the idealized case of a *collisionless* plasma (i.e., $\tau \rightarrow \infty$), we see that $\epsilon'' = 0$. Thus the complex index of refraction becomes

$$
\begin{array}{c|c|c}
 & \omega < \omega_p & \omega > \omega_p \\
\hline
n = & 0 & \sqrt{1 - \dfrac{\omega_p^2}{\omega^2}} \\
k = & \sqrt{\dfrac{\omega_p^2}{\omega^2}} & 0
\end{array}
\qquad (13\text{-}44)
$$

The idealized medium has $n = 0$ up to ω_p (we will see in Section 14-2 that this leads to total reflection), and abruptly becomes a lossless dielectric above ω_p. For a real plasma, Figure 13-7 shows that the transition is actually spread out over a range of frequency.

The plasma frequency for a typical metal was estimated in Problem 5-17 as about 10^{16} s^{-1}, which is in the ultraviolet region of the spectrum. For rarefied gaseous plasmas, the density will be many orders of magnitude smaller, so the plasma frequency will be correspondingly lower. In the ionosphere, for example, it is somewhat under 100 megahertz, which lies between the AM and FM broadcast bands.

Example 13-14

Discuss the complex index of refraction for a dielectric having a single resonance described by the classical oscillator model (see Section 4-6).

Solution Since there is no dc conductivity (no free carriers), $\epsilon_{\text{eff}}(\omega)$ is given by Eq. (4-38), namely

$$\epsilon_{\text{eff}} = \epsilon_0\left(1 + \frac{\omega_p^2}{\omega_r^2 - \omega^2 - i\gamma\omega}\right) \tag{13-45}$$

The real and imaginary parts are obtained from Eqs. (4-39) and (4-40) and are plotted in Figure 4-10. Now from Eqs. (13-30) we see that when $\epsilon' \ll \epsilon''$, n and k are essentially equal. This will be the case when $\omega \cong \omega_r$. When ϵ' is positive and $\gg \epsilon''$, n is large and k is small. This occurs when $\omega \ll \omega_r$ and when $\omega \gg \omega_p$. Finally, when ϵ' is negative but $|\epsilon'| \gg \epsilon''$, k is large and n is small. This occurs for a frequency range $\omega_r \ll \omega < \omega_p$.

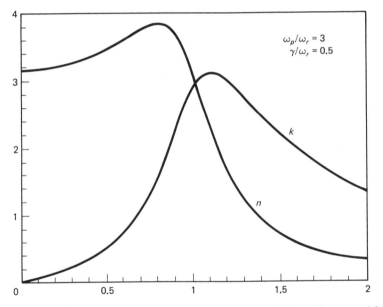

$\omega_p/\omega_r = 3$
$\gamma/\omega_r = 0.5$

Figure 13-9 Complex index of refraction for a classical oscillator model dielectric.

An example of the frequency dependence of n and k is shown in Figure 13-9. The curves shown in this figure are for cases where the damping is relatively weak, so that the resonance is well developed and sharp. The condition is $\omega_r\tau \gg 1$, so that at ω_r the particles execute many cycles of oscillation before the motion is interrupted by a collision. Quite different behavior occurs under the opposite condition, $\omega_r\tau \ll 1$. Then the damping inhibits any appreciable motion except at frequencies $\ll \omega_r$.

Example 13-15

Discuss the complex index of refraction of a strongly damped medium at low frequency.

Solution   At zero frequency, Eq. (13-45) gives

$$\epsilon(0) = \epsilon_0\left(1 + \frac{\omega_p^2}{\omega_r^2}\right) \quad \text{or} \quad \frac{\epsilon_0\omega_p^2}{\omega_r^2} = \epsilon(0) - \epsilon_0$$

Thus for frequencies $\omega \ll \omega_r$, Eq. (13-45) can be written (neglecting ω^2 in the denominator)

$$\epsilon_{\text{eff}}(\omega) - \epsilon_0 = \frac{\epsilon(0) - \epsilon_0}{1 - i\omega\tau'} \tag{13-46a}$$

where

$$\tau' = \frac{\gamma}{\omega_r^2} \tag{13-46b}$$

The real and imaginary parts are

$$\epsilon'(\omega) - \epsilon_0 = \frac{\epsilon(0) - \epsilon_0}{1 + (\omega\tau')^2} \tag{13-46c}$$

$$\epsilon''(\omega) = [\epsilon(0) - \epsilon_0]\frac{\omega\tau'}{1 + (\omega\tau')^2} \tag{13-46d}$$

The foregoing equations are known as the *Debye equations* and describe the process of *relaxation*. When displaced, the system tends to sag back to equilibrium with a characteristic time τ'. The frequency dependence of ϵ' and ϵ'' and the corresponding complex index of refraction are illustrated in Figure 13-10.

The condition $\gamma \gg \omega_r$ is commonly met in cases where the polarization is due to rotation of permanent dipoles (see Section 4-5). A good example is liquid water, whose room temperature dielectric response function is about 80 at zero frequency

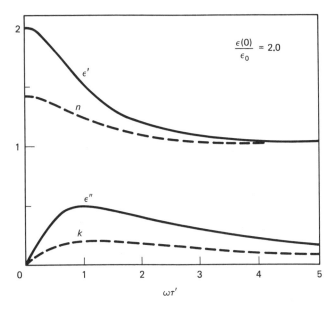

Figure 13-10   Effective dielectric response function and complex index of refraction for a strongly damped dielectric medium.

but only about 2 at optical frequencies, the latter value being due to intramolecular electronic resonances.

Relaxation can also occur in free-carrier systems (i.e., conductors). This can be seen from the close similarity between Eqs. (13-46) and (5-41) for the frequency-dependent conductivity,

$$g(\omega) = \frac{g(0)}{1 - i\omega\tau} = \frac{\epsilon_0 \omega_p^2 \tau}{1 - i\omega\tau} \qquad \text{(5-41) (13-47)}$$

Here the relaxation time is the reciprocal of the collision frequency γ. This is, however, *not* the same as the *dielectric relaxation time* defined in Eq. (5-20). The latter is the characteristic time for decay of an excess charge, and is

$$\tau_{\text{diel}} = \frac{\epsilon_0}{g(0)} = \frac{1}{\omega_p^2 \tau} = \frac{\gamma}{\omega_p^2} \qquad \text{(13-48)}$$

On comparison with Eq. (13-46b), this suggests that ω_p plays the role of a resonant frequency for free-carrier systems. In Section 13-5 we shall see that this is indeed the case.

13-4 ENERGY PROPAGATION

In this section we work out some of the energy relations for plane-wave fields.

Nonconductive Media

Example 13-16

Calculate the time-average energy density and energy flux in a plane-polarized plane wave in a dielectric medium, assuming a real frequency-independent dielectric response function.

Solution   The energy density is the sum of the electric and magnetic contributions [Eqs. (8-16) and (9-8)]. The power theorem, Eq. (8-46), gives the time average as

$$\bar{u} = \frac{1}{2} \operatorname{Re} \left(\frac{\mathbf{E} \cdot \mathbf{D}^*}{2} + \frac{\mathbf{H} \cdot \mathbf{B}^*}{2} \right)$$

With the aid of Eq. (13-15), this is found to be

$$\bar{u} = \frac{1}{4} E_0^2 \left(\epsilon + \frac{1}{\mu c^2} \right) = \frac{1}{2} \epsilon E_0^2 \qquad \text{(13-49)}$$

which shows that the electric and magnetic energy densities are equal. The energy flux is given by the Poynting vector, Eq. (10-3). Its time-average value is

$$\bar{\mathbf{S}} = \frac{1}{2} \operatorname{Re} (\mathbf{E} \times \mathbf{H}^*)$$

$$= \frac{1}{2} \left[\mathbf{E}_0 \times \left(\frac{\hat{\mathbf{u}} \times \mathbf{E}_0}{\mu c} \right) \right]$$

$$= \frac{1}{2} E_0^2 \sqrt{\frac{\epsilon}{\mu}} \, \hat{\mathbf{u}} \qquad \text{(13-50)}$$

Example 13-17

What is the velocity of energy propagation in Example 13-16?

Solution   Since the flux is the density multiplied by the appropriate velocity, we have

$$\mathbf{v}_E \equiv \frac{\bar{\bar{\mathbb{S}}}}{\bar{u}} = \frac{\hat{\mathbf{u}}}{\sqrt{\mu\epsilon}} = c\hat{\mathbf{u}} \tag{13-51}$$

Thus the energy propagates at the wave speed and in the wave direction.

Example 13-18

Calculate the time-average energy flux in a standing wave (Example 13-4).

Solution

$$\mathbf{B} = \frac{-1}{i\omega}\, \text{curl } \mathbf{E}$$

$$= \frac{2K}{i\omega}(\hat{\mathbf{k}} \times \mathbf{E}_0) \sin Kz\, e^{-i\omega t}$$

Thus Re $(\mathbf{E} \times \mathbf{H}^*) = 0$, since $\mathbf{B}$ is pure imaginary.

This is an obvious result, since the two equal and opposite traveling waves have equal and opposite energy fluxes. In a similar way it can be shown that in a circularly polarized (Example 13-6) or elliptically polarized plane wave, the two energy fluxes are additive, as are the densities.

Conductive Media

A calculation similar to that of Example 13-16 is not applicable to a *dispersive* medium (ϵ a function of frequency) for a very subtle reason: The energy density expressions were derived (Sections 8-1 and 9-1) for media with *linear* response, but a dispersive medium is *not truly linear*. It is only quasi-linear in the complex sense that the *steady-state* response amplitude is proportional to the driving field amplitude. But the energy content includes work done during the transient buildup of the response, and thus cannot be related solely to the final amplitude.

Pulse Propagation

For an infinite uninterrupted wave, there is no operational meaning to the flow velocity. It is like water going into a pipe at one end and coming out the other. The only way to ascertain the velocity of flow is to *identify* some point (e.g., by injecting a dye) and measure the transit velocity of this point.

In the case of an electromagnetic wave, the only way to identify a particular point is to *modulate* the wave amplitude (e.g., by sending out a short pulse of waves). Such a pulse cannot, of course, be truly monochromatic. It must contain a range of frequencies, so that the corresponding infinite waves can destructively interfere everywhere beyond the limits of the pulse. In fact, the shorter the pulse, the wider must be the frequency range included. Since in a dispersive medium each frequency travels with a different velocity, the pulse will evidently become distorted in shape as it moves along, so that no perfectly well-defined velocity exists.

Group Velocity under Normal Dispersion

A reasonably well-defined velocity does exist in regions of *normal dispersion*. These are the frequency ranges far from any resonances or relaxations, where the index of refraction is a *slowly* increasing *real* function of ω. To avoid the complication of dealing with an integral over a continuous range of frequencies (a *Fourier integral*), we consider a wave composed of just two frequencies, ω and $\omega + \Delta\omega$, with equal amplitudes E_0. The total electric field is then

$$\mathbf{E}(z, t) = \mathbf{E}_0[e^{iKz - i\omega t} + e^{i(K + \Delta K)z - i(\omega + \Delta\omega)t}$$

$$= \mathbf{E}_0 e^{i\left(K + \frac{\Delta K}{2}\right)z - i\left(\omega + \frac{\Delta\omega}{2}\right)t}\left[2\cos\left(\frac{\Delta K}{2}z - \frac{\Delta\omega}{2}t\right)\right] \qquad (13\text{-}52)$$

The term in brackets is the *modulation envelope* of the wave field described by the exponential factor. The velocity of motion of this envelope is obviously

$$v_g = \frac{\Delta\omega}{2} \div \frac{\Delta K}{2} = \frac{\Delta\omega}{\Delta K}$$

If a continuum of frequencies are included, all but one of the lobes are wiped out by destructive interference, leaving a single pulse. If the frequency range is not too large, the velocity of this pulse envelope is, by analogy with the foregoing,

$$v_g = \frac{d\omega}{dK} = \left(\frac{dK}{d\omega}\right)^{-1} \qquad (13\text{-}53)$$

This relation holds for all kinds of waves, not just electromagnetic ones.

Example 13-19

Express the group velocity in terms of the index of refraction.

Solution   By Eq. (13-8),

$$\frac{dK}{d\omega} = \frac{d}{d\omega}\left(\frac{\omega n}{c_0}\right) = \frac{1}{c_0}\left(n + \omega\frac{dn}{d\omega}\right)$$

so that

$$v_g = \frac{c_0}{n + \omega(dn/d\omega)} \qquad (13\text{-}54)$$

Thus in dispersionless ranges of frequency (i.e., where $dn/d\omega = 0$), v_g is the same as the wave velocity c. In normal dispersion ranges (i.e., where $dn/d\omega$ is positive but small), v_g is less than c and, in fact, less than c_0 even when $n < 1$.

In frequency ranges near resonances or relaxations, both n and k are large and rapidly varying. Thus the distortion of the pulse is extreme and a velocity cannot even be exactly defined, let alone specified by a simple formula such as Eq. (13-54). The problem has been solved exactly, but it is too complicated to discuss here. One result, though, is simple and very important: No part of the pulse ever travels at a speed greater than c_0. Thus no violation of relativity can ever occur, no matter how exotic the pulse shape or the medium's response may be.

13-5 PLASMA OSCILLATIONS AND WAVES

We have seen several indications that exceptional behavior may occur at the plasma frequency of a system. Most striking is the fact that ϵ_{eff} is zero in a collisionless plasma. Thus the wave equation for **E** [Eq. (13-26)] reduces simply to

$$\nabla^2 \mathbf{E}_0 = 0 \qquad \text{for } \mathbf{E} = \mathbf{E}_0(x, y, z)e^{-i\omega_p t}$$

This is the same as the electrostatic Laplace equation, so we may expect *static-like* fields to be permitted as solutions. Such fields are *irrotational* (i.e., curl-free), so by Faraday's law they are not accompanied by a magnetic field.

To investigate such fields let us consider a plasma in which the positive ion charge is rigidly fixed in place while the electrons are free to move. We start by giving the electron distribution a displacement $\zeta(z, t)$ relative to the fixed ions. There is then a polarization

$$P(z, t) = nq\zeta(z, t)$$

and a corresponding "bound" charge density*

$$\rho_p = -\text{div } P = nq\frac{d\zeta}{dz}$$

Thus there is an electric field obeying div $E = (1/\epsilon_0)\rho_p$, so that

$$\frac{\partial}{\partial z} E(z, t) = -nq\frac{d\zeta}{dz}$$

or

$$E(z, t) = -\frac{nq}{\epsilon_0}\zeta(z, t)$$

Thus the equation of motion for the moving charges is

$$m\frac{\partial^2\zeta}{\partial t^2} = qE = \frac{-nq^2}{\epsilon_0}\zeta$$

or

$$\zeta(z, t) = \zeta(z, 0)e^{-i\omega_p t} \tag{13-55}$$

where (recall) $\omega_p^2 = nq^2/m\epsilon_0$. This describes *plasma oscillations*. Any pattern of displacements simply oscillates at frequency ω_p. It may be made to look like a wave by taking $\zeta(z, 0) = \xi_0 e^{iKz}$, but there is no transport of energy, since the group velocity $d\omega/dK$ is zero.

---

*It is a matter of semantics whether we call this charge density "bound" or "free." By using the present designation, we avoid the necessity of rederiving the wave equation for a charged medium. Furthermore, we have $D = \epsilon_0 E + P = 0$, consistent with $\epsilon_{eff} = 0$.

Actual waves may, however, occur if the positive charge distribution is somewhat compressible. The compression forces then augment the electric restoring force to give real waves at frequencies greater than ω_p. These are longitudinal waves, quite analogous to sound waves except for the lower-limit frequency at ω_p. Still further types of wave propagation can occur in gaseous plamsas with a static magnetic field present. These include *magnetosonic waves*, which are longitudinal waves propagating perpendicular to the magnetic field, and *Alfven waves*, which are transverse waves propagating along the magnetic field. Unfortunately, the details of these interesting modes of propagation are a bit too complicated to consider here.

FURTHER REFERENCES

Material of this chapter is covered to some degree in all standard texts. Interesting further details, including wave propagation in anisotropic dielectrics, is found in books on optics, such as the excellent treatise by R. W. Ditchburn, *Light*, Interscience, New York, 1953. Plasma waves are discussed by Jackson and in many engineering-oriented texts, such as C. C. Johnson, *Field and Wave Electrodynamics*, McGraw-Hill, New York, 1965.

PROBLEMS

13-1. Derive Eq. (13-4) directly from the Maxwell equations.

13-2. Show that $\cos kx \cos \omega t$ is a solution of the wave equation. What kind of wave does it represent?

13-3. A wave field has $E_x = E_1 \cos (kz - \omega t)$, $E_y = E_2 \sin (kz - \omega t)$. What kind of wave is this? Write $\mathbf{E}$ and $\mathbf{B}$ in complex exponential form. Calculate the time-average Poynting vector.

13-4. $\psi = (\hat{\mathbf{i}} - i\hat{\mathbf{j}})e^{iKz}$ describes a right-circularly polarized wave propagating in the z direction. What does $(\hat{\mathbf{i}} - i\hat{\mathbf{j}})e^{-iKz}$ describe?

13-5. In both circularly polarized and unpolarized light, the field varies over all transverse directions. What is the difference?

13-6. Derive the value of E/H for a plane-wave field in vacuum.

13-7. Derive Eq. (13-30).

13-8. The Navy is interested in very low frequency radio broadcasting. Why?

13-9. Suppose in some medium $\epsilon_r = -\epsilon_i$ at frequency ω_1 and $\epsilon_r = \epsilon_i$ at ω_2. How do the optical constants compare at the two frequencies?

13-10. What would happen if k (the imaginary part of the complex index of refraction) were negative? Can this occur?

13-11. What is the numerical relation between Poynting vector and electric field amplitude in vacuum?

13-12. Solar energy impinges on the top of the atmosphere at the rate of about 1.3 kW/m². What is the electric field amplitude?

13-13. A small laboratory gas laser puts out about 1 mW of beam power. If the beam is focused down to spot 0.01 mm in diameter, what is the electric field amplitude?

13-14. A high-power pulsed laser amplifier delivers pulses of 10^{-4} J in 100 femtoseconds (femto = 10^{-15}). What is the pulse length? The peak power? If focused to 1 mm diameter, the electric field amplitude? How does this compare with the field acting on the electron in a hydrogen atom (orbit radius ~ 0.5 Å)? [*Note:* for a description of a laser system meeting these specifications, see R. L. Ford, C. V. Shank, and R. T. Yen, *Appl. Phys. Lett.* **41** (1982), 233.]

13-15. In a collisionless plasma, calculate the group velocity for $\omega > \omega_p$.

13-16. For the artificial "square" absorption band model of Example 4-18, let ω_a and ω_b ($\omega_b > \omega_2$) be the frequencies for which K_{re} vanishes. Discuss the optical constants in the five relevant frequency ranges.

chapter 14

Wave Fields with Boundaries

14-1 REFLECTION AND REFRACTION AT A DIELECTRIC INTERFACE

We now turn to the simplest class of boundary-value problems with wave fields. The boundary comprises an infinite planar interface between two (linear) media. A specified plane electromagnetic wave is present in medium 1, moving toward the interface at some angle of incidence. The problem is to deduce what other fields must be present, in terms of the properties of the two media. In this first section we consider only lossless dielectrics.

The field must satisfy certain boundary conditions at the interface (see Sections 11-1 and 12-1). To do this, it is generally necessary to have a second wave in medium 1 (the *reflected* wave) and a field of some sort in medium 2. Most commonly, the latter is also a plane-wave field (the *transmitted* wave), and we shall consider this situation first. The special case of total internal reflection will be taken up later.

Kinematical Relations

Considerable information can be gleaned simply from the fact that boundary relations of some sort must pertain.

1. *The incident, reflected, and transmitted waves must all have the same frequency.* This is because the boundary relations, whatever they may be, have to be satisfied *at all times*, hence the fields must all have the same time-variation.

2. *All three directions of propagation must be coplanar.* The incident wave direction $\hat{\mathbf{u}}_i$ and the interface normal $\hat{\mathbf{N}}$ define a unique direction $\hat{\mathbf{u}}_i \times \hat{\mathbf{N}}$ (unless $\hat{\mathbf{u}}_i$ and $\hat{\mathbf{N}}$ coincide). Any plane normal to this direction is called the *plane of incidence*. The

planes of constant phase (e.g., the electric field maxima) intersect the interface in a set of equally spaced parallel lines running perpendicular to the plane of incidence, as shown in Figure 14-1. Since the boundary relations must pertain *uniformly* over the interface plane, the constant-phase traces of the other waves must run parallel to these. Hence the other direction unit vectors must also lie in the plane of incidence.

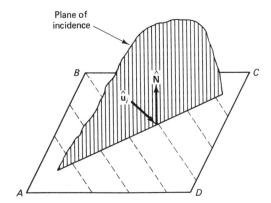

Figure 14-1   Traces of planes of constant phase (dashed lines) in the interface plane *ABCD*.

3. *The law of reflection* (*Figure 14-2*). The incident and reflected waves both lie in medium 1, hence are described by wave functions $\exp(iK_1\hat{\mathbf{u}}_i \cdot \mathbf{r})$ and $\exp(iK_1\hat{\mathbf{u}}_r \cdot \mathbf{r})$, respectively. For points in the interface plane, these functions must have the *same spatial periodicity* in order to satisfy boundary relations uniformly. Thus $\hat{\mathbf{u}}_i \cdot \mathbf{r}_s = \hat{\mathbf{u}}_r \cdot \mathbf{r}_s$ for any vector $\mathbf{r}_s$ lying in the interface plane. In other words, $(\hat{\mathbf{u}}_i - \hat{\mathbf{u}}_r)$ is perpendicular to the interface plane. Hence the two unit vectors must make the same angle with the normal $\hat{\mathbf{N}}$ to this plane, that is,

$$\theta_r = \theta_i \tag{14-1}$$

4. *The law of refraction* (*Figure 14-3*): The transmitted wave is described by the wave function $\exp(iK_2\hat{\mathbf{u}}_t \cdot \mathbf{r})$. Since this must also have the same periodicity in

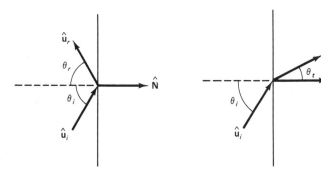

Figure 14-2   Geometry of reflection.

Figure 14-3   Geometry of refraction.

the interface plane, we have

$$(K_2\hat{\mathbf{u}}_t - K_1\hat{\mathbf{u}}_i) \cdot \mathbf{r}_s = 0$$

Since, by Eq. (13-8), K is proportional to the index of refraction n, this gives

$$n_2 \sin \theta_t = n_1 \sin \theta_i \qquad (14\text{-}2)$$

which is *Snell's law of refraction.*

It should be reiterated that up to this point we have not used the details of the boundary relations at all, just the fact that some sort of relations exist. Thus, much of the foregoing pertains to any sort of wave, not just an electromagnetic one. We shall now see that the detailed boundary relations lead to the determination of the *relative amplitudes* of the three waves.

Dynamical Relations

For definiteness in writing out the field components, we establish the following standard geometry:

Plane $z = 0$ is the interface plane.

Plane $y = 0$ (or any plane parallel thereto) is the plane of incidence.

Thus the three wave-direction unit vectors are

$$\hat{\mathbf{u}}_i = \cos \theta_i \, \hat{\mathbf{k}} + \sin \theta_i \, \hat{\mathbf{i}} \qquad (14\text{-}3\text{a})$$

$$\hat{\mathbf{u}}_r = -\cos \theta_i \, \hat{\mathbf{k}} + \sin \theta_i \, \hat{\mathbf{i}} \qquad (14\text{-}3\text{b})$$

$$\hat{\mathbf{u}}_t = \cos \theta_t \, \hat{\mathbf{k}} + \sin \theta_t \, \hat{\mathbf{i}} \qquad (14\text{-}3\text{c})$$

From Eq. (13-15) the fields in the incident wave are

$$\mathbf{E}_i = \mathbf{E}_{0i} e^{iK_1\hat{\mathbf{u}}_i \cdot \mathbf{r} - i\omega t}$$

$$= \mathbf{E}_{0i} e^{iK_1(x \sin \theta_i + z \cos \theta_i) - i\omega t} \qquad (14\text{-}4\text{a})$$

$$\mathbf{H}_i = \frac{n_1}{\mu_1 c_0} \hat{\mathbf{u}}_i \times \mathbf{E}_{0i} e^{iK_1(x \sin \theta_i + z \cos \theta_i) - i\omega t} \qquad (14\text{-}4\text{b})$$

The fields in the other waves can be written in similar form with the obvious changes of signs and subscripts.

The next step is to consider the *polarization* of the incident wave. Examples 13-5 and 13-6 suggest that *any* polarization (plane, circular, or elliptical) can be obtained as a superposition of two plane-polarized waves. Accordingly, we need consider only the latter. The principal orientations are designated as *p-polarization* (electric field *parallel* to plane of incidence) and *s-polarization* (electric field perpendicular, or *senkrecht* in German, to plane of incidence). These have, of course, to be treated separately.

Case 1 : s-Polarization (Figure 14-4)

$$\mathbf{E}_{0i} = E_{0i}\hat{\mathbf{j}} \tag{14-5}$$

It seems reasonable to assume that the other two waves will also have their **E** fields perpendicular to the plane of incidence, and we shall see that all conditions can, indeed, be satisfied in this way. (*Note:* We are considering only *isotropic* media, that is, media whose properties are the same in all directions. The more complicated case of *anisotropic* media will not be taken up.) Thus we write $\mathbf{E}_0 = E_0\hat{\mathbf{j}}$ for each of the waves.

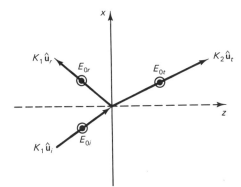

Figure 14-4 Wave vectors and electric field amplitude vectors for s-polarization.

In other words, the amplitude factors for the electric and magnetic fields of the three waves are

$$E_{0i}\hat{\mathbf{j}}: \qquad \frac{n_1}{\mu_1 c_0}E_{0i}(-\cos\theta_i\,\hat{\mathbf{i}} + \sin\theta_i\,\hat{\mathbf{k}})$$

$$E_{0r}\hat{\mathbf{j}}: \qquad \frac{n_1}{\mu_1 c_0}E_{0r}(\cos\theta_i\,\hat{\mathbf{i}} + \sin\theta_i\,\hat{\mathbf{k}}) \tag{14-6}$$

$$E_{0t}\hat{\mathbf{j}}: \qquad \frac{n_2}{\mu_2 c_0}E_{0t}(-\cos\theta_t\,\hat{\mathbf{i}} + \sin\theta_t\,\hat{\mathbf{k}})$$

The boundary relations apply to the *total* fields in the two media at the interface. Explicitly, these are the sum of the incident and reflected wave fields at $z = 0$ in medium 1, and the transmitted wave field at $z = 0$ in medium 2.

The continuity of tangential components of **E** [Eq. (11-1)] gives

$$(E_{0i} + E_{0r})e^{iK_1 x\sin\theta_i} = E_{0t}e^{iK_2 x\sin\theta_t}$$

or, in view of Snell's law [Eq. (14-2)],

$$E_{0i} + E_{0r} = E_{0t} \tag{14-7a}$$

Since neither medium is an ideal conductor, Eq. (12-1) states that the tangential components (x components in the present geometry) of the total **H** field must also be continuous. This leads to

$$\frac{n_1}{\mu_1}(-E_{0i} + E_{0r})\cos\theta_i = \frac{n_2}{\mu_2}(-E_{0t})\cos\theta_t \tag{14-7b}$$

Solving the two equations for the ratios gives

$$\frac{E_{0r}}{E_{0i}} = \frac{(n_1/\mu_1)\cos\theta_i - (n_2/\mu_2)\cos\theta_t}{(n_1/\mu_1)\cos\theta_i + (n_2/\mu_2)\cos\theta_t} \qquad (14\text{-}8a)$$

$$\frac{E_{0t}}{E_{0i}} = \frac{2(n_1/\mu_1)\cos\theta_i}{(n_1/\mu_1)\cos\theta_i + (n_2/\mu_2)\cos\theta_t} \qquad (14\text{-}8b)$$

Note that the boundary condition on the normal component of D does not pertain in the present case since $\mathbf{D}$ is purely tangential. The condition on the normal component of $\mathbf{B}$ does pertain, but gives no additional information. (The reader should check this.)

With the use of Snell's law for θ_t, the amplitude ratios are completely determined. An example of the behavior is shown for the case $n_2/n_1 = 1.5$, $\mu_1 = \mu_2$ (an air-to-glass interface) in Figure 14-5. The negative sign of E_{0r} means simply that the reflected wave suffers a 180-degree phase shift. From Eq. (14-8a), it is seen that this will occur whenever the second medium is more "optically dense."

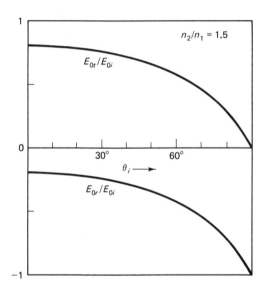

Figure 14-5 Amplitude ratios for s-polarization.

Case 2 : p-Polarization (Figure 14-6)

The field amplitude vectors now are conventionally chosen as

$$E_{0i}(\hat{\mathbf{i}}\cos\theta_i - \hat{\mathbf{k}}\sin\theta_i): \qquad \frac{n_1}{\mu_1 c_0}E_{0i}\hat{\mathbf{j}}$$

$$E_{0r}(-\hat{\mathbf{i}}\cos\theta_i - \hat{\mathbf{k}}\sin\theta_i): \qquad \frac{n_1}{\mu_1 c_0}E_{0r}\hat{\mathbf{j}} \qquad (14\text{-}9)$$

$$E_{0t}(\hat{\mathbf{i}}\cos\theta_t - \hat{\mathbf{k}}\sin\theta_t): \qquad \frac{n_2}{\mu_2 c_0}E_{0t}\hat{\mathbf{j}}$$

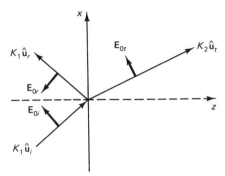

Figure 14-6 Wave vectors and electric field amplitude vectors for *p*-polarization.

Thus the boundary relations for the tangential components are

$$(E_{0i} - E_{0r}) \cos \theta_i = E_{0t} \cos \theta_t \tag{14-10a}$$

$$\frac{n_1}{\mu_1}(E_{0i} + E_{0r}) = \frac{n_2}{\mu_2} E_{0t} \tag{14-10b}$$

Solving this pair of equations results in

$$\frac{E_{0r}}{E_{0i}} = \frac{(n_2/\mu_2) \cos \theta_i - (n_1/\mu_1) \cos \theta_t}{(n_2/\mu_2) \cos \theta_i + (n_1/\mu_1) \cos \theta_t} \tag{14-11a}$$

$$\frac{E_{0t}}{E_{0i}} = \frac{2(n_1/\mu_1) \cos \theta_i}{(n_2/\mu_2) \cos \theta_i + (n_1/\mu_1) \cos \theta_t} \tag{14-11b}$$

The amplitude ratios are shown for $n_2/n_1 = 1.5$ in Figure 14-7. Equations (14-8) and (14-11) are called *Fresnel's equations*.

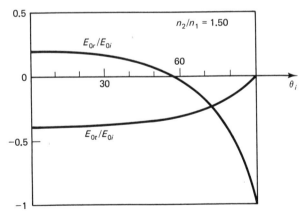

Figure 14-7 Amplitude ratios for *p*-polarization.

Example 14-1

When both media are nonmagnetic, find the conditions under which there is no reflected wave.

Solution   For the s-polarization case, it looks as if the numerator of Eq. (14-8a) can vanish. However, if we include Snell's law, the condition for this to occur becomes

$$n_1 \cos \theta_i = n_2 \sqrt{1 - \left(\frac{n_1}{n_2}\right)^2 \sin^2 \theta_i}$$

or

$$\cos^2 \theta_i = \left(\frac{n_2}{n_1}\right)^2 - \sin^2 \theta_i$$

which can hold only if $n_2 = n_1$ (i.e., if there is no optical interface).

For the p-polarization case, it is best to include Snell's law explicitly. We multiply the first and second terms of the numerator and denominator of Eq. (14-11a) by $2n_1 \sin \theta_i$ and $2n_2 \sin \theta_t$ (which are equal), respectively, to obtain (remembering that we are assuming $\mu_1 = \mu_2 = \mu_0$)

$$\begin{aligned}
\frac{E_{0r}}{E_{0i}} &= \frac{\sin 2\theta_i - \sin 2\theta_t}{\sin 2\theta_i + \sin 2\theta_t} \\
&= \frac{\cos (\theta_i + \theta_t) \sin (\theta_i - \theta_t)}{\sin (\theta_i + \theta_t) \cos (\theta_i - \theta_t)} \\
&= \frac{\tan (\theta_i - \theta_t)}{\tan (\theta_i + \theta_t)} \quad\quad (14\text{-}12)
\end{aligned}$$

It is seen that the numerator again can vanish only if the indexes of refraction are equal (then $\theta_t = \theta_i$). However, the amplitude ratio can also vanish if the denominator becomes infinite. This occurs when $\theta_i + \theta_t = 90°$, or $\sin \theta_t = \cos \theta_i$; then Snell's law gives

$$\tan \theta_i = \frac{n_2}{n_1} \quad\quad (14\text{-}13)$$

This value of θ_i is called the *Brewster angle*. It is the angle at which the E_{0r}/E_{0i} curve in Figure 14-7 crosses the axis.

Example 14-2

Write the boundary relation for **D** in the case just considered (p-polarization, nonmagnetic media).

Solution   From Eqs. (14-9) we can obtain the normal components of **D**, which must by Eq. (11-2) be equal on the two sides of the interface. Thus

$$\epsilon_1 (E_{0i} + E_{0r}) \sin \theta_i = \epsilon_2 (E_{0t}) \sin \theta_t$$

But Snell's law is $\sqrt{\epsilon_1} \sin \theta_i = \sqrt{\epsilon_2} \sin \theta_t$, so the relation is

$$n_1 (E_{0i} + E_{0r}) = n_2 (E_{0t})$$

which is seen to be exactly the same as was obtained from the tangential components of **H** in Eq. (14-10b).

This provides a more physical understanding of the Brewster effect; when Eq. (14-13) is satisfied, both boundary conditions can be taken care of by the incident and transmitted waves alone, without the necessity of a reflected wave. Furthermore, it is clear that this cannot occur in the *s*-polarized case.

The Brewster effect is widely used in making polarizers for light (Figure 14-8). If a beam of "mixed" light is incident on a glass plate at the Brewster angle, the reflected beam is *s*-polarized (but weak). After passing through a stack of such plates, the transmitted beam is almost completely *p*-polarized and quite strong. Windows at the ends of gas laser discharge tubes are commonly oriented at the Brewster angle to minimize reflection losses of the *p*-polarized light.

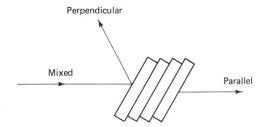

Figure 14-8   Production of polarized light via the Brewster effect.

Reflection and Transmission Coefficients

These coefficients are defined in terms of the ratios of energy fluxes per unit area of *interface*. The energy flux per unit area of *wavefront* is given by the Poynting vector. From Eq. (13-17) we obtain

$$R = \frac{\mathcal{S}_r \cos \theta_r}{\mathcal{S}_i \cos \theta_i} = \left(\frac{E_{0r}}{E_{0i}}\right)^2 \tag{14-14}$$

$$T = \frac{\mathcal{S}_t \cos \theta_t}{\mathcal{S}_i \cos \theta_i} = \sqrt{\frac{\epsilon_1 \mu_2}{\epsilon_2 \mu_1}} \left(\frac{E_{0t}}{E_{0i}}\right)^2 \frac{\cos \theta_t}{\cos \theta_i} \tag{14-15}$$

Example 14-3

Show that conservation of energy holds in reflection and refraction.

Solution   For each polarization, substitute the expressions for the amplitude ratios from Eqs. (14-8) and (14-11). Then it is straightforward to verify that in each case

$$R + T = 1 \tag{14-16}$$

14-2 REFLECTION WITH COMPLEX WAVE VECTORS

Since the propagation constant K is proportional to the index of refraction of the medium, Eq. (13-27) shows that K may be complex. Thus the *wave vector* $K\hat{\mathbf{u}}$ may be described as a *complex vector*. It is, however, a specialized sort of complex vector, since both the real and the imaginary parts have the *same direction*. We now investigate the more general case in which the two parts have *different* directions. Such waves are called *inhomogeneous plane waves*. The fields are described by wave function of the form

$$f(r, t) = f_0 e^{i\mathbf{K} \cdot \mathbf{r} - i\omega t} \tag{14-17}$$

where

$$\mathbf{K} = K_{re}\hat{\mathbf{u}} + K_{1m}\hat{\mathbf{v}} \tag{14-18}$$

On inserting form (14-17) into the wave equation (13-18), we find, instead of Eqs. (13-29),

$$K_{re}^2 - K_{im}^2 = \omega^2 \mu \epsilon' \tag{14-19a}$$

$$2K_{re}K_{im}\hat{\mathbf{u}} \cdot \hat{\mathbf{v}} = \omega^2 \mu \epsilon'' \tag{14-19b}$$

One immediate consequence is that in a lossless dielectric $\hat{\mathbf{u}}$ and $\hat{\mathbf{v}}$ must be perpendicular to each other (unless $K_{im} = 0$, in which case $\hat{\mathbf{v}}$ is meaningless).

Inhomogeneous plane waves may be set up when an ordinary plane wave is incident on an interface. To handle such cases, we have to reexamine the kinematical relations at the interface, since in the preceding section these were developed only for real wave vectors. To keep the discussion reasonably tractable, we consider only situations in which the wave vectors in medium 1 are real. Then $(K_{re}\hat{\mathbf{u}} + K_{im}\hat{\mathbf{v}})$ designates the complex wave vector in medium 2.

The kinematical conditions on frequency and coplanarity of the wave vectors are obviously unchanged. Hence the law of reflection remains valid. What requires reexamination is the law of refraction. There are now two kinematical interface relations bearing on this.

1. Since propagation is unattenuated in medium 1, any vector $\mathbf{r}_s$ lying in the interface must give

$$K_{im}\hat{\mathbf{v}} \cdot \mathbf{r}_s = 0 \tag{14-20}$$

Thus the attenuation direction is *normal to the interface*, which means that the planes of constant amplitude are parallel to the interface (i.e., $\hat{\mathbf{v}} = \hat{\mathbf{k}}$ in our chosen geometry).

2. The periodicity along the interface is the same in both media. If direction $\hat{\mathbf{u}}$ makes angle α with the normal (so that $\hat{\mathbf{u}} = \cos \alpha \, \hat{\mathbf{k}} + \sin \alpha \, \hat{\mathbf{i}}$) this says

$$K_{re} \sin \alpha = K_1 \sin \theta_i \tag{14-21a}$$

Hence

$$K_{re}\hat{\mathbf{u}} \cdot \hat{\mathbf{v}} = K_{re} \cos \alpha = \sqrt{K_{re}^2 - K_1^2 \sin^2 \theta_i} \tag{14-21b}$$

Either of these forms comprises the law of refraction for this situation.

Combining these results, we write the complex wave vector in medium 2 as

$$\mathbf{K} = K_{re}(\sin \alpha \, \hat{\mathbf{i}} + \cos \alpha \, \hat{\mathbf{k}}) + iK_{im} \, \hat{\mathbf{k}} \tag{14-22a}$$

$$= K_1 \sin \theta_i \, \hat{\mathbf{i}} + (\sqrt{K_{re}^2 - K_1^2 \sin^2 \theta_i} + iK_{im})\hat{\mathbf{k}} \tag{14-22b}$$

Equations (14-19) then become

$$K_{re}^2 - K_{im}^2 = \omega^2 \mu_2 \epsilon_2' \qquad (14\text{-}23a)$$

$$2K_{im}\sqrt{K_{re}^2 - K_1^2 \sin^2 \theta_i} = \omega^2 \mu_2 \epsilon_2'' \qquad (14\text{-}23b)$$

These two equations can be solved straightforwardly in much the same manner as Example 13-7, but the expressions are complicated and not very perspicuous. Instead of writing them out, we shall consider two important special cases.

Total Internal Reflection

We take medium 2 (as well as medium 1) to be a lossless dielectric, so that $\epsilon_2'' = 0$, $\epsilon_2' = \epsilon_2$. Then the solution of Eqs. (14-23) is

$$K_{re} = K_1 \sin \theta_i = \frac{\omega n_1}{c_0} \sin \theta_i \qquad (14\text{-}24)$$

$$K_{im} = \sqrt{K_{re}^2 - \omega^2 \mu_2 \epsilon_2}$$

$$= \frac{\omega}{c_0}\sqrt{n_1^2 \sin^2 \theta_i - n_2^2} \qquad (14\text{-}25)$$

The latter demands $n_1 \sin \theta_i \geq n_2$, or $\theta_i \geq \theta_c$, where

$$\sin \theta_c = \frac{n_2}{n_1} \qquad (14\text{-}26)$$

Thus

$$K_{im} = \frac{\omega n_1}{c_0}\sqrt{\sin^2 \theta_i - \sin^2 \theta_c} \qquad (14\text{-}27)$$

Obviously, this can occur only if $n_2 < n_1$, that is, in reflection from an optically *less dense* medium, such as a glass-to-air interface.

θ_c is called the *critical angle*. For angles of incidence $\theta_i < \theta_c$, we simply have ordinary refraction. As θ_i is increased to θ_c, the angle of refraction θ_t increases to 90 degrees. For still larger angles of incidence $(\theta_i > \theta_c)$ there is an inhomogeneous plane wave in medium 2. This is also called an *evanescent* wave. To obtain the amplitude of this wave, we proceed just as in ordinary refraction to match the total tangential field components at the interface in the two media.

s-Polarization

The electric field in medium 2 is described by

$$\mathbf{E}_t(x, z, t) = E_{0t}\hat{\mathbf{j}}e^{iK_{re}x - K_{im}z - i\omega t} \qquad (14\text{-}28)$$

For this the magnetic field is given by

$$i\omega\mu_2\mathbf{H}_t = \text{curl } \mathbf{E}_t$$

$$= (iK_{re}\hat{\mathbf{k}} + K_{im}\hat{\mathbf{i}})E_{0t}e^{iK_{re}x - K_{im}z - i\omega t} \qquad (14\text{-}29)$$

The matching equations thus are, instead of Eqs. (14-7),

$$E_{0i} + E_{0r} = E_{0t} \tag{14-30a}$$

$$\frac{n_1}{\mu_1 c_0}(-E_{0i} + E_{0r}) \cos \theta_i = \frac{-i}{\omega \mu_2} K_{im} E_{0t}$$

or

$$-E_{0i} + E_{0r} = -i\Gamma E_{0t} \tag{14-30b}$$

where

$$\Gamma = \frac{\mu_1 c_0}{n_1} \frac{K_{im}}{\omega \mu_2} = \frac{\mu_1}{\mu_2} \frac{\sqrt{\sin^2 \theta_i - \sin^2 \theta_c}}{\cos \theta_i} \tag{14-31}$$

The solution of Eqs. (14-30) is

$$\frac{E_{0r}}{E_{0i}} = \frac{1 - i\Gamma}{1 + i\Gamma} \tag{14-32a}$$

$$\frac{E_{0t}}{E_{0i}} = \frac{2}{1 + i\Gamma} \tag{14-32b}$$

Example 14-4

Show that the reflection is total.

Solution

$$R = \left| \frac{E_{0r}}{E_{0i}} \right|^2 = 1$$

Example 14-5

For an air–glass interface ($n = 1$ for air, 1.5 for glass) compare the reflection coefficients as function of the angle of incidence for the two directions air-to-glass and glass-to-air.

Solution In the subcritical range of incidence angles, Eq. (14-8a) holds. When Snell's law is incorporated, this gives

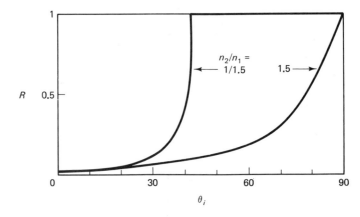

Figure 14-9 Reflection coefficients in both directions for an air–glass interface, *s*-polarization.

$$R = \left[\frac{\cos \theta_i - \sqrt{(n_2/n_1)^2 - \sin^2 \theta_i}}{\cos \theta_i + \sqrt{(n_2/n_1)^2 - \sin^2 \theta_i}}\right]^2$$

Calculated values for the two index ratios, 1.5 and 1/1.5, are shown in Figure 14-9.

Example 14-6

Calculate the electric field amplitude in medium 2 for the case $n_2 < n_1$. Take $\mu_1 = \mu_2$ for simplicity.

Solution Part 1   For the subcritical range Eq. (14-8b) gives the amplitude ratio. Since

$$\sin \theta_t = \frac{n_1}{n_2} \sin \theta_i = \frac{\sin \theta_i}{\sin \theta_c} \qquad (14\text{-}33)$$

this may be written

$$\frac{E_{0t}}{E_{0i}} = \frac{2 \cos \theta_i}{\cos \theta_i + \sin \theta_c \cos \theta_t}$$

$$= \frac{2 \cos \theta_i}{\cos \theta_i + \sqrt{\sin^2 \theta_c - \sin^2 \theta_i}} \qquad (\theta_i < \theta_c) \qquad (14\text{-}34a)$$

Solution Part 2   For incidence angles above critical, Eq. (14-32b) gives

$$\left|\frac{E_{0t}}{E_{0i}}\right| = \sqrt{\frac{2}{1 + i\Gamma} \frac{2}{1 - i\Gamma}} = \frac{2}{\sqrt{1 + \Gamma^2}}$$

$$= \frac{2}{\sqrt{1 + \dfrac{\sin^2 \theta_i - \sin^2 \theta_c}{\cos^2 \theta_i}}}$$

$$= \frac{2 \cos \theta_i}{\cos \theta_c} \qquad (\theta_i > \theta_c) \qquad (14\text{-}34b)$$

The behavior is shown in Figure 14-10.

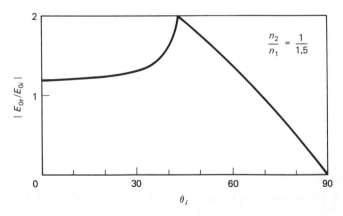

Figure 14-10   Amplitude of transmitted wave for *s*-polarization. For $\theta > \theta_c$, the curve gives the value just beyond the interface plane. The amplitude then decays exponentially with distance into medium 2.

Example 14-7

Calculate the time-average Poynting vector of the evanescent wave in medium 2.

Solution   The electric and magnetic fields are given by Eqs. (14-28) and (14-29). Thus

$$\bar{\bar{S}} = \frac{1}{2} \text{Re} \, (\mathbf{E} \times \mathbf{H}^*)$$

$$= \frac{1}{2} \text{Re} \left(E_{0t} \hat{\mathbf{j}} \times \frac{-iK_{\text{re}}\hat{\mathbf{k}} + K_{\text{im}}\hat{\mathbf{i}}}{-i\omega\mu_2} \right) E_{0t} e^{-2K_{\text{im}}z}$$

$$= \frac{1}{2} E_{0t}^2 \frac{K_{\text{re}}}{\omega\mu_2} e^{-2K_{\text{im}}z} \hat{\mathbf{i}} \tag{14-35}$$

Thus the time-average energy flow is entirely parallel to the interface, with no steady-state energy flow into the interior. This is, of course, consistent with the total reflection.

Example 14-8

Calculate the phase shift of the reflected wave under total reflection conditions.

Solution   Equation (14-32a) may be written

$$\frac{E_{0r}}{E_{0t}} = \frac{\sqrt{1 + \Gamma^2}e^{-i\phi}}{\sqrt{1 + \Gamma^2}e^{i\phi}} = e^{-2i\phi} \tag{14-36}$$

where $\phi = \tan^{-1}\Gamma$. Since, by Eq. (14-31), Γ ranges from 0 to ∞ as θ_i increases from θ_c to 90 degrees, the phase shift -2ϕ ranges from 0 to -180 degrees, as shown in Figure 14-11.

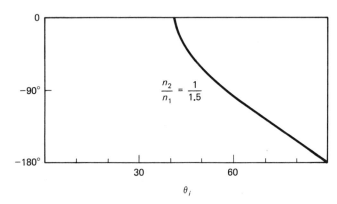

Figure 14-11   Phase shift of internally reflected wave for s-polarization.

Example 14-9

Calculate the attenuation length of the evanescent wave in medium 2.

Solution   From Eq. (14-28) the field amplitude varies as $e^{-K_{\text{im}}z}$. Hence the penetration depth is

$$\delta = K_{\text{im}}^{-1} = \frac{c_0}{\omega\sqrt{n_1^2 \sin^2 \theta_i - n_2^2}}$$

$$= \frac{c_0/n_1}{\omega\sqrt{\sin^2 \theta_i - \sin^2 \theta_c}} = \frac{\lambda_1/2\pi}{\sqrt{\sin^2 \theta_i - \sin^2 \theta_c}} \qquad (14\text{-}37)$$

where λ_1 is the incident wavelength. The penetration depth is seen to be infinite just above the critical angle and to decrease as θ_i increases.

Frustrated Total Reflection

Even though the evanescent wave carries no energy flux into medium 2, it possesses some energy content, as Example 14-7 clearly shows. This energy can be absorbed by an absorbing material placed in medium 2 reasonably close to the interface (i.e., within the penetration depth of the evanescent wave). The reflection can then, of course, no longer be complete. The total reflection is said to be *frustrated*.

There are some useful applications of this effect. One is to measure absorption spectra on extremely minute samples in cases where transmission measurements are not practicable. Another is a device for inkless fingerprinting, as illustrated in Figure 14-12.

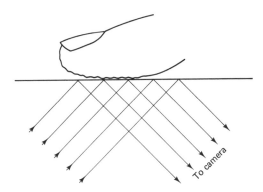

Figure 14-12 Inkless fingerprinting by frustrated total internal reflection. (Harrick Scientific Corp., Ossining, N.Y., with permission.)

p-Polarization

The calculation follows much the same lines. $\mathbf{E}_t$ will now have x and z components, which must satisfy div $\mathbf{E} = 0$. $\mathbf{H}_t$ has only a y component, obtained from the Maxwell curl $\mathbf{E}$ equation. The details are left for a problem. The result is again total reflection, but with a different phase shift.

Metallic Reflection

Another important case that can be handled quite easily is when medium 2 is a good conductor, and ω is low enough so that ϵ_2 and g_2 are real. Then, by Eqs. (13-25), $\epsilon_2' = \epsilon_2$, $\epsilon_2'' = g_2/\omega$, and Eqs. (14-23) become

$$K_{\text{re}}^2 - K_{\text{im}}^2 \cong 0 \tag{14-38a}$$

$$K_{\text{im}}^2(K_{\text{re}}^2 - K_1^2 \sin^2 \theta_i) = \left(\frac{\omega \mu_2 g_2}{2}\right)^2 \tag{14-38b}$$

After substitution, the latter is a biquadratic whose solution is

$$K_{\text{re}}^2 = \frac{1}{2}[K_1^2 \sin^2 \theta_i + \sqrt{K_1^4 \sin^4 \theta_i + (\omega \mu_2 g_2)^2}]$$

$$= \frac{1}{2} K_1^2 \sin^2 \theta_i \left[1 + \sqrt{1 + \left(\frac{\omega \mu_2 g_2}{K_1^2 \sin^2 \theta_i}\right)^2}\right]$$

$$\cong \frac{\omega \mu_2 g_2}{2}$$

or

$$K_{\text{re}} \cong K_{\text{im}} \cong \sqrt{\frac{\omega \mu_2 g_2}{2}} \tag{14-39}$$

Thus, from Eq. (14-21a),

$$\sin \alpha = \frac{K_1 \sin \theta_i}{K_{\text{re}}} \ll 1$$

so that the propagation direction in a good conductor is very nearly perpendicular to the surface, irrespective of the angle of incidence. In other words, the complex wave vector in medium 2 is (very nearly)

$$\mathbf{K}_2 = \sqrt{i\omega \mu_2 g_2}\,\hat{\mathbf{k}} \tag{14-40}$$

The reflected amplitude is again obtained by matching tangential field components at the interface. For the s-polarization case we have, following the lines of Eqs. (14-7) but remembering that $\cos \theta_t \cong 1$,

$$E_{0i} + E_{0r} = E_{0t} \tag{14-41a}$$

$$\frac{K_1}{\omega \mu_1}(-E_{0i} + E_{0r}) \cos \theta_i = \frac{K_2}{\omega \mu_2}(-E_{0t}) \tag{14-41b}$$

These give

$$\frac{E_{0r}}{E_{0i}} = \frac{Y - 1}{Y + 1} \tag{14-42}$$

where

$$Y = \frac{\mu_2 K_1 \cos \theta_i}{\mu_1 K_2}$$

Thus the reflection coefficient is

$$R \equiv \left|\frac{E_{0r}}{E_{0i}}\right|^2 = \frac{(1 - Y)(1 - Y^*)}{(1 + Y)(1 + Y^*)}$$

$$\cong 1 - 2(Y + Y^*)$$

Now, in view of Eq. (14-40),

$$Y + Y^* = \frac{\mu_2 \omega \sqrt{\mu_1 \epsilon_1} \cos \theta_i}{\mu_1 \sqrt{\omega \mu_2 g_2}} \left(\frac{1}{\sqrt{i}} + \frac{1}{\sqrt{-i}} \right)$$

$$= \sqrt{\frac{\mu_2}{\mu_1} \frac{\omega \epsilon_1}{g_2}} \cos \theta_i \left(\frac{\sqrt{2}}{1+i} + \frac{\sqrt{2}}{1-i} \right)$$

$$= \sqrt{2 \frac{\mu_2}{\mu_1} \frac{\omega \epsilon_1}{g_2}} \cos \theta_i$$

Thus

$$R = 1 - 2\sqrt{2} \sqrt{\frac{\mu_2}{\mu_1} \frac{\omega \epsilon_1}{g_2}} \cos \theta_i \tag{14-43}$$

This is known as the *Hagen–Rubens relation*, after the experimenters who first tested it on metallic reflection in the far infrared. It is *not* valid at visible frequencies, since g is then no longer purely real (see Section 5-6).

Example 14-10

Estimate the reflection coefficient of a typical good metal at a wavelength of 100 μm.

Solution   We may take copper at room temperature as typical example; then $g_2 \cong 6 \times 10^7$ mho/m. The angular frequency is

$$\omega = \frac{2\pi c_0}{\lambda} \cong 2 \times 10^{13} \text{ s}^{-1}$$

Then

$$\frac{\omega \epsilon_1}{g_2} \cong \frac{2 \times 10^{13} \times 8.85 \times 10^{-12}}{6 \times 10^7} \cong 3 \times 10^{-6}$$

$$R = 1 - 2\sqrt{2} \sqrt{3 \times 10^{-6}} \cong 0.995$$

Thus metals are *strong reflectors* of electromagnetic waves. This property persists up to visible frequencies (although the calculation is more complicated because g is complex) and accounts for the typical shiny appearance of metals.

Reflection at Normal Incidence

If $\theta_i = 0$, Eqs. (14-23) simplify to

$$K_{re}^2 - K_{im}^2 = \omega^2 \mu_2 \epsilon_2'$$

$$2K_{im}K_{re} = \omega^2 \mu_2 \epsilon_2''$$

These are equivalent to Eqs. (13-29), which is natural since the two parts of the complex wave vector now have the same direction, namely $\hat{\mathbf{k}}$. Thus, as in an unbounded conductive medium, K is proportional to the complex index of refraction $n + ik$. The Fresnel relations, found as usual from the boundary relations, turn out just like Eqs. (14-8) or (14-11) with $\theta_i = \theta_t = 0$ and with the complex indexes instead of the real ones.

Example 14-11

Write the reflection coefficient for normal incidence from vacuum on a lossy but nonmagnetic dielectric.

Solution   Equation (14-8a) becomes

$$\frac{E_{0r}}{E_{0i}} = \frac{1 - (n_2 + ik_2)}{1 + (n_2 + ik_2)} = \frac{(-n_2 + 1) - ik_2}{(n_2 + 1) + ik_2}$$

$$R = \left| \frac{E_{0r}}{E_{0i}} \right|^2 = \frac{(n_2 - 1)^2 + k_2^2}{(n_2 + 1)^2 + k_2^2} \tag{14-44}$$

This shows that either large n or large k (or both) lead to R nearly 1. In other words, either highly lossy or highly refractive materials are good reflectors. In the case of metals, both factors are operative.

14-3 GUIDED WAVES

In many important applications, electromagnetic waves are propagated not in unbounded media but in regions confined by various types of boundaries. When the region is infinitely long in one dimension and of uniform cross section in the other two, the structure is called a *waveguide*. A simple example is just a metal pipe or duct. Shining a flashlight through a very narrow duct is an example of guided-wave propagation.

The fields must, of course, satisfy certain boundary conditions (depending on the type of material) at the guide surface. Ordinary plane waves cannot do this because the fields are uniform over the transverse planes. In some cases, combinations of plane waves traveling in oblique directions and reflecting off the walls will suffice, but rather than attempt to work with these it is better to start over from the beginning and investigate the fields more generally. This can be done either in terms of the potentials or the fields themselves. The former is somewhat more concise and elegant, but the latter is more straightforward, and we shall follow it. For simplicity, we shall deal with *ideal conductors* as boundaries.

The most general form possible for **E** and **H** fields in a wave propagating in a *single direction* (taken to be the z direction) is

$$\mathbf{E} = [e_x(x, y)\hat{\mathbf{i}} + e_y(x, y)\hat{\mathbf{j}} + e_z(x, y)\hat{\mathbf{k}}]e^{i(Kz-\omega t)} \tag{14-45a}$$

$$\mathbf{H} = [h_x(x, y)\hat{\mathbf{i}} + h_y(x, y)\hat{\mathbf{j}} + h_z(x, y)\hat{\mathbf{k}}]e^{i(Kz-\omega t)} \tag{14-45b}$$

Each of the six components must satisfy the wave equation, and they must be related to each other by the Maxwell equations. For fields of this form, the x and y components of the curl equations, $-\dot{\mathbf{B}} = \text{curl } \mathbf{E}$ and $\dot{\mathbf{D}} = \text{curl } \mathbf{H}$, become

$$i\omega\mu h_x = \frac{\partial e_z}{\partial y} - iKe_y \tag{14-46a}$$

$$i\omega\mu h_y = \frac{-\partial e_z}{\partial x} + iKe_x \tag{14-46b}$$

$$-i\omega\epsilon e_x = \frac{\partial h_z}{\partial y} - iKh_y \tag{14-46c}$$

$$-i\omega\epsilon e_y = \frac{-\partial h_z}{\partial x} + iKh_x \tag{14-46d}$$

By combining these equations, each of the transverse functions may be expressed in terms of the longitudinal ones. Equations (14-46a) and (14-46d) contain h_x and e_y; Eqs. (14-46b) and (14-46c) contain h_y and e_x. By substituting from one to the other in these pairs, we obtain

$$(\mu\epsilon\omega^2 - K^2)e_x = iK\frac{\partial e_z}{\partial x} + i\omega\mu\frac{\partial h_z}{\partial y} \tag{14-47a}$$

$$(\mu\epsilon\omega^2 - K^2)e_y = iK\frac{\partial e_z}{\partial y} - i\omega\mu\frac{\partial h_z}{\partial x} \tag{14-47b}$$

$$(\mu\epsilon\omega^2 - K^2)h_x = iK\frac{\partial h_z}{\partial x} - i\omega\epsilon\frac{\partial e_z}{\partial y} \tag{14-47c}$$

$$(\mu\epsilon\omega^2 - K^2)h_y = iK\frac{\partial h_z}{\partial y} + i\omega\epsilon\frac{\partial e_z}{\partial x} \tag{14-47d}$$

The coefficient on the left in these equations can also be cast in terms of the wavelength of the guided wave and the wavelength of a wave of the same frequency in the same medium without the boundaries. From Eqs. (14-45) the guided wavelength is

$$\lambda_g = \frac{2\pi}{K} \tag{14-48a}$$

The "free space" wavelength, on the other hand, is

$$\lambda_0 = \frac{c}{v} = \frac{2\pi c}{\omega} = \frac{2\pi}{\omega\sqrt{\mu\epsilon}} \tag{14-48b}$$

Thus

$$\mu\epsilon\omega^2 - K^2 = \left(\frac{2\pi}{\lambda_0}\right)^2 - \left(\frac{2\pi}{\lambda_g}\right)^2 \tag{14-48c}$$

Many textbooks cast the formulas in terms of these wavelengths rather than ω and K. Some also use the symbol $\lambdabar$, called *lambda-bar*, for $\lambda/2\pi$.

The next task is to write the wave equation for e_z and h_z, which, according to Eqs. (14-45), are functions of the two transverse coordinates x and y. From Eqs. (13-3) and (13-4), with the form of fields we are considering we get

$$\left(\frac{\partial^2}{\partial x^2} + \frac{\partial^2}{\partial y^2} - K^2 + \mu\epsilon\omega^2\right)f(x, y) = 0$$

or

$$\left(\frac{\partial^2}{\partial x^2} + \frac{\partial^2}{\partial y^2}\right)f = -(\mu\epsilon\omega^2 - K^2)f \tag{14-49}$$

This equation states that the sum of the second partial derivatives reproduces a multiple of the same function. Possible solutions are trigonometric and/or exponential (or hyperbolic) functions. However, since boundary conditions at the guide surface will have to be satisfied, functions that *return to the same value* on going across the guide are needed. Hence the appropriate form for our purpose is

$$f(x, y) = (A_1 \sin \alpha x + A_2 \cos \alpha x)(A_3 \sin \beta y + A_4 \cos \beta y) \qquad (14\text{-}50)$$

To narrow this down further, we need the explicit boundary conditions on e_z and h_z. Remember that we are assuming the guide wall to be an ideal conductor, so the fields vanish within it. Thus the continuity of tangential components of E demands that

$$e_z = 0 \qquad \text{at the walls} \qquad (14\text{-}51)$$

For the magnetic field, things are not as simple. The tangential components do not have to be continuous since there may well be surface currents in the wall. However, by Eq. (12-2) the normal component must be zero. The normal component lies in the xy plane. Perpendicular to it in this plane is a tangential component of E, which must also be zero. From Eqs. (14-46c) and (14-46d) we see that these require the derivative of h_z in the direction normal to the surface to vanish. Thus

$$\frac{\partial h_z}{\partial n} = 0 \qquad \text{at the walls} \qquad (14\text{-}52)$$

TE and TM Modes

Since the boundary conditions on e_z and h_z are different, it is convenient to divide the field into two classes:

$$TE \ (transverse \ electric): \quad e_z = 0 \ \text{everywhere}$$

$$TM \ (transverse \ magnetic): \quad h_z = 0 \ \text{everywhere}$$

Since solutions of the wave equation can be superimposed, and since the equations relating the various components are linear, the general solution may be written as a sum of the two types.

Rectangular Waveguides

In the following we consider the guide to be a hollow rectangular cylinder with faces at $x = 0$, $x = a$, $y = 0$, and $y = b$ (see Figure 14-13). We are, of course, interested in fields in the *interior* of such a guide, and we assume this to be filled with a uniform linear isotropic dielectric medium (usually air in practice). We shall use the term "free space" to describe the same medium but of infinite extent.

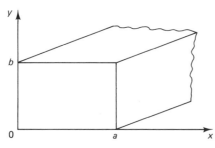

Figure 14-13 Geometry of a rectangular waveguide.

Example 14-12

Find the explicit form of e_z for a TM wave.

Solution   Apply the boundary condition Eq. (14-51) to the general form of Eq. (14-50). The condition at $x = 0$ requires that $A_2 = 0$. For the condition at $x = a$, we cannot simply set $A_1 = 0$, for then the entire function would vanish. Instead, we require that $\sin \alpha a = 0$, hence αa must be an integer multiple of π, $\alpha a = n\pi$. Similarly, the conditions at $y = 0$ and $y = b$ require $A_4 = 0$, $\beta b = m\pi$. Thus

$$e_z(x, y) = e_{z0} \left(\sin \frac{n\pi}{a} x \right) \left(\sin \frac{m\pi}{b} y \right) \tag{14-53}$$

where e_{z0} is a constant and n and m are *integers*. A specific pair of integers gives the "TM$_{nm}$ mode."

Example 14-13

Find the explicit form of h_z for a TE wave.

Solution   Equation (14-52) states that $\partial h_z/\partial x = 0$ at $x = 0$ and $x = a$, and $\partial h_z/\partial y = 0$ at $y = 0$ and $y = b$. Now

$$\frac{\partial h_z}{\partial x} = \alpha (A_1 \cos \alpha x - A_2 \sin \alpha x)(A_3 \sin \beta y + A_4 \cos \beta y)$$

so the condition at $x = 0$ requires $A_1 = 0$, then that at $x = a$ again requires $\alpha a = n\pi$. Similarly,

$$\frac{\partial h_z}{\partial y} = \beta \left(A_2 \cos \frac{n\pi}{a} x \right)(A_3 \cos \beta y - A_4 \sin \beta y)$$

so the condition at $y = 0$ requires $A_3 = 0$, then that at $y = b$ again requires $\beta b = m\pi$. Thus

$$h_z(x, y) = h_{z0} \left(\cos \frac{n\pi}{a} x \right) \left(\cos \frac{m\pi}{b} y \right) \tag{14-54}$$

where h_{z0} is a constant and n and m are again integers. A specific pair of integers gives the "TE$_{nm}$ mode."

Cutoff Frequency

By putting e_z and h_z back into the wave equation, we now show that there is a *minimum frequency* for the propagation of any specific mode. For both types, Eq. (14-49) becomes

$$-\left(\frac{n\pi}{a} \right)^2 - \left(\frac{m\pi}{b} \right)^2 = -(\mu\epsilon\omega^2 - K^2)$$

or

$$\mu\epsilon\omega^2 = K^2 + \left(\frac{n\pi}{a} \right)^2 + \left(\frac{m\pi}{b} \right)^2 \tag{14-55}$$

Now for a propagating wave, K must be real, so K^2 must be nonnegative. Thus the *cutoff frequency* for an (n, m) mode is given by

$$\mu\epsilon\omega_c^2 = \left(\frac{n\pi}{a}\right)^2 + \left(\frac{m\pi}{b}\right)^2 \tag{14-56a}$$

$$\omega_c = c\sqrt{\left(\frac{n\pi}{a}\right)^2 + \left(\frac{m\pi}{b}\right)^2} \tag{14-56b}$$

The relation of K to ω for the guided wave, Eq. (14-55), then can be written

$$K^2 = \mu\epsilon(\omega^2 - \omega_c^2) = \frac{\omega^2 - \omega_c^2}{c^2}$$

$$K\left(\equiv \frac{2\pi}{\lambda_g}\right) = \frac{1}{c}\sqrt{\omega^2 - \omega_c^2} \tag{14-57}$$

For the unbounded medium, on the other hand, the relation is $K = \omega/c$ [Eq. (13-8)]. It is seen that the guided-wave relation becomes asymptotically "free-space-like" when the frequency is far above cutoff (Figure 14-14). From Eqs. (14-48) it can be seen that this corresponds to a free-space wavelength much smaller than the guide dimensions.

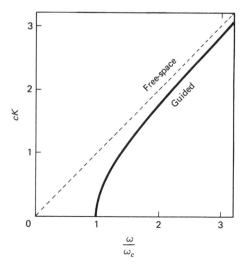

Figure 14-14 Relation of wave number to frequency for guided and free-space waves.

The wavelength of the guided wave is

$$\lambda_g = \frac{2\pi}{K} = \frac{2\pi c}{\sqrt{\omega^2 - \omega_c^2}} = \frac{2\pi c/\omega}{\sqrt{1 - (\omega_c/\omega)^2}} = \frac{\lambda_0}{\sqrt{1 - (\omega_c/\omega)^2}} \tag{14-58}$$

Thus the guided wavelength is *larger* than the free-space wavelength at the same frequency, and becomes *infinite* at cutoff. Since the propagation is described by the factor $e^{i(Kz-\omega t)}$, the speed is

$$v = \frac{\omega}{K} = \frac{\omega c}{\sqrt{\omega^2 - \omega_c^2}} = \frac{c}{\sqrt{1 - (\omega_c/\omega)^2}} \tag{14-59}$$

Thus the speed is also *greater* than the free-space wave speed, and it too becomes infinite at cutoff.

This may be disturbing to some readers, as there is a common misconception that the speed of light in vacuum is the fastest that anything can travel. This is simply not true. We saw one counterexample in Chapter 13 in the wave velocity at very high frequency in a dispersive medium. For another example, if we shine a flashlight at a wall and rotate the flashlight, the spot on the wall can be made to move as fast as we wish by standing far enough away from the wall. Yet another example is found in the breaking of waves on a beach: If the waves come in at a slight angle, the "break-point" moves along the beach much faster than the waves through the water, as indicated in Figure 14-15. As we shall see later, this is actually a very good analog of the waveguide situation. (The correct statement about the speed of light is that it cannot be exceeded by any *material body* or by any *signal* of any sort. In the examples just given, clearly the spot of light cannot be used to signal along the wall, nor can the breakpoint be used to signal along the beach.)

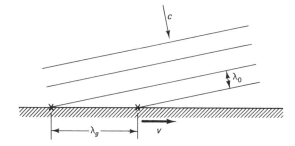

Figure 14-15   Analogy of waves in guide to breakers on a beach.

Example 14-14

Which modes of each type have the lowest cutoff frequencies?

Solution   From Eq. (14-56b) it is clear that we want n and m to be as small as possible. For TE, this would be $n = 1$, $m = 0$ (assuming that $a > b$); they cannot both be zero since, by Eq. (14-54), h_z would then be simply a constant and the other field components would vanish. For TM, neither of them can be zero since, by Eq. (14-53), e_z would vanish; so the lowest possibility is $n = 1$, $m = 1$.

Example 14-15

Calculate the "cutoff wavelength" (free-space wavelength at the cutoff frequency) for each of the lowest modes.

Solution

$$\text{TE}_{10}: \quad \omega_c = \frac{\pi c}{a} \qquad\qquad\qquad \lambda_c = 2a \qquad\qquad (14\text{-}60\text{a})$$

$$\text{TM}_{11}: \quad \omega_c = \pi c \sqrt{\frac{1}{a^2} + \frac{1}{b^2}} \qquad \lambda_c = \frac{2a}{\sqrt{1 + (a/b)^2}} \qquad (14\text{-}60\text{b})$$

Thus we see that cutoff occurs when the free-space wavelength becomes about as large as the transverse dimensions of the guide.

Example 14-16

Calculate the fields in the TE_{10} mode.

Solution   On inserting Eq. (14-54) into Eqs. (14-47) we obtain (remembering that $e_z = 0$ for TE modes)

$$h_z = h_{z0} \cos \frac{\pi x}{a}$$

$$e_x = 0$$

$$e_y = \frac{(-i\omega\mu)(-\pi/a)}{\mu\epsilon\omega^2 - K^2} h_{z0} \sin \frac{\pi x}{a}$$

$$h_x = \frac{iK(-\pi/a)}{\mu\epsilon\omega^2 - K^2} h_{z0} \sin \frac{\pi x}{a}$$

$$h_y = 0$$

The coefficients may be simplified by the use of Eqs. (14-56) and (14-57) to give

$$e_y = i\mu c \frac{\omega}{\omega_c} h_{z0} \sin \frac{\pi x}{a}$$

$$h_x = -i\sqrt{\left(\frac{\omega}{\omega_c}\right)^2 - 1}\, h_{z0} \sin \frac{\pi x}{a}$$

It is customary to express all components in terms of the electric field amplitude, so we let $E_0 = i\mu c(\omega/\omega_c)h_{z0}$; then

$$\mathbf{E} = E_0 \hat{\mathbf{j}} \sin \frac{\pi x}{a} e^{i(Kz-\omega t)} \tag{14-61a}$$

$$\mathbf{H} = \frac{-E_0}{\mu c}\left[\sqrt{1 - \left(\frac{\omega_c}{\omega}\right)^2} \sin \frac{\pi x}{a} \hat{\mathbf{i}} + i\frac{\omega_c}{\omega} \cos \frac{\pi x}{a} \hat{\mathbf{k}}\right] e^{i(Kz-\omega t)} \tag{14-61b}$$

A sketch of the field-line patterns is shown in Figure 14-16.

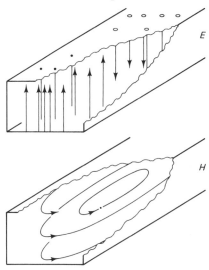

Figure 14-16   Field patterns of TE_{10} mode in a rectangular waveguide.

Example 14-17

Show that the TE_{10} field is equivalent to two plane waves traveling in oblique directions.

Solution   Note that

$$\sin \frac{\pi x}{a} = \frac{e^{i\pi x/a} - e^{-i\pi x/a}}{2i}$$

Thus **E** can be written as the sum of two terms, the first of which varies as

$$\exp\left[i\left(\frac{\pi x}{a} + Kz - \omega t\right)\right] = \exp i\left[\frac{\omega_c}{c}x + \frac{\omega}{c}\sqrt{1 - \left(\frac{\omega c}{\omega}\right)^2}z - \omega t\right]$$

$$= \exp i\left\{\frac{\omega}{c}\left[\frac{\omega_c}{\omega}x + \sqrt{1 - \left(\frac{\omega_c}{\omega}\right)^2}z\right] - \omega t\right\}$$

$$(14\text{-}62)$$

This is a plane wave traveling at an angle θ to the z axis given by

$$\sin \theta = \frac{\omega_c}{\omega} \quad (\leq 1)$$

The second term of **E** obviously corresponds to a wave going at the negative of this angle. Thus the field consists of a pair of plane waves reflecting back and forth from the vertical faces of the guide, as shown in Figure 14-17. Other modes can be similarly decomposed into plane waves, but more terms are required and the expressions get a bit messy.

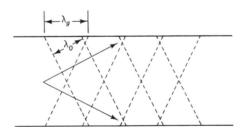

Figure 14-17   Pair of oblique plane waves comprising the TE_{10} field.

This example illustrates the analogy with the breaking of water waves on a beach, which we discussed after Eq. (14-59), except that in the present case the waves simply reflect off the boundary. The intersection point of a given plane in one of the two waves moves along the wall at a speed $c/\cos \theta$, which agrees with Eq. (14-59). On the other hand, if the wave is used to carry a signal (say by "chopping" it in a dot-dash fashion) the latter will progress along the guide at a speed

$$v_{\text{sig}} = c \cos \theta = \frac{c^2}{v} \qquad (14\text{-}63)$$

because it zigzags back and forth with the individual waves. Thus the signal velocity is *always less* than the free-space wave speed. It is, in fact, just equal to the *group velocity*, defined in Eq. (13-53).

Example 14-18

Calculate the current flowing in the walls of a waveguide carrying a TE_{10} wave.

Solution   According to Eq. (12-1), the surface current density is related to the discontinuity in the tangential component of **H**. Taking region 2 to be the interior of the guide and region 1 to be within the wall material, we have $\mathbf{H}_1 = 0$ (in the ideal conductor), hence $\mathbf{j} = \mathbf{H} \times \hat{\mathbf{n}}$, with **H** given in Eq. (14-61b). On the vertical walls, $\hat{\mathbf{n}} = \pm\hat{\mathbf{i}}$, so the current flow is in the y direction. On the horizontal walls, $\hat{\mathbf{n}} = \pm\hat{\mathbf{j}}$, so there are both x and z components of current. The converging lines of current flow provide the concentrations of surface charge needed to terminate the electric field lines.

Example 14-19

Calculate the time-average power flow along the guide in a TE_{10} wave (Figure 14-18).

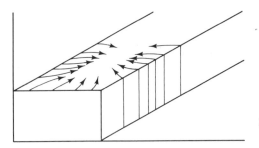

Figure 14-18   Pattern of current flow in walls of guide carrying a TE_{10} wave.

Solution   At any point the time-average Poynting vector is, from the fields given in Eqs. (14-61),

$$\bar{\mathcal{S}} = \frac{1}{2}\operatorname{Re}(\mathbf{E} \times \mathbf{H}^*) = \frac{1}{2}\frac{E_0^2}{\mu c}\sqrt{1 - \left(\frac{\omega_c}{\omega}\right)^2}\sin^2\frac{\pi x}{a}\,\hat{\mathbf{k}}$$

(Note that H_z is imaginary and does not contribute to the real part of $\mathbf{E} \times \mathbf{H}^*$.)

The total power flow is obtained by integrating $\bar{\mathcal{S}}$ over the cross-sectional area of the guide, that is,

$$\mathcal{P} = \frac{1}{2}\frac{E_0^2}{\mu c}\sqrt{1 - \left(\frac{\omega_c}{\omega}\right)^2}\int_0^b dy \int_0^a dx \sin^2\frac{\pi x}{a}\,\hat{\mathbf{k}}$$

$$= \frac{1}{2}\frac{E_0^2}{\mu c}\sqrt{1 - \left(\frac{\omega_c}{\omega}\right)^2}\left(\frac{ab}{2}\right)\hat{\mathbf{k}}$$

The power is zero at cutoff and increases as the frequency increases above the cutoff value. On comparing with Example 13-17, we see that the area-average energy velocity is diminished by the factor $\sqrt{1 - (\omega_c/\omega)^2}$ and thus is the same as the signal velocity, Eq. (14-63).

Attenuation in Waveguides

In real metallic waveguides the walls are good but of course not ideal conductors. A rigorous solution of the Maxwell equations becomes very much more complicated since the fields penetrate into the skin depth and the boundary conditions are much harder to handle. We can, however, obtain a fair idea of the behavior by starting from the ideal solution and making corrections for the finite conductivity.

We first show qualitatively that a wave in a real guide must be attenuated. The current in the walls (Example 14-16) will not be a true surface current but will be spread out through the skin depth. This current leads to ohmic losses which extract energy from the wave and thus attenuate it. To get an idea of the frequency dependence, we first note that at cutoff there is no energy flow; hence the loss is total and the attenuation coefficient is infinite. As the frequency increases, the power flow increases rapidly, so the attenuation obviously decreases rapidly. When the frequency is well above cutoff, the power flow levels off. However, the skin depth continues to decrease and this causes the loss again to increase, as may be seen from the Hagen–Rubens relation of Eq. (14-43). Hence the attenuation passes through a broad minimum and then increases slowly. In practice, one chooses the size of waveguide such that the attenuation is near minimum at the desired frequency of operation.

TEM Waves

Up to this point we have been considering guided waves in which *one* of the fields is transverse to the direction of propagation. In a free-space wave, *both* fields are transverse, and it is natural to ask whether such waves (so-called TEM or *transverse electric and magnetic* waves) can exist in a guide. It turns out that they can, but only for a certain class of guide configurations. From Eqs. (14-47) we see that if e_z and h_z are both identically zero, the only way any transverse components can exist is to have

$$\mu\epsilon\omega^2 - K^2 = 0 \quad \text{or} \quad K = \omega\sqrt{\mu\epsilon} = \frac{\omega}{c} \tag{14-64}$$

This is the same as the free-space relation (13-8). In particular, there is *no cutoff condition*. The latter feature gives a hint as to the type of guide structure required for a TEM wave. For the frequency to go right down to zero, the guide must be able to sustain a *static* field, and this requires at least two separate conductors. Familiar examples are the two-wire transmission line and the coaxial cable.

From a consideration of the wave equation, we can show that the fields in a TEM wave are closely related to static fields. For in view of the relation of K to ω, all field components obey

$$\left(\frac{\partial^2}{\partial x^2} + \frac{\partial^2}{\partial y^2}\right)f = 0 \tag{14-65}$$

Thus in any transverse plane the electric field looks like an electrostatic field subject to the boundary condition that each conductor is an equipotential. There is a potential difference between the conductors, and the entire pattern is modulated by the wave factor $e^{i(Kz-\omega t)}$. This results in longitudinal currents in the conductor surfaces, which gave rise to the transverse magnetic field.

Example 14-20

A *strip line* consists of a pair of parallel conductive ribbons separated by a spacing that is small compared with their width. Describe the TEM wave in such a structure.

Solution   Except for the fringe field near the edges (which we neglect) the static electric field is uniform and perpendicular to the ribbon planes, that is,

$$\mathbf{E}_{\text{static}} = E_0\hat{\mathbf{j}}$$

Thus, for the TEM wave,

$$\mathbf{E}(z, t) = E_0\hat{\mathbf{j}}e^{iKz-i\omega t} \tag{14-66a}$$

Thus, from the Maxwell equation,

$$\mathbf{B}(z, t) = \frac{1}{i\omega}\,\text{curl }\mathbf{E}$$

$$= \frac{K}{\omega}(-\hat{\mathbf{i}})E_0 e^{iKz-i\omega t} \tag{14-66b}$$

Strip lines are extensively used for signal transmission between elements of high-speed semiconductor chips. They have the advantages of easy fabrication by film deposition methods and of maximum signal propagation speed. The major disadvantage is that the open sides permit some "leakage" of energy and interference of extraneous signals.

Example 14-21

Describe the TEM wave on a *coaxial line*, which consists of two coaxial cylindrical conductors.

Solution   The static electric field is radial (Figure 14-19):

$$\mathbf{E}_{\text{static}} = \frac{E_0}{s}\hat{\mathbf{a}}_s \qquad \text{for } a \le s \le b$$

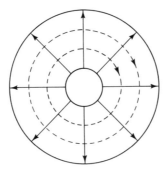

Figure 14-19   Transverse field pattern for TEM wave in a coaxial line. Solid lines: **E**; dashed lines: **H**.

Thus the TEM fields are

$$\mathbf{E}(s, z, t) = \frac{E_0 \hat{\mathbf{a}}_s}{s} e^{iKz - i\omega t} \tag{14-67a}$$

$$\mathbf{B}(s, z, t) = \frac{K}{\omega} \frac{E_0 \hat{\mathbf{a}}_\phi}{s} e^{iKz - i\omega t} \tag{14-67b}$$

Coaxial cables are very widely used for signal transmission at frequencies below the microwave range. They can be made quite flexible by using wire braid for the outer conductor, and are then nearly as convenient as ordinary two-wire power lines. They offer the advantage that the signals are very thoroughly shielded from outside interference. At higher frequencies, however, losses in the dielectric spacers become significant, and waveguides are generally preferred.

14-4 RESONANT CAVITIES

We now consider the wave fields that can exist in a finite region of space completely surrounded by an equipotential (i.e., ideal conductor) boundary. In principle, the region may be of any shape whatever, but we shall deal only with rectangular prisms. These may be considered as sections of hollow rectangular waveguide closed off by conductive end plates. There are then additional boundary conditions at the end plates (say, $z = 0$ and $z = d$), namely that E_x and E_y must be zero. Hence the fields can no longer have an e^{iKz} dependence, since this factor cannot vanish. The only other possibility consistent with the wave equation is a sinusoidal dependence, $\sin \gamma z$ or $\cos \gamma z$. This means that Eqs. (14-47) need modification: γ^2 instead of K^2 on the left and $\partial/\partial z$ instead of iK on the right. For example, instead of Eq. (14-47a) we now have

$$(\mu\epsilon\omega^2 - \gamma^2)E_x = \frac{\partial^2 E_z}{\partial z \, \partial x} + i\omega\mu\frac{\partial H_z}{\partial y} \tag{14-68}$$

Thus it can be seen that the additional boundary conditions require the following:

TE modes ($E_z = 0$): $\dfrac{\partial H_z}{\partial y} = 0$ at $z = 0, d$

TM modes ($H_z = 0$): $\dfrac{\partial E_z}{\partial z} = 0$ at $z = 0, d$

Since the other boundary conditions are unchanged, we may combine these with Eqs. (14-53) and (14-54) to obtain the wave functions.

$$\text{TE:}\quad H_z = H_{z0}\left(\cos\frac{n\pi}{a}x\right)\left(\cos\frac{m\pi}{b}y\right)\left(\sin\frac{p\pi}{d}z\right)e^{-i\omega t} \tag{14-69a}$$

$$\text{TM:}\quad E_z = E_{z0}\left(\sin\frac{n\pi}{a}x\right)\left(\sin\frac{m\pi}{b}y\right)\left(\cos\frac{p\pi}{d}z\right)e^{-i\omega t} \tag{14-69b}$$

where n, m, and p are integers.

Resonant Frequencies

On putting each of the foregoing forms back into the wave equation, we find in both cases:

$$-\left(\frac{n\pi}{a}\right)^2 - \left(\frac{m\pi}{b}\right)^2 - \left(\frac{p\pi}{d}\right)^2 - \mu\epsilon\omega^2 = 0$$

or

$$\omega_{nmp} = c\sqrt{\left(\frac{n\pi}{a}\right)^2 + \left(\frac{m\pi}{b}\right)^2 + \left(\frac{p\pi}{d}\right)^2} \tag{14-70}$$

Thus for each mode (n, m, p), the field can exist at *one single frequency* only. This is characteristic of any undamped resonant system, such as a frictionless pendulum or tuning fork. For the cavity fields, there are many different resonant modes, depending on the choice of integers n, m, and p, and on the choice of directions to designate as x, y, and z in Eqs. (14-69).

Field Patterns

As in waveguides, all transverse field components can be obtained from the longitudinal one. The modified forms of Eqs. (14-47), as exemplified by Eq. (14-68), must, of course, be used.

Example 14-22

Write the equations for the transverse components of a TE mode in a rectangular cavity.

Solution   We write Eq. (14-69a) concisely as

$$H_z = (\cos \alpha x)(\cos \beta y)(\sin \gamma z) \tag{14-71a}$$

Then Eq. (14-68) and its three companions give

$$(\mu\epsilon\omega^2 - \gamma^2)E_x = i\omega\mu(\cos \alpha x)(-\beta \sin \beta y)(\sin \gamma z) \tag{14-71b}$$

$$(\mu\epsilon\omega^2 - \gamma^2)E_y = -i\omega\mu(-\alpha \sin \alpha x)(\cos \beta y)(\sin \gamma z) \tag{14-71c}$$

$$(\mu\epsilon\omega^2 - \gamma^2)H_x = (-\alpha \sin \alpha x)(\cos \beta y)(\gamma \cos \gamma z) \tag{14-71d}$$

$$(\mu\epsilon\omega^2 - \gamma^2)H_y = (\cos \alpha x)(-\beta \sin \beta y)(\gamma \cos \gamma z) \tag{14-71e}$$

Example 14-23

What is the lowest resonant frequency in a cavity $3 \times 4 \times 5$ cm?

Solution   We first have to find the smallest set of integers that give a nonvanishing field. First consider TE modes. From Eq. (14-71a), γ cannot be zero, otherwise H_z, and with it all the other components, will vanish. Furthermore, from Eqs. (14-71b) and (14-71c), we cannot have both α and β equal to zero; otherwise, all components of E will vanish. Thus $(1, 0, 1)$ or $(0, 1, 1)$ are the lowest possible TE modes. For TM, similar reasoning discloses that none of the integers can be zero.

To find the lowest frequency, Eq. (14-70) shows that we should associate the nonzero integers with the largest dimensions. Thus

$$\omega_{0,1,1} = c\sqrt{\left(\frac{\pi}{4}\right)^2 + \left(\frac{\pi}{5}\right)^2}$$

$$\cong 3.02 \times 10^{10} \text{ s}^{-1}$$

Example 14-24

Write the field components of the $TE_{0,1,1}$ mode of Example 14-23.

Solution   In a very abbreviated notation, Eqs. (14-71) boil down to

$$H_z = (1)(\cos)(\sin)$$
$$E_x = (1)(\sin)(\sin)$$
$$E_y = 0$$
$$H_x = 0$$
$$H_y = (1)(\sin)(\cos)$$

A rough sketch of this field pattern is shown in Figure 14-20.

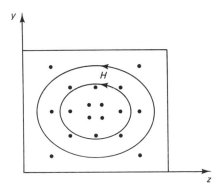

Figure 14-20   Field pattern in TE_{011} cavity mode (schematic).

Quality Factor Q

The perfectly sharp resonances that we have been discussing were the result of the use of ideal conductor boundary relations. In a cavity with real metal walls, there will be field penetration and energy loss just as in waveguides. To describe the effects of this energy loss, it is convenient to use the analogy of a damped mechanical oscillator, which we discussed at length in Section 4-6. If we define

$$Q \equiv \omega_r \tau \tag{14-72}$$

where τ is the collision time, then the equation of motion, Eq. (4-36), can be written

$$\ddot{x} + \omega_r^2 x + \frac{\omega_r}{Q}\dot{x} = \frac{f_0}{m}e^{-i\omega t} \tag{14-73}$$

Here f_0 represents the amplitude of the "driving force"; in the case of a cavity, this would be the amplitude of an electromagnetic wave fed in through a small aperture. The meaning of Q is thus seen to be

$$Q = \frac{\text{amplitude of restoring force}}{\text{amplitude of damping force}} \qquad (14\text{-}74a)$$

$$= \frac{\text{stored energy}}{\text{energy dissipated per radian of oscillation}} \qquad (14\text{-}74b)$$

The amplitude of oscillation is given by

$$x_0 = \frac{f_0}{m} \frac{1}{\omega_r^2 - \omega^2 - i\omega\omega_r/Q}$$

$$|x_0|^2 = \left(\frac{f_0}{m}\right)^2 \frac{1}{(\omega_r^2 - \omega^2)^2 + \omega^2\omega_r^2/Q^2} \qquad (14\text{-}75)$$

If Q is large, $|x_0|$ has its maximum at a frequency ω_m very near to ω_r. By differentiation we find that

$$\omega_m^2 = \omega_r^2\left(1 - \frac{1}{2Q^2}\right) \qquad (14\text{-}76)$$

Thus, for frequencies close to the resonance, Eq. (14-75) may be written approximately as

$$|x_0|^2 = \left(\frac{f_0}{m}\right)^2 \frac{1}{(\omega_r - \omega)^2(\omega_r + \omega)^2 + \omega^2\omega_r^2/Q^2}$$

$$\cong \left(\frac{f_0}{m}\right)^2 \frac{1}{(\omega_r - \omega)^2(2\omega_r)^2 + \omega_r^4/Q^2}$$

$$= \left(\frac{f_0}{2m\omega_r}\right)^2 \frac{1}{(\Delta\omega)^2 + \omega_r^2/4Q^2} \qquad (14\text{-}77)$$

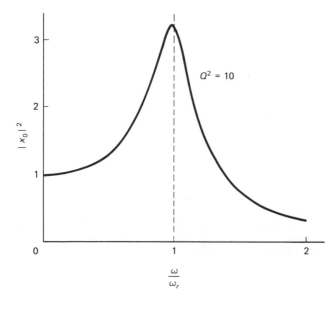

$Q^2 = 10$

Figure 14-21 Shape of a damped resonance. A rather low value of Q has been used to emphasize the shift of the peak to a frequency slightly below the undamped resonant frequency.

The half-maximum points occur at

$$\Delta\omega = \pm\frac{\omega_r}{2Q} \tag{14-78}$$

and Q is a measure of the *relative narrowness* of the resonance,

$$Q = \frac{\omega_r}{2\,\Delta\omega} \cong \frac{\text{resonant frequency}}{\text{full width at half maximum}} \tag{14-79}$$

The shape of the resonance line is indicated in Figure 14-21.

In practical cavities for microwave frequencies, Q values of several thousand are common.

14-5 DIELECTRIC WAVEGUIDES (FIBER OPTICS)

A method that has developed strongly in recent years is the use of dielectric rods to guide waves. Conditions are chosen so that the wave is evanescent outside the rod, and therefore no energy is radiated away. When the frequency is in or near the visible range, the method is called *fiber optics*.

To illustrate the principle, we shall simplify the geometry to a single slab or strip line with boundaries at $x = \pm a/2$. The wave functions must thus be

1. Sinusoidal in region 1 ($|x| < a/2$)
2. Exponentially decaying with respect to $|x|$ in region 2 ($|x| > a/2$)

It is convenient to further subdivide the modes into two more classes: even (cosine) and odd (sine) dependence on x. One example would be TE (even) modes

$$H_z^{(1)} = \cos \alpha x \, e^{iKz} \tag{14-80a}$$

$$H_z^{(2)} = Ce^{-l|x|}e^{iKz} \tag{14-80b}$$

Since both of these must satisfy the wave equations for their own domains,

$$-\alpha^2 - K^2 + \omega^2\mu_1\epsilon_1 = 0 \tag{14-81a}$$

$$l^2 - K^2 + \omega^2\mu_2\epsilon_2 = 0 \tag{14-81b}$$

Subtracting these gives

$$\alpha^2 + l^2 = \omega^2\Delta \tag{14-82}$$

where

$$\Delta \equiv \mu_1\epsilon_1 - \mu_2\epsilon_2$$

The other field components are obtained from Eqs. (14-47) which, remember, are valid whenever we have e^{iKz} dependence. Thus

$$E_y^{(1)} = \frac{-i\omega\mu}{\alpha^2}(-\alpha \sin \alpha x)e^{iKz} \tag{14-83a}$$

$$E_y^{(2)} = \frac{-i\omega\mu}{-l^2}(-lCe^{-l|x|})e^{iKz} \tag{14-83b}$$

The boundary relations at $x = a/2$ then are

$$\cos \frac{\alpha a}{2} = Ce^{-la/2} \qquad (H_z \text{ continuous}) \qquad (14\text{-}84\text{a})$$

$$\frac{1}{\alpha} \sin \frac{\alpha a}{2} = -\frac{C}{l} e^{-la/2} \qquad (E_y \text{ continuous}) \qquad (14\text{-}84\text{b})$$

Dividing these, we get

$$-\alpha \cot \frac{\alpha a}{2} = l = \sqrt{\omega^2 \Delta - \alpha^2} \qquad (14\text{-}84\text{c})$$

This is a transcendental equation for α as function of ω. A graphical solution, plotting both sides as functions of α, is illustrated in Figure 14-22.

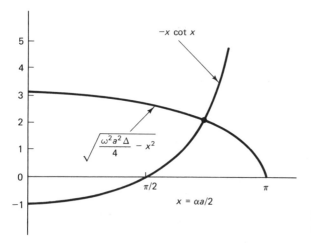

Figure 14-22 Graphical solution of Eq. (14-84c).

The other three mode types are handled in much the same way, and are left for the problems. The overall results may be stated quite concisely. Let

$$X = \frac{\alpha a}{2} \qquad A = a\omega \frac{\sqrt{\Delta}}{2}$$

Then the four transcendental equations are

$$\frac{\sqrt{A^2 - X^2}}{X} = \begin{cases} -\cot X & \text{for TE (even)} \\ \tan X & \text{for TE (odd)} \\ -\dfrac{\epsilon_2}{\epsilon_1} \cot X & \text{for TM (even)} \\ \dfrac{\epsilon_2}{\epsilon_1} \tan X & \text{for TM (odd)} \end{cases} \qquad (14\text{-}85)$$

For each class, there may be 0, 1, 2, . . . solutions, depending on the parameter A. There is always a cutoff frequency below which no solutions are possible, then a range with only one, then two, and so on. In this respect dielectric waveguides behave much like hollow metal ones.

For fiber optics, circular filaments are used, so the detailed mathematics is rather more complicated, but the principles are the same. A tremendous effort has gone into the development of low-loss glass filaments, and these are coming into wide use in communication systems.

PROBLEMS

14-1. If a plane-polarized wave of amplitude E_0 in vacuum is incident normally on a dielectric whose index of refraction is n, what are the electric and magnetic field amplitudes in the dielectric?

14-2. If a wave of power density 1 kW/m² is incident normally on a sheet of glass (index of refraction 1.5), what is the magnitude of the time-average Poynting vector in the glass?

14-3. Can the Brewster effect occur on reflection from a less-dense medium (i.e., $n_2 < n_1$)?

14-4. Work out the Fresnel relations for total internal reflection of a p-polarized incident wave.

14-5. A wave of angular frequency ω and amplitude E_0 in vacuum is incident normally on a sheet of metal (conductivity g, permeability μ). Calculate the (complex) electric and magnetic field amplitudes just inside the surface.

14-6. A wave polarized perpendicular to the plane of incidence is incident at angle θ from vacuum onto a sheet of ideal conductor. Calculate the total electric and magnetic fields and the time-average Poynting vector in the space, and the induced charge and current densities in the conductive surface. Take the direction of incidence to lie in the xz plane with positive components in both the x and z directions.

14-7. Repeat Problem 14-6 for the case of polarization parallel to the plane of incidence.

14-8. A wave is incident normally on an ideal conductive plane. Calculate the force per unit area on the plane and compare it with the momentum transferred from the incident to the reflected wave. *Hint:* The force is exerted by the magnetic field on the induced current.

14-9. A beam of light of angular frequency ω is incident normally on a dielectric film of index n and thickness d. Calculate the reflection coefficient. *Hint:* There are waves traveling in both directions within the film, as well as reflected and transmitted ones. Thus there are five waves to satisfy the two sets of boundary conditions.

14-10. A beam of light of angular frequency ω is incident normally from a medium of index n_1 onto a dielectric film of index n_2 and thickness d backed by an infinite medium of index n_3. Find the values of n_2 and d such that there is no reflected wave.

14-11. For a rectangular waveguide of transverse dimensions 5 cm × 4 cm, enumerate all modes of which the cutoff frequency ω_c does not exceed $1.5 \times 10^{10}\,\pi$ rad/s.

14-12. For a rectangular cavity of dimensions 5 cm × 4 cm × 3 cm, enumerate all modes for which the resonant frequency does not exceed $(3\pi/4) \times 10^{10}$ Hz.

14-13. Verify the last three parts of Eq. (14-85).

chapter 15

Radiation

15-1 RETARDED POTENTIALS

Once again it would be a good idea for the reader to review Sections 7-1 and 7-2, especially the derivations of Eqs. (7-19) and (7-20). In Chapters 13 and 14 we considered the homogeneous form of these equations, in which the source terms on the right were set either equal to zero or proportional to the wave field. We now deal with the inhomogeneous form; the terms on the right are to be regarded as some *specified* functions of the coordinates and the time.

We start with the simplest imaginable geometry, namely a point charge fixed in position, but of varying magnitude: $q = q(t)$. In view of the current continuity relation, Eq. (2-6), this is not physically realizable, but it gives us a good starting point for the theory. We seek the Lorentz gauge scalar potential as a function of field point position and time. From the symmetry of the problem U can at any instant depend only on the radial distance from the charge; that is, $U = U(r, t)$. Thus it is natural to write Eq. (7-20) in spherical coordinates, namely

$$\frac{1}{r^2}\frac{\partial}{\partial r}\left(r^2 \frac{\partial U}{\partial r}\right) - \mu\epsilon \frac{\partial^2 U}{\partial t^2} = \frac{1}{\epsilon}\rho(r, t)$$

$$= 0 \qquad \text{for } r \neq 0 \qquad (15\text{-}1)$$

Now if q is constant, the second term vanishes and the solution is the static point-charge potential $U(r) = q/4\pi\epsilon_0 r$. Since the general time-dependent solution must include this case, we are led to try a solution of the form

$$U(r, t) = r^{-1}f(r, t) \qquad (15\text{-}2a)$$

$$\frac{\partial U}{\partial r} = -r^{-2}f + r^{-1}\frac{\partial f}{\partial r} \qquad (15\text{-}2b)$$

$$\nabla^2 U = r^{-2}\frac{\partial}{\partial r}\left(r^2 \frac{\partial U}{\partial r}\right) = r^{-1}\frac{\partial^2 f}{\partial r^2} \qquad (15\text{-}2c)$$

Thus the equation for f becomes

$$r^{-1}\left(\frac{\partial^2 f}{\partial r^2} - \mu\epsilon\frac{\partial^2 f}{\partial t^2}\right) = 0 \qquad \text{for } r \neq 0 \tag{15-3}$$

This has the form of a homogeneous one-dimensional wave equation, Eq. (13-5). The general solution was worked out in the few lines following that equation. We write it in the form

$$f(r, t) = f_1\left(t - \frac{r}{c}\right) + f_2\left(t + \frac{r}{c}\right) \tag{15-4}$$

Consider the first term. For the *static* case, in which $q(t) = q_0$, we know that f must reduce to $q_0/4\pi\epsilon$. Hence the appropriate form for the time-varying point charge is

$$f(r, t) = \frac{q(t_r)}{4\pi\epsilon} \qquad U(r, t) = \frac{q(t_r)}{4\pi\epsilon r} \tag{15-5}$$

where

$$t_r = t - \frac{r}{c} \tag{15-6}$$

[It might be thought that terms involving time derivatives of q could also be present. However, at points very close to the charge, where $t_r \cong t$, the solution must reduce to $q(t)/4\pi\epsilon_0 r$, so there can be no other terms.] Thus the potential at any point at time t depends on the source strength at an *earlier* time (the so-called *retarded time*) t_r. The amount of retardation is just the *transit time of a light signal from the source point to the field point.*

Now what about the second term of Eq. (15-4)? This would make the potential appear to depend on the source strength at a *later* ("advanced") time $t + r/c$. Since the field point has no way of "knowing" what the source is going to be doing at a later time, this term violates the principle of casuality and is ruled out.

We can easily generalize this solution to an arbitrary time-varying charge distribution $\rho(\mathbf{r}, t)$. Since each source element contributes its own retarded potential at any field point,

$$U(\mathbf{r}, t) = \frac{1}{4\pi\epsilon} \iiint \frac{\rho(\mathbf{r}', t_r)}{|\mathbf{r} - \mathbf{r}'|} d^3r' \tag{15-7}$$

where

$$t_r(t, \mathbf{r}, \mathbf{r}') = t - \frac{|\mathbf{r} - \mathbf{r}'|}{c} \tag{15-8}$$

Note that each source element is counted at its own retarded time.

In a similar way we determine the Lorentz gauge vector potential for an arbitrary time-varying current distribution $\mathbf{J}(\mathbf{r}, t)$. Since each component of Eq. (7-19) has the same form as Eq. (7-20),

$$\mathbf{A}(\mathbf{r}, t) = \frac{\mu}{4\pi} \iiint \frac{\mathbf{J}(\mathbf{r}', t_r)}{|\mathbf{r} - \mathbf{r}'|} d^3r' \tag{15-9}$$

The last two results are, of course, closely related. The charge and current distributions are not independent but are related by the equation of current continuity, Eq. (2-6). As a consequence, **A** and U are related by the *Lorentz condition*, Eq. (7-18). In many cases, once U is known it is easier to find **A** from the latter than by direct evaluation of the integral.

Although the retarded potentials are easy to write, the calculation of the field from them is considerably more complicated than in the static case. The fields involve derivatives of the potentials with respect to the field-point coordinates. In the static case these operate only on the *distance function* $1/R$, but in the retarded case, the *time* t_r is also a function of the field-point coordinates. For example, we would have

$$\frac{\partial}{\partial x}\left[\frac{\rho(\mathbf{r}', t_r)}{R}\right] = \rho(\mathbf{r}', t_r)\frac{\partial}{\partial x}\left(\frac{1}{R}\right) + \frac{1}{R}\left(\frac{\partial \rho}{\partial t}\right)_{t=t_r}\frac{\partial t_r}{\partial R}\frac{\partial R}{\partial x} \tag{15-10}$$

The first term is just the same as in static problems. The second, however, is unique to time-varying situations and leads to characteristically different forms of fields, as we shall illustrate in the next sections.

15-2 ELECTRIC DIPOLE RADIATION

The isolated time-varying point charge is, as we noted, not a physically realistic source since it violates current continuity. The simplest "realistic" source is an oscillating dipole which consists of a pair of harmonically varying point charges $\pm q_0 e^{-i\omega t}$, connected by a wire of length l, as indicated in Figure 15-1. For this, Eq. (15-7) becomes

$$U = \frac{q_0 e^{-i\omega(t-r_b/c)}}{4\pi\epsilon r_b} - \frac{q_0 e^{-i\omega(t-r_a/c)}}{4\pi\epsilon r_a} \tag{15-11}$$

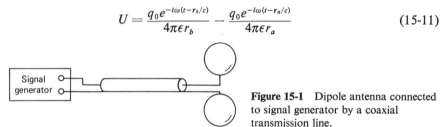

Figure 15-1 Dipole antenna connected to signal generator by a coaxial transmission line.

Signal generator

To approximate this for large distances from the source, we note by Figure 15-2 that

$$r_a \cong r + \tfrac{1}{2}lC \qquad r_b \cong r - \tfrac{1}{2}lC \qquad \text{(where } C \equiv \cos\theta)$$

Thus

$$U \cong \frac{q_0 e^{-i\omega(t-r/c)}}{4\pi\epsilon r}\left[\frac{e^{-(1/2)iKlC}}{1 - \tfrac{1}{2}lC/r} - \frac{e^{(1/2)iKlC}}{1 + \tfrac{1}{2}lC/r}\right]$$

To simplify further, each factor in the brackets may be expressed as the first two terms of a Taylor series: $e^x \cong 1 + x$, $(1 + x)^{-1} \cong 1 - x$, giving

$$U = \frac{q_0 e^{-i\omega(t-r/c)}}{4\pi\epsilon r} l \cos\theta\left(\frac{1}{r} - \frac{2\pi i}{\lambda}\right) \tag{15-12}$$

where $\lambda = 2\pi/K = 2\pi c/\omega = $ free-space wavelength.

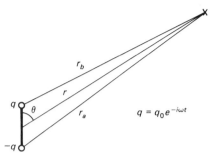

Figure 15-2 Geometry for dipole antenna, Eq. (15-11).

Note that if $\omega = 0$ this reduces to the electrostatic dipole potential, the second term of Eq. (1-26). Note also that the time retardation from the origin to the field point appears explicitly in the potential.

For the retarded vector potential we note that in the short straight wire connecting the two charges there is a current $I = \dot{q}$, so there is an element of chargeamentum at the origin of amount equal to

$$-i\omega q_0 e^{-i\omega t}\hat{\mathbf{k}} = I_0 e^{-i\omega t}\hat{\mathbf{k}} \qquad (15\text{-}13)$$

Inserting this into Eq. (15-9) gives

$$\mathbf{A} = \frac{I_0 l\mu e^{-i\omega(t-r/c)}}{4\pi r}\,\hat{\mathbf{k}} \qquad (15\text{-}14)$$

Since we neglected the length of the current line and assumed that the current was constant along it, the potential is exact only for a point dipole.

Having the potentials, we now proceed to calculate the fields. It is best to carry out the calculations in spherical coordinates; hence we put $\hat{\mathbf{k}} = \hat{\mathbf{a}}_r \cos\theta - \hat{\mathbf{a}}_\theta \sin\theta$. Then

$$\mathbf{B} = \text{curl}\,\mathbf{A} = \left[\frac{1}{r}\frac{\partial}{\partial r}(rA_\theta) - \frac{1}{r}\frac{\partial A_r}{\partial \theta}\right]\hat{\mathbf{a}}_\phi$$

$$= \frac{I_0 l\mu}{4\pi}\left\{\frac{1}{r}\frac{\partial}{\partial r}[-\sin\theta\, e^{-i(\cdot)}] - \frac{1}{r}\frac{\partial}{\partial \theta}\left[\frac{1}{r}\cos\theta\, e^{-i(\cdot)}\right]\right\}\hat{\mathbf{a}}_\phi$$

On carrying out the differentiations, we obtain

$$\mathbf{B} = \frac{I_0 l\mu}{4\pi r}\sin\theta\left[\frac{1}{r} - \frac{2\pi i}{\lambda}\right]e^{-i\omega(t-r/c)}\,\hat{\mathbf{a}}_\phi \qquad (15\text{-}15)$$

Note that very close to the dipole the first term in the brackets is dominant and the field looks like that due to a steady current element. But this term dies off as $1/r^2$ and at distances of many wavelengths the second term is dominant. This term came, of course, from the derivative of the retardation factor $e^{i\omega r/c}$.

For the calculation of **E** we use Eq. (7-11).

$$-\frac{\partial \mathbf{A}}{\partial t} = i\omega \mathbf{A} = \frac{\omega^2 q_0 l \mu}{4\pi r} e^{-i\omega(t-r/c)}(\hat{\mathbf{a}}_r \cos\theta - \hat{\mathbf{a}}_\theta \sin\theta)$$

$$-\text{grad } U = -\hat{\mathbf{a}}_r \frac{\partial U}{\partial r} - \frac{\hat{\mathbf{a}}_\theta}{r}\frac{\partial U}{\partial \theta}$$

$$= -\frac{q_0 l}{4\pi\epsilon}\left\{\hat{\mathbf{a}}_r \cos\theta \frac{\partial}{\partial r}\left[\left(\frac{1}{r^2} - \frac{2\pi i}{\lambda r}\right)e^{-i\omega(t-r/c)}\right]\right.$$

$$\left. + \hat{\mathbf{a}}_\theta \frac{1}{r}\left(\frac{1}{r^2} - \frac{2\pi i}{\lambda r}\right)(-\sin\theta)e^{-i\omega(t-r/c)}\right\}$$

$$= -\frac{q_0 l}{4\pi\epsilon}\left\{\hat{\mathbf{a}}_r \cos\theta\left[\frac{-2}{r^3} + \frac{2\pi i}{\lambda r^2} + \frac{2\pi i}{\lambda}\left(\frac{1}{r^2} - \frac{2\pi i}{\lambda r}\right)\right]\right.$$

$$\left. - \hat{\mathbf{a}}_\theta \sin\theta\left[\frac{1}{r^3} - \frac{2\pi i}{\lambda r^2}\right]\right\}e^{-i\omega(t-r/c)}$$

Putting these together [recalling from Eq. (13-6) that $\mu = 1/\epsilon c^2$], we get

$$\mathbf{E} = \frac{q_0 l}{4\pi\epsilon}\left\{\hat{\mathbf{a}}_r \cos\theta\left[\frac{2}{r^3} - \frac{4\pi i}{\lambda r^2}\right] + \hat{\mathbf{a}}_\theta \sin\theta\left[\frac{1}{r^3} - \frac{2\pi i}{\lambda r^2} - \frac{4\pi^2}{\lambda^2 r}\right]\right\}e^{-i\omega(t-r/c)} \quad (15\text{-}16)$$

Here there are terms falling off as r^{-1}, r^{-2}, and r^{-3}. For points in the *near zone* ($r \ll \lambda$), the latter dominate, so

$$\mathbf{E} \simeq \frac{q_0 l}{4\pi\epsilon}\frac{1}{r^3}(2\hat{\mathbf{a}}_r \cos\theta + \hat{\mathbf{a}}_\theta \sin\theta)e^{-i\omega t} \quad (15\text{-}17)$$

This is exactly the form of an electrostatic dipole field, Eq. (1-27b).

For the *far zone* ($r \gg \lambda$), we retain only the terms that fall off the most slowly. For convenience, we write the coefficients in terms of the amplitude of the oscillating dipole moment, $p_0 = q_0 l = I_0 l/(-i\omega)$. Then

$$\mathbf{H} = -\frac{\omega p_0}{2\lambda r}\sin\theta\, e^{i(Kr-\omega t)}\,\hat{\mathbf{a}}_\phi \quad (15\text{-}18a)$$

$$\mathbf{E} = -\frac{\pi p_0}{\epsilon\lambda^2 r}\sin\theta\, e^{i(Kr-\omega t)}\,\hat{\mathbf{a}}_\theta \quad (15\text{-}18b)$$

Both fields contain a wave factor propagating in the outward radial direction. The fields are perpendicular to this direction and to each other. Furthermore, the quotient of their amplitudes at any point is

$$\frac{H_0}{E_0} = \frac{\omega p_0}{2\lambda}\frac{\epsilon\lambda^2}{\pi p_0} = \frac{\epsilon\omega\lambda}{2\pi} = \epsilon c = \sqrt{\frac{\epsilon}{\mu}} \quad (15\text{-}19)$$

which is the same as that in Eq. (13-15). Thus in any small region the fields look exactly like those of an electromagnetic plane wave. The only difference is that the amplitudes do not remain constant in the direction of propagation but fall off slowly as $1/r$. This is because the wavefronts are spherical rather than truly planar.

Example 15-1

Calculate the energy flow at any point in the far zone.

Solution The time-average Poynting vector is

$$\bar{\mathbb{S}} = \tfrac{1}{2}\,\text{Re}\,(\mathbf{E} \times \mathbf{H}^*)$$

$$= \frac{\pi\omega}{4\epsilon}\,\frac{p_0^2}{\lambda^3}\,\frac{\sin^2\theta}{r^2}\,\hat{\mathbf{a}}_r$$

$$= \frac{1}{32\pi^2}\sqrt{\frac{\mu}{\epsilon}}\,\frac{\omega^4}{c^2}\,p_0^2\,\frac{\sin^2\theta}{r^2}\,\hat{\mathbf{a}}_r \qquad (15\text{-}20)$$

Thus the radiation goes radially outward and its intensity depends on direction through the factor $\sin^2\theta$. It is strongest in the equatorial plane and zero along the direction of the dipole.

Example 15-2

Calculate the total power radiated. Cast the result in terms of the current amplitude.

Solution Integrate the Poynting vector magnitude over a sphere:

$$\mathcal{P} = \frac{1}{32\pi^2}\sqrt{\frac{\mu}{\epsilon}}\,\frac{\omega^4}{c^2}\,p_0^2 \int_0^\pi \frac{\sin^2\theta}{r^2}\,2\pi r^2 \sin\theta\,d\theta$$

$$= \frac{1}{12\pi}\sqrt{\frac{\mu}{\epsilon}}\,\frac{\omega^4}{c^2}\,p_0^2 \qquad (15\text{-}21a)$$

$$= \frac{\pi}{3}\sqrt{\frac{\mu}{\epsilon}}\left(\frac{l}{\lambda}\right)^2 I_0^2 \qquad (15\text{-}21b)$$

This result has an interesting physical interpretation. Dimensionally, l/λ is dimensionless and $\sqrt{\mu/\epsilon}$ has the dimensions of a *resistance*. Thus Eq. (15-21b) may be written as

$$\mathcal{P} = \tfrac{1}{2}R_{\text{rad}}I_0^2$$

where

$$R_{\text{rad}} = \frac{2\pi}{3}\left(\frac{l}{\lambda}\right)^2\sqrt{\frac{\mu}{\epsilon}} \qquad (15\text{-}22)$$

The factor of $\tfrac{1}{2}$ comes from the power theorem, Eq. (8-46), since we used the complex exponential form $I_0 e^{-i\omega t}$ for the time-varying current. R_{rad} is the *radiation resistance* of the dipole. It is the resistance presented to the terminals of a transmission line from a signal generator driving the oscillating current, as illustrated in Figure 15-1. As far as measurements at the generator are concerned, there is no way of telling whether the load is an actual resistor dissipating the energy or the dipole radiating it away.

15-3 MAGNETIC DIPOLE RADIATION

The next step in complexity of source distributions is an oscillating magnetic dipole, which consists of a small loop of harmonically varying current $I_0 e^{-i\omega t}$. For simplicity, we take the loop to be planar and circular and choose axes so that the field point has

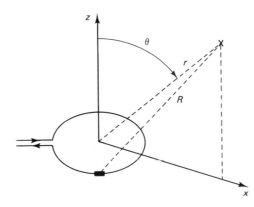

Figure 15-3   Magnetic dipole antenna and coordinate axes used in calculations of vector potential.

coordinates $x, 0, z$, as shown in Figure 15-3. Then everything proceeds as in Section 2-6 except that we use the retarded value of the time-varying current at every source point. Thus Eq. (15-9) becomes

$$A(x, 0, z) = \frac{\mu I_0}{4\pi} \int_0^{2\pi} \frac{e^{-i\omega(t - R/c)} a \, d\phi' \cos \phi'}{R} \hat{j}$$

where

$$R = (r^2 + a^2 - 2ax \cos \phi')^{1/2}$$

$$\cong r\left(1 - ax \frac{\cos \phi'}{r^2}\right)$$

$$A(x, 0, z) = \frac{\mu I_0 a}{4\pi r} e^{-i\omega(t - r/c)} \hat{j} \int_0^{2\pi} \frac{e^{-i \frac{2\pi}{\lambda} \frac{ax \cos \phi'}{r}}}{1 - \frac{ax \cos \phi'}{r^2}} \cos \phi' \, d\phi'$$

To evaluate the integral within the spirit of the approximations we again expand both factors in Taylor series, so the integral becomes

$$\int_0^{2\pi} \left[1 - i\left(\frac{2\pi a}{\lambda}\right) \frac{x}{r} \cos \phi'\right] \left(1 + \frac{ax}{r^2} \cos \phi'\right) \cos \phi' \, d\phi'$$

$$= \frac{ax}{r}\left(\frac{1}{r} - \frac{2\pi i}{\lambda}\right) \int_0^{2\pi} \cos^2 \phi' \, d\phi'$$

$$= \frac{\pi ax}{r}\left(\frac{1}{r} - \frac{2\pi i}{\lambda}\right)$$

Thus, now dropping the particular choice of axes that was used only to simplify the calculation,

$$A(r, \theta, \phi) = \frac{\mu}{4\pi r}(\pi a^2 I_0)e^{-i\omega(t - r/c)} \sin \theta \left(\frac{1}{r} - \frac{2\pi i}{\lambda}\right) \hat{a}_\phi \qquad (15\text{-}23)$$

For zero frequency this reduces to the vector potential of the static magnetic dipole. Since there is no charge density at any point the scalar potential is zero.

Example 15-3

Calculate the radiation fields due to the oscillating magnetic dipole.

Solution   Since $U = 0$, $\mathbf{E} = -\dot{\mathbf{A}} = i\omega\mathbf{A}$. For $r \gg \lambda$, this becomes

$$\mathbf{E} = \frac{\mu m_0}{4\pi r} e^{-i\omega(t-r/c)} \sin\theta \left(\frac{-2\pi i}{\lambda}\right)(i\omega) \,\hat{\mathbf{a}}_\phi$$

$$= \frac{\mu m_0}{4\pi r} e^{-i\omega(t-r/c)} \sin\theta \, \frac{\omega^2}{c} \,\hat{\mathbf{a}}_\phi \tag{15-24}$$

$$\mathbf{H} = \mathrm{curl}\, \frac{\mathbf{A}}{\mu}$$

$$= \frac{1}{\mu r^2 \sin\theta} \left(\hat{\mathbf{a}}_r \frac{\partial}{\partial\theta} - r\hat{\mathbf{a}}_\theta \frac{\partial}{\partial r}\right)\left(r\sin\theta\, A_\phi\right)$$

$$= \frac{m_0 e^{-i\omega t}}{4\pi r^2 \sin\theta} \left(\hat{\mathbf{a}}_r \frac{\partial}{\partial\theta} - r\hat{\mathbf{a}}_\theta \frac{\partial}{\partial r}\right)\left[\left(\frac{1}{r} - \frac{2\pi i}{\lambda}\right) e^{2\pi i r/\lambda} \sin^2\theta\right]$$

$$= \frac{m_0}{4\pi} \left[\hat{\mathbf{a}}_r \left(\frac{1}{r^3} - \frac{2\pi i}{\lambda r^2}\right) 2\cos\theta - \hat{\mathbf{a}}_\theta \left(\frac{-1}{r^3} + \frac{2\pi i}{\lambda r^2} + \frac{4\pi^2}{\lambda^2 r}\right)\sin\theta\right] e^{-i\omega(t-r/c)}$$

$$\tag{15-25}$$

For zero frequency (infinite wavelength) this reduces to the field of a static magnetic dipole. On the other hand, for $r \gg \lambda$ it becomes

$$\mathbf{H} = -\frac{m_0}{4\pi r} \frac{\omega^2}{c^2} \sin\theta\, e^{-i\omega(t-r/c)} \,\hat{\mathbf{a}}_\theta \tag{15-26}$$

Thus at large distances $\mathbf{E}$ and $\mathbf{H}$ are again perpendicular to each other and have the proper relationship for an outward-going electromagnetic wave.

Example 15-4

Calculate the radiation resistance of a magnetic dipole.

Solution   From the preceding results the time-average Poynting vector is

$$\bar{\mathbb{S}} = \frac{\mu}{c} \frac{m_0^2}{4\pi^2 r^2} \frac{\omega^4}{c^2} \sin^2\theta\, \hat{\mathbf{a}}_r \tag{15-27}$$

Putting $m_0 = \pi a^2 I_0$, integrating over a sphere, and simplifying, we obtain

$$\mathcal{P} = \frac{2\pi}{3} \sqrt{\frac{\mu}{\epsilon}} \left(\frac{2\pi a}{\lambda}\right)^4 I_0^2 \tag{15-28a}$$

$$R_{\mathrm{rad}} = \frac{4\pi}{3} \sqrt{\frac{\mu}{\epsilon}} \left(\frac{2\pi a}{\lambda}\right)^4 \tag{15-28b}$$

Note that the radiation resistance of the magnetic dipole goes as the fourth power of the ratio of dimension of the structure to the wavelength, whereas for the electric dipole [Eq. (15-22)], it is the second power. Thus for radiation of fairly long wavelength compared with the source dimensions (e.g., radiation of visible light by atoms) the electric dipole is much more effective. However, both types of dipole give the same radiation *pattern*, having a $\sin^2 \theta$ dependence.

It is obvious that there exists an endless hierarchy of elementary "multipole" radiation sources. The next one after the magnetic dipole would be the electric quadrupole, such as a pair of equal but opposite electric dipoles slightly displaced from each other. For this, the radiation resistance is comparable to that of an equal-size magnetic dipole, but the angular pattern is more complicated. Then comes the magnetic quadrupole, weaker by another factor of $(a/\lambda)^2$, and so on. These multipole fields play an important role in the radiation emitted by molecules, atoms, and nuclei, but are too complicated for us to go into here.

15-4 RADIATION FROM ANTENNAS

In the preceding sections we dealt with radiating structures whose dimensions were small compared with the free-space wavelength at the frequency of the oscillating currents. For this reason it was permissible to treat the currents as *uniform* over the structures. We turn now to larger structures, where this simplifying assumption cannot be made. The problem is then an extremely difficult one since the radiated fields depend on the currents and they, in turn, depend in part on the fields. The usual approach is to break the problem into two parts: (1) find the approximately correct current distribution by neglecting the radiation, then (2) calculate the radiation produced by that current distribution. For simplicity, we shall confine ourselves to straight wires fed by a signal generator connected across a gap at the middle of the wire, as in Figure 15-4.

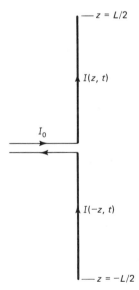

Figure 15-4   Current flow in a linear center-fed antenna.

We start by seeking the form of the variation of current along an *infinitely long* center-fed straight-wire antenna under conditions such that no radiation occurs. To prevent radiation, we may imagine the wire to be surrounded by a conductive cylinder. The complete structure is then a coaxial transmission line so the current variation is

$$I(z, t) = I_0 e^{i(Kz - \omega t)} \tag{15-29}$$

where, as usual,

$$K = \frac{\omega}{c} = \frac{2\pi}{\lambda}$$

Now suppose that the antenna is of finite length L. Since the current must be zero at the ends of the wire, we need a standing-wave rather than the traveling-wave form of Eq. (15-29). This can be made up of two oppositely moving waves. Since the current is fed in from a generator at the center, we must allow for a discontinuity of form there. Hence

$$I(z, t) = (A_1 e^{iKz} + B_1 e^{-iKz})e^{-i\omega t} \quad \text{for} \quad z > 0$$

$$I(z, t) = (A_2 e^{iKz} + B_2 e^{-iKz})e^{-i\omega t} \quad \text{for} \quad z < 0$$

For the current to vanish at $z = L/2$ and $-L/2$,

$$A_1 e^{iKL/2} + B_1 e^{-iKL/2} = 0 \qquad B_1 = -A_1 e^{iKL} \tag{15-30a}$$

and also

$$B_2 = -A_2 e^{-iKL} \tag{15-30b}$$

Now the current at the feed-in point is

$$I_0 e^{-i\omega t} = I(0, t) = (A_1 + B_1)e^{-i\omega t} = (A_2 + B_2)e^{-i\omega t}$$

Thus

$$I_0 = A_1 + B_1 = A_1(1 - e^{iKL})$$

$$= A_1 e^{iKL/2}\left(-2i \sin \frac{KL}{2}\right) \tag{15-31a}$$

and also

$$I_0 = A_2 e^{-iKL/2}\left(2i \sin \frac{KL}{2}\right) \tag{15-31b}$$

This can all be put together concisely in the form

$$I(z, t) = \frac{I_0}{\sin KL/2} \sin K\left(\frac{L}{2} - |z|\right) e^{-i\omega t} \tag{15-32}$$

Examples of this form of current distribution are shown in Figure 15-5.

We can now calculate the vector potential at large distances from the current line. The current is of the form $I(z, t) = I(z)e^{-i\omega t}$, so Eq. (15-9) becomes

$$\mathbf{A}(\mathbf{r}, t) = \frac{\mu}{4\pi} \int \frac{I(z')e^{-i\omega(t - R/c)}}{R} \hat{\mathbf{k}} \, dz'$$

$$= \frac{\mu}{4\pi} e^{-i\omega t} \hat{\mathbf{k}} \int \frac{I(z')e^{iKR}}{R} dz' \tag{15-33}$$

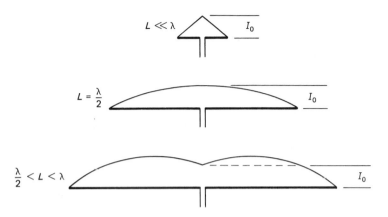

Figure 15-5 Current distributions in center-fed linear antennas, Eq. 15-32.

For points whose distance is large compared with the size of the antenna we may use the approximation

$$R = r - z' \cos \theta + \cdots$$

$$\mathbf{A}(\mathbf{r}, t) = \frac{\mu}{4\pi} e^{i(Kr - \omega t)} \hat{\mathbf{k}} \int I(z') \frac{e^{-iKz' \cos \theta}}{r[1 - (z'/r) \cos \theta + \cdots]} dz'$$

$$\cong \frac{\mu e^{i(Kr - \omega t)}}{4\pi r} \hat{\mathbf{k}} \int I(z') e^{-iKz' \cos \theta} dz' \qquad (15\text{-}34)$$

For the current distribution of Eq. (15-32), the vector potential becomes

$$\mathbf{A}(\mathbf{r}, t) = \frac{\mu I_0}{\sin (KL/2)} \frac{e^{i(Kr - \omega t)}}{4\pi r} \hat{\mathbf{k}} \int_{-L/2}^{L/2} \sin K\left(\frac{L}{2} - |z'|\right) e^{-iKz' \cos \theta} dz' \qquad (15\text{-}35)$$

The integral is explicitly

$$\int_{-L/2}^{0} \sin K\left(\frac{L}{2} + z'\right) e^{-iKz' \cos \theta} dz' + \int_{0}^{L/2} \sin K\left(\frac{L}{2} - z'\right) e^{-iKz' \cos \theta} dz'$$

In the first term we change the name of the dummy variable of integration from z' to $-z'$ to obtain *in toto* (using the shorthand notation $C = \cos \theta$)

$$\int_{0}^{L/2} \sin K\left(\frac{L}{2} - z'\right)(2 \cos [KCz']) dz'$$

$$= \int_{0}^{L/2} \left[\sin K\left(\frac{L}{2} - z' + Cz'\right) + \sin K\left(\frac{L}{2} - z' - Cz'\right)\right] dz'$$

$$= \frac{2[\cos (KL \cos \theta/2) - \cos (KL/2)]}{K \sin^2 \theta} \qquad (15\text{-}36)$$

Having obtained the vector potential, we may now proceed to calculate the magnetic field, that is,

$$\mathbf{B} = \text{curl } \mathbf{A}$$

$$= \frac{\mu I_0 e^{-i\omega t}}{2\pi K \sin (KL/2)} \text{curl} \left[\frac{f(\theta) e^{iKr}}{r} \hat{\mathbf{k}} \right] \tag{15-37}$$

where

$$f(\theta) = \frac{\cos [(KL/2) \cos \theta] - \cos (KL/2)}{\sin^2 \theta} \tag{15-38}$$

The complete calculation of the curl would obviously be a fairly formidable task. However, if we want only the *radiation* field, that is, the part that falls off only as $1/r$, things are not so bad. On looking into the calculations involved, we see that a term of this form can come only from the derivative of the e^{iKr} factor; the other factors may be treated as constants for this purpose. In other words, the term we are interested in is

$$\frac{f(\theta)}{r} \text{curl} (e^{iKr} \hat{\mathbf{k}}) = \frac{f(\theta)}{r} iKe^{iKr} \left(\hat{\mathbf{i}} \frac{\partial r}{\partial y} - \hat{\mathbf{j}} \frac{\partial r}{\partial x} \right)$$

$$= \frac{f(\theta)}{r} iKe^{iKr} \sin \theta \, \hat{\mathbf{a}}_\phi \tag{15-39}$$

Putting Eq. (15-39) back into Eq. (15-37) gives the explicit expression for the magnetic field. The electric field is then proportional (and perpendicular) to it in the usual plane-wave relation, that is,

$$\mathbf{E} = \frac{f(\theta)}{r} i\omega^{iKr} \sin \theta \, \hat{\mathbf{a}}_\theta \tag{15-40}$$

Example 15-5

Sketch the angular distribution of the radiated power for a "half-wave" antenna, that is, one whose length is exactly one-half wavelength.

Solution $L = \lambda/2$; $KL = (2\pi/\lambda)(\lambda/2) = \pi$;

$$f(\theta) = \frac{\cos [(\pi/2) \cos \theta]}{\sin^2 \theta}$$

According to Eq. (15-39), $\mathbf{B}$ and $\mathbf{E}$ are each proportional to $f(\theta) \sin \theta$, and the Poynting vector will therefore be proportional to the square of this, namely

$$\mathcal{S} \propto \left\{ \frac{\cos [(\pi/2) \cos \theta]}{\sin \theta} \right\}^2 \tag{15-41}$$

This function is readily seen to have its maximum at $\theta = \pi/2$ and to be zero at $\theta = 0$ and π. In general shape it is not too different from the function $\sin^2 \theta$ of Eq. (15-20) which describes the angular distribution from a dipole. However, the present one is somewhat more sharply peaked as indicated in Figure 15-6; in other words, more of the power goes out in the equatorial plane.

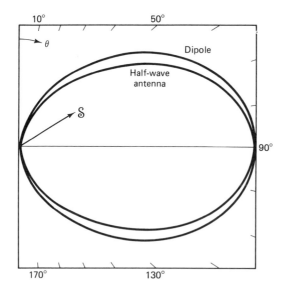

Figure 15-6 Polar plots of radiation patterns from a dipole (outer curve) and a center-fed half-wave antenna.

As the antenna length is increased, the radiation pattern becomes multilobed and quite complicated. However, concentrated single-lobe emission can be obtained by dividing a long antenna into half-wavelength segments with suitable inductors in between to keep all the currents in phase. A parallel array of such antennas can concentrate the energy into a single directional beam, which is obviously useful for point-to-point communication. The detailed calculations are quite messy, for the basic principles are the same as we have been discussing.

15-5 SCATTERING

When an electromagnetic wave impinges on a small body, the charged particles within the body experience forces due to the wave fields and are set into oscillatory motion. They therefore radiate, in accordance with the oscillating multipole moments set up in the body. The result is that energy is extracted from the incident wave and sent out in some angular distribution over the entire 4π solid angle around the body. This is called *scattering*. The effectiveness of a given body as a scatterer is described by its *scattering cross section*, which is defined as

$$\sigma \equiv \frac{\text{total scattered power}}{\text{incident energy flux}} \qquad (15\text{-}42)$$

σ is seen to have the dimensionality of an *area*. It may be thought of as the area of incident wave front required to deliver the power that gets scattered out.

Example 15-6

Calculate the scattering cross section of a free point charge q in vacuum.

Solution   Let the rest position of the particle be at the origin, and the wave be incident along the z direction polarized in the xz plane. The force is mainly due to the electric field of the wave, and the resulting displacement of the particle is

$$x(t) = \frac{qE_0}{m\omega^2} e^{-i\omega t}$$

There is thus an oscillating electric dipole moment whose amplitude is

$$p_0 = qx_0 = \frac{q^2}{m\omega^2} E_0$$

and from Eq. (15-21a) the total power radiated is

$$\mathcal{P} = \frac{1}{12\pi} \sqrt{\frac{\mu_0}{\epsilon_0}} \frac{\omega^4}{c^2} \left(\frac{q^2}{m\omega^2}\right)^2 E_0^2$$

The incident energy flux is, by Eq. (13-50),

$$\bar{\mathbb{S}} = \frac{1}{2} \sqrt{\frac{\epsilon_0}{\mu_0}} E_0^2$$

Hence the cross section is

$$\sigma \equiv \frac{\mathcal{P}}{\bar{\mathbb{S}}} = \frac{1}{6\pi} \frac{\mu_0}{\epsilon_0} \frac{q^4}{m^2 c^2}$$

With the aid of the identity $\mu\epsilon c^2 = 1$ this is readily put into the more common form

$$\sigma \equiv \frac{8\pi}{3} \left(\frac{q^2}{4\pi\epsilon_0 mc^2}\right)^2 \tag{15-43}$$

Equation (15-43) is called the *Thomson scattering formula*. The quantity in parentheses is called the *classical radius* of the scattering particle. It is 2.81×10^{-15} m for an electron, and is, of course, even smaller for more massive particles.

Example 15-7

Calculate the scattering cross section of a small dielectric sphere in vacuum.

Solution   If the radius is much smaller than the wavelength of the incident wave, the field may be regarded as uniform at all times. Hence, according to Example 11-26, the field in the sphere is $3E_0/(\kappa + 2)$, and the resulting dipole moment amplitude is

$$p_0 = \frac{4\pi a^3}{3} \frac{3}{\kappa + 2} (\kappa - 1)\epsilon_0 E_0$$

$$= 4\pi\epsilon_0 a^3 \frac{\kappa - 1}{\kappa + 2} E_0 \tag{15-44}$$

Thus, as in Example 15-6,

$$\mathcal{P} = \frac{1}{12\pi}\sqrt{\frac{\mu_0}{\epsilon_0}}\frac{\omega^4}{c^2}\left(4\pi\epsilon_0 a^3\frac{\kappa-1}{\kappa+2}\right)^2 E_0^2$$

and

$$\sigma = \frac{\mathcal{P}}{\mathcal{S}} = \frac{8\pi}{3}\frac{\omega^4}{c^4}a^6\left(\frac{\kappa-1}{\kappa+2}\right)^2 \qquad (15\text{-}45\text{a})$$

$$= \frac{8\pi}{3}\left(\frac{2\pi a}{\lambda}\right)^4\left(\frac{\kappa-1}{\kappa+2}\right)^2 a^2 \qquad (15\text{-}45\text{b})$$

The cross section is seen to vary as ω^4, so that higher frequencies are much more strongly scattered. This is known as *Rayleigh scattering*. Molecules in the atmosphere are, approximately, tiny dielectric spheres, so Rayleigh scattering explains the main features of why the sky is blue and sunsets are red. Since the molecular polarizability plays the key role in both scattering and refraction, it is not surprising to find that the cross section may be related to the index of refraction. We leave it to one of the problems to work this out.

Polarization by Scattering

If the incident beam is *unpolarized* (which means that the plane of polarization varies randomly over all orientations containing the incident wave vector), the induced dipole moment will vary randomly over all orientations in the perpendicular plane. Let us consider the scattered wave at a particular deflection angle, say 90 degrees. To be specific about it, we take the incident direction along z and observation point somewhere out along x, say $R\hat{\mathbf{i}}$, and we consider only the electric dipole radiation. According to Eqs. (15-18) the dipole radiation fields vary as $\sin\theta$, where θ is the angle between the dipole direction and the radius vector to the field point; furthermore, the electric field lies in the plane defined by these two vectors. Applying these relations to the present case, we see that the radiation reaching point $R\hat{\mathbf{i}}$ is due to the y components of the oscillating dipole moment and is completely polarized in the xy plane. At deflection angles other than 90 degrees, the same basic considerations apply but result in only partial polarization of the scattered radiation.

There are several important applications of this effect. Historically, Barkla used it to prove the electromagnetic nature of x-rays shortly after their discovery. He placed a second scatterer at $R\hat{\mathbf{i}}$ and measured the variation in the doubly scattered intensity as function of direction in a plane perpendicular to x. A more current application is the use by photographers of "polarizing" filters to obtain beautiful sky and cloud effects. The blue sky-light, having been Rayleigh scattered, is somewhat polarized and can therefore be partly blocked by a suitably oriented filter. The light from clouds, on the other hand, was scattered by much larger particles (water droplets) and is therefore white and unpolarized. Thus the contrast between clouds and sky can be enhanced by the filter.

15-6 QUANTUM EFFECTS

The reader is probably well aware that the classical radiation formulas we have been discussing are only approximations, valid under restricted conditions. We emphasized in earlier sections that the simple multipole source structures had to be much smaller in size than the wavelength of the radiation they were emitting. What we did not mention was that for the validity of the classical concepts, they also had to be *larger* than a certain other characteristic size that depends on their structure and energy. Atoms and simple molecules, it turns out, generally satisfy the first criterion but not the second.

A major crisis in classical physics arose late in the nineteenth century in connection with atomic structure. Atom sizes were by then known to be about an angstrom (e.g., from measurements of ion masses and condensed matter densities). Atoms were therefore imagined to consist of a ball of positively charged "jelly" with electrons embedded in it. Measurements of the x-ray scattering power of atoms, in comparison with the Thomson cross section [Eq. (15-43)] of an electron, showed that the number of electrons is only of the order of the relative atomic weight, so that practically all the mass of the atom is due to the positive part. The crisis arose when Rutherford measured the scattering of α particles (small energetic charged particles emitted spontaneously by certain radioactive substances) from heavy atoms such as gold. The unexpectedly high proportion of large-deflection-angle scattering events proved conclusively that the massive positive part of the atom is concentrated in a very tiny volume (the *nucleus*); otherwise, it could not produce a strong enough electrostatic field to give the large deflections.

The dilemma, then, was as follows: The electrons clearly had to be outside the nucleus. The only way they could avoid falling straight in was to be in orbit, like planets around the sun. But an orbiting electron is equivalent to a two-component oscillating dipole moment, and will thus radiate its energy away and fall in anyway, taking only a tiny fraction of a second to do so! Thus classical physics was utterly unable to account for the obvious long-term stability of atoms.

The resolution of the dilemma came from the invention of quantum mechanics. In the earliest version of this theory, Niels Bohr simply hypothesized that certain orbits (i.e., *states of motion*) were stable and would not radiate. Only when an atom made a *quantum jump* from one state to another could radiation be emitted or, for that matter, absorbed. Furthermore, the frequency of the radiation was related not to any frequency of internal motion in the atom but simply to the energy difference between the two states according to

$$\omega = \frac{2\pi}{h} \Delta E \qquad (15\text{-}46)$$

where h is a universal constant, called *Planck's constant*, 6.06×10^{-34} joule-second. Thus electromagnetic energy at frequency ω can be emitted or absorbed only in discrete amounts (*quanta*) $h\omega/2\pi$, or $h\nu$. Such a unit of electromagnetic energy is called a *photon*, a concept first introduced by Einstein.

Although the quantum picture of radiation is radically different from the classical picture, there are, nevertheless, some connections. In regard to rates, Bohr's *correspondence principle* states that under certain conditions the statistical probability of photon emission is such that the rate of energy release is correctly given by the classical multipole formulas. The condition is that the total energy of the emitting system be vastly greater than that of a single photon. This is amply met by all large-scale systems, such as antennas, wire loops, and so on. A second connection has to do with angular distributions. During the process of a quantum jump the charge distribution within an atom or molecule changes, and can be described by a transient multipole distribution. Statistically, the angular distribution of photon emission corresponds to the multipole pattern, such as the electric dipole [Eq. (15-20)], magnetic dipole [Eq. (15-27)], and so on. Finally, the polarization of the emitted radiation must be such that the *angular momentum* (moment of the electromagnetic momentum about the emitting atom) carried away be the photon is just equal to the change in angular momentum of the atom.

We have noted that photon emission occurs when an atom (or molecule or nucleus) jumps from a higher to a lower-energy state. This process can occur in two ways: (1) *spontaneous emission* takes place statistically without regard to external conditions, and (2) if a field of the correct frequency is already present, *stimulated emission* occurs in addition. The latter is just the inverse of the process of absorption, and in both these processes the rate is proportional to the external field intensity. Stimulated emission is the central process in *lasers*. A majority of the active atoms are put into the higher-energy state by some separate energy input, which can take a variety of forms. Then once a field is initiated by a spontaneous transition, it very rapidly builds up by stimulated emission. Since the stimulated photon is *coherent* (in phase) with the stimulating field, the field builds up to tremendous strength.

Quantum principles also play a role in scattering processes, as well as emission and absorption. As we saw in Section 10-2, the momentum density of an electromagnetic field is proportional to the energy flux. Therefore a photon, energy of amount $h\nu$ moving at speed c, must have a definite momentum

$$p_{photon} = \frac{E}{c} = \frac{h\nu}{c} \tag{15-47}$$

If the photon is deflected, there is a change of momentum, which must of course come from recoil of the scatterer. The latter thus undergoes a *change of kinetic energy*, which, in turn, can only come from the photon. Therefore, the scattered photon is *shifted slightly lower in frequency*, by an amount that increases with the deflection angle. This is called the *Compton effect*. It has been experimentally verified in x-ray scattering, and provides prima facie evidence for the reality of photons.

FURTHER REFERENCES

Most of the material in this chapter is covered in all the standard texts as well as in many electrical engineering texts. Antenna theory is treated extensively by Stratton, and scattering by Jackson. More information on quantum effects can be found in texts on modern physics, such as A. Beiser, *Concepts of Modern Physics*, 3rd ed., McGraw-Hill, New York, 1981; K. Krane, *Modern Physics*, Wiley, New York, 1983; and R. B. Leighton, *Principles of Modern Physics*, McGraw-Hill, New York, 1959. For a bit more depth, a good start is D. Park, *Introduction to the Quantum Theory*, McGraw-Hill, New York, 1964.

PROBLEMS

15-1. A point charge moves along the x axis at speed v. For a field point at the origin, calculate the retarded time t_r as function of t.

15-2. Repeat Problem 15-1 except that field point is $(0, b, 0)$.

15-3. An oscillating electric dipole lies parallel to an ideal conductive plane at distance l. Calculate the radiation field at a point straight out from the dipole along the normal to the plane, at distance R. (*Hint:* There is an image dipole.)

15-4. Repeat Problem 15-3 for an arbitrary field point.

15-5. Is a very short center-fed linear antenna the same as a dipole? Explain.

15-6. By taking the limit of Eq. (15-36), compare the vector potential of a short center-fed linear antenna with that of a dipole for a given current amplitude I_0.

15-7. An electric dipole 1 cm long is radiating 1 W of power at a wavelength of 1 m. Calculate (a) the current amplitude, and (b) the Poynting vector at a distance of 1 km in the equatorial plane.

15-8. An electric dipole p_0 lies in the xy plane and rotates about the z axis with angular velocity ω. Calculate the Poynting vector at a distant point (r, θ, ϕ).

15-9. A point charge of mass m and charge q moves in a circular orbit of radius a at angular velocity ω. If $\omega \ll c/a$, what fraction of its kinetic energy does it radiate away in one revolution?

15-10. A bar magnet rotates about an axis perpendicular to its length. Will it radiate?

15-11. A uniform spherical shell of charge oscillates in radius (a "breathing" oscillation). Will it radiate?

15-12. Two equal circular wire loops are centered at the origin. Loop 1 lies in the xz plane and carries current $I_1 = I_0 e^{-i\omega t}$, loop 2 lies in the yz plane and carries current $I_2 = \alpha I_0 e^{-i\omega t}$. Calculate the radiation pattern in the xy plane if (a) $\alpha = 1$, and (b) $\alpha = i$.

15-13. A plane wave propagating through a gas of scatterers is attenuated. Find the distance for the Poynting vector to be reduced by factor $1/e$ in terms of the number density n and cross section σ of the scatterers.

15-14. In Problem 15-13, relate the attenuation length to the index of refraction of the gas.

15-15. In the Barkla experiment (Section 15-5) on double 90-degree scattering of x-rays, in what directions are intensity maxima observed?

15-16. With the sun directly overhead, what is the polarization of sky light near the horizon?

chapter 16

Special Relativity Theory

16-1 PRINCIPLE OF RELATIVITY

Theories of relativity deal with an exceedingly simple question: How are physical measurements affected by the *reference frames* of the measuring instruments? By a reference frame we mean simply the *system coordinates* upon which the measurements are based. To take a trivial example, suppose that two surveyors measure the length of a building, setting up their transits in different places as in Figure 16-1. Then, although they will obtain *different coordinates* of the end points of the building, they will (assuming the measurements to be accurate) determine the *same length*. The situation may be stated in overly formalized terms as follows:

1. Coordinate measurements by the two observers are related to each other by "transformation equations"

$$x' = x - a$$
$$y' = y - b \qquad (16\text{-}1)$$
$$z' = z - c$$

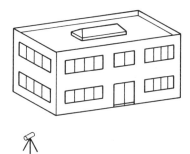

Figure 16-1 Two sets of surveying instruments measuring a building.

where x', y', and z' are coordinates measured by surveyor S', x, y, and z are coordinates measured by surveyor S, and a, b, and c are the coordinates of the instruments of S', relative to those of S.

2. Lengths of objects are *invariant* (unchanged) under this transformation. This follows from the fact that lengths involve *differences* of coordinates. If $L_{AB} = x_A - x_B$ and $L'_{AB} = x'_A - x'_B$, then obviously $L'_{AB} = L_{AB}$.

Another example seems to be nearly as trivial (but really isn't!): Suppose that the same surveyors again measure the building but this time S' sets up his instruments on a railroad flat car moving at a uniform speed v past the building (Figure 16-2).

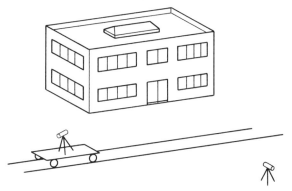

Figure 16-2   Two sets of surveying instruments, one of them on a moving railroad car, measuring a building.

He would have to use some specialized techniques (perhaps photographic) to be sure that he obtained coordinates of the two end points at the same time, but other than this technical difficulty we would intuitively expect L and L' again to be equal. (This is, in fact, wrong, but let us follow our intuition for a while.) In formal terms, we would replace Eqs. (16-1) by

$$x' = x - vt - a$$
$$y' = y - b$$
$$z' = z - c \qquad (16\text{-}2)$$
$$t' = t$$

where we have chosen the x and x' directions along the motion and have included the "obvious" fourth relation only because time is involved in the measurements by S'. These equations are called the *Galilean transformation*. Not only are lengths invariant under this transformation, but so also are *accelerations*. Thus, assuming force and mass to be invariant, Newton's laws of motion are *equally valid* in two reference frames connected by the Galilean transformation. This leads to a fundamental postulate, the *principle of relativity for mechanics*. It states that for mechanical phenomena *only relative motions* can be of significance. In other words, any unaccelerated reference frame (so-called *inertial frame*) is as good as any other for the statement of the laws. There is *no preferred inertial frame*.

The last few sentences have to be read very carefully. They pertain to statements of laws, not descriptions of phenomena. For the latter, there may indeed be preferred frames. A description of water waves, for example, is best done in a frame fixed relative to the body of water. Motion relative to the water could then be related to motion relative to such waves via the Galilean transformation, which says that the velocities should be *additive*. For example, a submarine navigator could ascertain the ship's velocity through the water by measuring the speed of sound in various directions from a source on a movable extended arm as suggested in Figure 16-3.

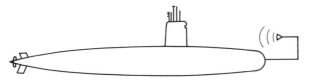

Figure 16-3 Submerged submarine measuring its speed relative to the water by means of sound velocities in various directions.

Once the principle of relativity for mechanics was recognized, several physicists proposed that a more wide-ranging principle of relativity might be valid for *all* physical laws, including in particular the laws of electromagnetism. It was recognized that this would entail a modification of the Galilean transformation. The reason is that the Maxwell equations contain a "built-in" speed, the speed of light, and were therefore incompatible with the idea of additivity of velocities in different equivalent reference frames.

16-2 THE ETHER

When the wave nature of light was established, it was natural to think of it as an analog of the more familiar mechanical types of waves, such as sound, water, elastic, and so on. All the latter types exist and propagate in a *medium* (air, water, etc.), so it was imagined that there existed throughout all space and throughout all matter a special medium for the propagation of light waves. This was called the *luminiferous ether* (*ether* for short). Much effort was devoted to ascertaining its properties. One of the early questions concerned the motion of matter through the ether. Does matter move freely through it or does the ether get "dragged along" by moving matter? Fizeau's 1851 measurement of the speed of light in moving water provided an answer. It appeared that the ether was *partially* dragged, such that in water of speed v the light speed was

$$c' = \frac{c}{n} + \alpha v \tag{16-3}$$

where the drag coefficient α was equal to $1/n^2$, n being the index of refraction.

Later efforts centered on the question of the earth's velocity through the ether. This is exactly analogous to the submarine navigation method of Figure 16-3, except that there is a major technical difficulty. Owing to the tremendous speed of light, only "round-trip" methods of measurement are possible in terrestrial laboratories.

Starting in 1881, Albert A. Michelson (later joined by E. W. Morley) made a series of ultraprecise interferometric measurements that compared the round-trip speeds of light in two perpendicular directions. The results were remarkable: No matter what the orientation of the apparatus, or what time of day, night, month, or year the observation was made, there was *no detectable difference* between the two speeds. This meant that if any motion of earth relative to ether existed it was *undetectably small.*

Several explanations of this unexpected result were offered, since it was utterly inconceivable that the whole vast sea of ether could be rigidly tied to the earth. Some of these were as follows:

1. *Fluid motion of the ether with total drag near massive bodies.* Aside from contradicting Fizeau's *partial* drag results, this would have contradicted the astronomical observation of stellar aberration (see Examples 16-6 and 16-14).

2. *"Emission theories"* of light, in which Maxwell's equations were modified so that the speed of light was fixed relative to its source rather than relative to the propagating medium. Repetitions of the Michelson–Morley experiment using extraterrestrial light sources (sun and stars) were thought to have ruled these out, but objections were raised based on the "extinction theorem" which shows that a light wave gradually acquires the velocity characteristic of the medium. However, still more recent experiments with very high energy photons, called *gamma rays*, have overcome the objections and firmly buried the emission theories.

3. *The "Fitzgerald contraction."* In 1892, G. F. Fitzgerald and H. A. Lorentz independently proposed that *all matter* contracts in the direction of its motion relative to the ether. A contraction ratio of $\sqrt{1 - v^2/c^2}$ would exactly cancel the expected velocity effect out of the Michelson–Morley experiment and thus account for the null results.

Einstein's Postulates

Although Lorentz and Poincaré were grappling with these problems, it remained for Einstein, in a brilliant paper of 1905, to cut through all the difficulty with a single stroke. Einstein built his entire edifice upon two simple postulates:

Postulate 1.   The principle of relativity holds; that is, no physical effect can distinguish between reference frames in uniform motion relative to each other. The restriction to *uniform* relative motion was later removed, leading to the *general* theory of relativity (1916). We shall, however, retain the restriction and thus confine ourselves to the *special* theory of relativity.

Postulate 2.   The speed of light (in vacuum) is a universal constant *no matter what inertial reference frame* it is measured from. This is simply a statement of the prima facie evidence furnished by Michelson–Morley. The situation is illustrated in Figure 16-4.

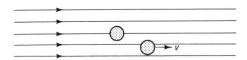

Figure 16-4 Two identical pieces of apparatus (symbolized by shaded circles), in motion relative to each other, measuring the speed of the same beam of light.

16-3 LORENTZ TRANSFORMATION

Postulate 2 is obviously at odds with the Galilean transformation, and so the latter has to be replaced. The correct transformation equations had been derived earlier by Lorentz, but Einstein rederived them on more general grounds, and we shall follow a later adaptation of this by McCrea.* In order to be quite clear about what we are doing it is helpful to define a new term: An *event* is a point in space-time. A good physical picture of an event is a spark occurring at some instant of time at some point in space. The event has coordinates (t, x, y, z) in some reference frame S and (t', x', y', z') in another frame S' that is moving relative to S. Our objective is to develop the relations between the two sets of coordinates. To simplify matters as much as possible, we shall take the x and x' axes along the direction of motion of S' relative to S, as shown in Figure 16-5, so that the transformation relations take the form

$$t' = t'(t, x; v) \tag{16-4a}$$

$$x' = x'(t, x; v) \tag{16-4b}$$

$$y' = y \tag{16-4c}$$

$$z' = z \tag{16-4d}$$

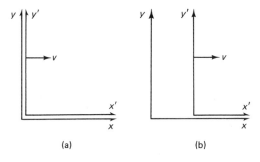

Figure 16-5 Axes of reference frames S and S': (a) at $t = t' = 0$; (b) at $t > 0, t' > 0$.

The relations must satisfy the following criteria:

1. They must be linear; otherwise, uniform motion relative to one frame would be accelerated relative to the other.

2. They must satisfy postulate 2.

3. They must reduce to the Galilean relations in the limit of small relative velocity v.

*W. H. McCrea, *Relativity Physics*, Methuen, London, 1947.

By criterion 1 we write

$$ct' = A(ct) + Bx \qquad \text{(16-5a)}$$

$$x' = C(ct) + Dx \qquad \text{(16-5b)}$$

$$y' = y \qquad z' = z \qquad \text{(16-5c)}$$

where (ct) is used as the time coordinate to keep all dimensionalities the same, and where we have dropped additive constants by appropriate choice of zero points.

For criterion 2, we consider a sequence of two events:

Event 1: A spark fires at the origins at the instant when they coincide (Figure 16-6a). Thus the coordinates of this event are $(0, 0, 0, 0)$ in both frames.

Event 2: Light from the spark actuates a photocell at some point (Figure 16-6b). The coordinates are (t_2, x_2, y_2, z_2) in S and (t'_2, x'_2, y'_2, z'_2) in S'.

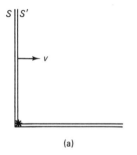

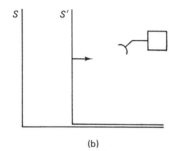

Figure 16-6 Events selected for analysis: (a) light flash at common origin, $t = x = y = z = 0$, $t' = x' = y' = z' = 0$; (b) light is registered by photocell at (t, x, y, z), (t', x', y', z').

(a)   (b)

According to postulate 2, these coordinates correspond to speed c in *both* frames.

$$ct_2 = \sqrt{x_2^2 + y_2^2 + z_2^2}$$

$$ct'_2 = \sqrt{x'^2_2 + y'^2_2 + z'^2_2}$$

In other words,

$$(ct'_2)^2 - (x'^2_2 + y'^2_2 + z'^2_2) = (ct_2)^2 - (x_2^2 + y_2^2 + z_2^2) \qquad \text{(16-6)}$$

On inserting Eqs. (16-5), this demands that

$$A^2 - C^2 = 1$$

$$B^2 - D^2 = -1 \qquad \text{(16-7)}$$

$$AB - CD = 0$$

A solution in terms of a single parameter γ (yet to be determined) is

$$A = D = \gamma \qquad \text{(16-8a)}$$

$$B = C = \pm\sqrt{\gamma^2 - 1} \qquad \text{(16-8b)}$$

To determine γ, we consider the point $x' = 0$, the origin of the frame S'. This point moves relative to S according to $x = vt$. Thus Eq. (16-5b) for this point reads

$$0 = \pm\sqrt{\gamma^2 - 1}\, ct + \gamma vt \qquad \text{(16-9)}$$

Thus (1) the minus sign is required if v is positive, and (2)

$$\gamma = \frac{1}{\sqrt{1 - \beta^2}} \tag{16-10}$$

where

$$\beta = \frac{v}{c} \tag{16-11}$$

Thus the complete Lorentz transformation is

$$ct' = \gamma ct - \beta \gamma x \tag{16-12a}$$

$$x' = -\beta \gamma ct + \gamma x \tag{16-12b}$$

$$y' = y \qquad z' = z \tag{16-12c}$$

It is clear that as $\beta \to 0$, $\gamma \to 1$, and Eqs. (16-12) reduce to (16-2) with $a = b = c = 0$. For the inverse transformation, we may simply reverse the sign of β, since the origin of S moves relative to S' with velocity $-v$. The relation of γ to β is plotted in Figure 16-7.

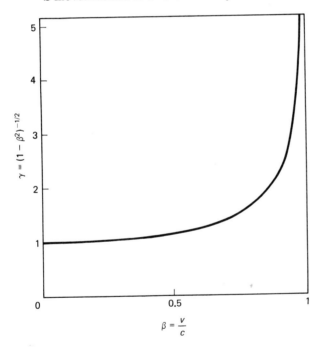

Figure 16-7   Relation of γ to β.

16-4 SIMULTANEITY

Let us consider two events whose coordinates in S' are $(t_1' = 0, x_1' = 0)$ and $(t_2' = 0, x_2' = L)$. These events are *simultaneous* in S', since they have the same time coordinates in this frame. By the inverse form of Eqs. (16-12) the coordinates of these same events in S are

$$ct_1 = 0 \qquad \text{and} \qquad ct_2 = \beta \gamma L \tag{16-13}$$

Thus the events are *not* simultaneous in *S*. Event 1 occurs *earlier* than event 2, when observed from frame *S*. Thus simultaneity of events is not absolute but is a relative concept, dependent on the observer's frame of reference.

This is one of the simplest yet most profound of the results of Einstein's postulates. Einstein himself repeatedly emphasized its overwhelming importance. The point is that there is no such thing as "absolute time" in the Newtonian sense. The *time of an event* is simply the reading of a (correctly set) clock located at the position of the event. If the clock is some distance away, the reading must be corrected for the transit time of the signal (e.g., a light flash) by which the event is sensed.

It is very useful in analyzing event sequences to picture each reference frame as being densely covered with a set of synchronized clocks. The synchronization must, of course, be done with signals. For example, to synchronize a clock located at a point one light-minute away, I might instruct an assistant stationed there to "set your clock to 12:00 when you see my light flash"; I would then flash the light at 11:59 on my clock. Clearly, these clocks would be synchronous *only* in that frame in which my assistant and I were at rest. In any relatively moving frame these clock readings, *after correction for transit times*, would not agree. The discrepancy would depend on distance as in Eq. (16-13). To help visualize the relation, it may help to keep in mind that *trailing clocks run late* (Figure 16-8).

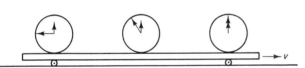

Figure 16-8 Illustration of trailing clocks running late. The clock readings symbolize values obtained by a ground-based observer *after* correcting the signal transit times. The clocks are synchronized relative to the train.

Considering again the same pair of events, we saw that event 2 was later in *S*, a frame moving at $-v$ relative to *S'*. Clearly, then, event 2 would be earlier in a frame *S''* moving at $+v$ relative to *S'*. Thus we see that not only simultaneity but even *temporal ordering* may be relative. The quantity of importance is the *space-time* interval Δs, defined by

$$(\Delta s)^2 = c^2(\Delta t)^2 - (\Delta r)^2 \tag{16-14}$$

By postulate 2, this is a *Lorentz invariant*, meaning that it has the same value in all inertial frames. We can then classify event pairs according to the sign of $(\Delta s)^2$:

$(\Delta s)^2 > 0$: timelike interval, definite "before/after" order

$(\Delta s)^2 > 0$: spacelike interval, cannot connect events by a light signal, indefinite order

$\Delta s = 0$: "on the light cone"

The term *light cone* refers to a four-dimensional space-time diagram, where a 45-degree cone about the *ct* axis separates the timelike region (inside the cone) from the spacelike region (outside) (Figure 16-9). Plotting event coordinates on such a diagram is often very helpful in figuring out the relationships.

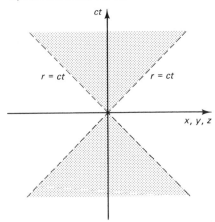

Figure 16-9 Space-time diagram showing the light cone (dashed lines). The upper shaded region is the "absolute future," the lower is the "absolute past" (both timelike intervals), and outside the cone is the "absolute elsewhere" (spacelike intervals).

16-5 RELATIVISTIC KINEMATICS

Many of the well-known relativistic effects follow as straightforward consequences of the Lorentz transformations. The following series of examples will illustrate them.

Example 16-1: Time Dilation

Show that in a timelike interval the time difference has its *minimum* value in that frame wherein the events occur at the same position.

Solution Let the events 1 and 2 have the coordinates $(0, x)$ and $(c\tau, x)$, respectively, in S. Then the time coordinates in S' are

$$ct_1' = -\beta\gamma x \qquad ct_2' = \gamma c\tau - \beta\gamma x \qquad (16\text{-}15)$$

so that

$$\Delta t' = \gamma\tau = \gamma\,\Delta t \qquad (16\text{-}16)$$

Since $\gamma \geq 1$, $\Delta t'$ is greater than Δt in any S'. The minimum time interval $\Delta t_{\min} = \tau$ is called the *proper time* interval. It is, of course, closely related to the invariant space-time interval

$$\Delta s = c\tau \qquad (16\text{-}17)$$

Example 16-2: Length Contraction

Show that any object has a *maximum* length in its own rest frame (i.e., that reference frame that is moving with the object).

Solution This goes back to the two-surveyor situation discussed in Section 16-1. The length of an object is defined as the distance between its end points *at the same time*. But now we realize that "same time" is different for different frames. Let the object be a rod fixed in S' with ends at $x_1' = 0$ and $x_2' = L$. These two equations describe the *world lines* (trajectories in space-time) of the rod ends. Written in terms of the S coordinates, they read

$$-\beta\gamma ct + \gamma x_1 = 0 \qquad (16\text{-}18\text{a})$$

$$-\beta\gamma ct + \gamma x_2 = L \qquad (16\text{-}18\text{b})$$

Subtracting at a fixed time t gives

$$\gamma(x_2 - x_1) = L$$

$$\Delta x = \gamma^{-1}L = \gamma^{-1}\Delta x' \qquad (16\text{-}19)$$

This result is exactly the same as the Fitzgerald contraction, but now it is a natural consequence of space-time geometry rather than an ad hoc hypothesis about the structure of matter.

It is important to note the distinction between the time-dilation and length-contraction effects. In time dilation, we compare coordinates of the *same pair* of events. In length contraction, we contrast two *different* event pairs, namely (x'_1, t') and (x'_2, t') in S' versus (x_1, t) and (x_2, t) in S. Both of these intervals are, incidentally, spacelike.

Example 16-3: Velocity Transformation

By differentiating the Lorentz transformation relations, deduce the transformation equations for velocity components.

Solution   Consider a particle having velocity components u_x, u_y, u_z relative to S, where $u_x = dx/dt$, etc. (Figure 16-10). By taking differentials of Eqs. (16-12), we find

$$dt' = \gamma\, dt - \frac{\beta\gamma}{c}\, dx \qquad (16\text{-}20\text{a})$$

$$dx' = -\beta\gamma c\, dt + \gamma\, dx \qquad (16\text{-}20\text{b})$$

$$dy' = dy \qquad (16\text{-}20\text{c})$$

Thus

$$u'_x \equiv \frac{dx'}{dt'} = \frac{\gamma(dx - \beta c\, dt)}{\gamma[dt - (\beta/c)\, dx]} = \frac{u_x - v}{1 - \dfrac{u_x v}{c^2}} \qquad (16\text{-}21\text{a})$$

$$u'_y \equiv \frac{dy'}{dt'} = \frac{u_y}{\gamma\left(1 - \dfrac{u_x v}{c^2}\right)} \qquad (16\text{-}21\text{b})$$

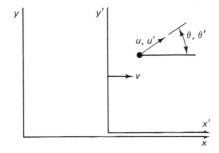

Figure 16-10   Moving particle having vector velocity **u** relative to S, **u**' relative to S'.

Note the asymmetry! The transverse components (transverse to the relative motions of the two frames) have an extra factor of γ in the denominator. Of course, for $\beta \ll 1$, these relations reduce to simple additivity as in the Galilean transformation.

The inverse relations are again easily obtained by reversing the sign of v, for example,

$$u_x = \frac{u'_x + v}{1 + \frac{u'_x v}{c^2}} \quad \text{etc.} \tag{16-22}$$

Example 16-4

Show that the combination of two speeds each less than c can never give a result greater than c.

Solution   The maximum resultant will be obtained if the two velocities are in the same direction. Accordingly let $u'_x = (1 - \epsilon_1)c$, $v = (1 - \epsilon_2)c$. Then Eq. (16-22) gives

$$\frac{u_x}{c} = \frac{(1 - \epsilon_1) + (1 - \epsilon_2)}{1 + (1 - \epsilon_1)(1 - \epsilon_2)} = \frac{2 - (\epsilon_1 + \epsilon_2)}{2 - (\epsilon_1 + \epsilon_2) + \epsilon_1 \epsilon_2} < 1 \tag{16-23}$$

Thus it is clear that no amount of "addition" of subluminal ($< c$) speeds can ever give a superluminal speed. This is in marked contrast to the Galilean result. For example, two speeds of $0.99c$ in the same direction combine to $0.99995c$ instead of $1.98c$.

Example 16-5

Show that if one of the speeds is equal to c, the resultant is also equal to c regardless of the relative directions.

Solution   Take $u' = c$, and suppose that this velocity makes angle θ' with the x' axis. Then

$$u_x = \frac{c \cos \theta' + v}{1 + \frac{vc \cos \theta'}{c^2}} \tag{16-24a}$$

$$u_y = \frac{c \sin \theta'}{\gamma_v \left(1 + \frac{vc \cos \theta'}{c^2}\right)} \tag{16-24b}$$

$$u^2 \equiv u_x^2 + u_y^2$$

$$= \frac{(c \cos \theta' + v)^2 + \left(1 - \frac{v^2}{c^2}\right)c^2 \sin^2 \theta'}{\left[1 + \left(\frac{v}{c}\right)\cos \theta'\right]^2}$$

$$= c^2 \tag{16-24c}$$

This simply verifies that the Lorentz transformation complies with postulate 2, which is just what it was invented for.

Example 16-6

For Example 16-5, find the angle of the velocity vector relative to the x axis in S.

Solution

$$\tan \theta = \frac{u_y}{u_x} = \frac{c \sin \theta'}{\gamma_v(c \cos \theta' + v)} = \frac{\sqrt{1 - \beta^2}}{1 + \beta \sec \theta'} \tan \theta' \qquad (16\text{-}25)$$

This gives the formula for *stellar aberration*, the apparent shift in direction of a star as the earth moves around the sun, illustrated in Figure 16-11.

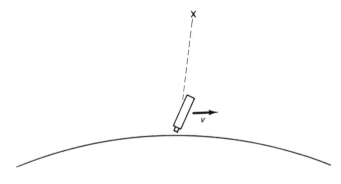

Figure 16-11 Stellar aberration. Telescope must be aimed ahead of true direction to star to compensate for velocity of earth.

Example 16-7: Acceleration Transformation

Derive the transformation equations for acceleration components.

Solution The differential of Eq. (16-21a) is

$$du'_x = \frac{\left(1 - \dfrac{u_x v}{c^2}\right) du_x - (u_x - v)\left(\dfrac{-v}{c^2} du_x\right)}{(1 - u_x v/c^2)^2}$$

$$= \frac{(1 - \beta^2)\, du_x}{\left(1 - \dfrac{u_x v}{c^2}\right)^2} = \frac{du_x}{\gamma_v^2 \left(1 - \dfrac{u_x v}{c^2}\right)^2}$$

Dividing this by Eq. (16-20a) gives

$$\frac{du'_x}{dt'} = \frac{1}{\gamma_v^3 \left(1 - \dfrac{u_x v}{c^2}\right)^3} \frac{du_x}{dt} \qquad (16\text{-}26a)$$

In the same way, from Eq. (16-21b) we obtain

$$\frac{du'_y}{dt'} = \frac{\left(1 - \dfrac{u_x v}{c^2}\right)\left(\dfrac{du_y}{dt}\right) + \left(\dfrac{u_y v}{c^2}\right)\left(\dfrac{du_x}{dt}\right)}{\gamma_v^2 \left(1 - \dfrac{u_x v}{c^2}\right)^3} \qquad (16\text{-}26b)$$

Example 16-8

Derive the transformation equations for the partial derivative operators.

Solution   From the chain rule of differential calculus

$$\frac{\partial}{\partial t} = \frac{\partial t'}{\partial t}\frac{\partial}{\partial t'} + \frac{\partial x'}{\partial t}\frac{\partial}{\partial x'}$$

The coefficients can be read off from Eqs. (16-12),

$$\frac{\partial}{\partial t} = \gamma\frac{\partial}{\partial t'} + (-\beta\gamma c)\frac{\partial}{\partial x'} = \gamma\left(\frac{\partial}{\partial t'} - v\frac{\partial}{\partial x'}\right) \tag{16-27a}$$

Similarly,

$$\frac{\partial}{\partial x} = \frac{\partial t'}{\partial x}\frac{\partial}{\partial t'} + \frac{\partial x'}{\partial x}\frac{\partial}{\partial x'} = \gamma\left(\frac{\partial}{\partial x'} - \frac{v}{c^2}\frac{\partial}{\partial t'}\right) \tag{16-27b}$$

Example 16-9: D'Alembertian Operator

Show that the operator $\square$, defined as $\nabla^2 - \frac{1}{c^2}\frac{\partial^2}{\partial t^2}$, is an *invariant*.

Solution   Substitute the derivative operators directly.

$$\square = \gamma^2\left(\frac{\partial}{\partial x'} - \frac{v}{c^2}\frac{\partial}{\partial t'}\right)\left(\frac{\partial}{\partial x'} - \frac{v}{c^2}\frac{\partial}{\partial t'}\right) + \frac{\partial^2}{\partial y'^2} + \frac{\partial^2}{\partial z'^2}$$

$$- \frac{1}{c^2}\gamma^2\left(\frac{\partial}{\partial t'} - v\frac{\partial}{\partial x'}\right)\left(\frac{\partial}{\partial t'} - v\frac{\partial}{\partial x'}\right)$$

$$= \frac{\partial^2}{\partial x'^2} + \frac{\partial^2}{\partial y'^2} + \frac{\partial^2}{\partial z'^2} - \frac{1}{c^2}\frac{\partial^2}{\partial t'^2} = \square' \tag{16-28}$$

16-6 RELATIVISTIC ELECTRODYNAMICS

A good starting point for the investigation of the transformation properties of electrodynamic quantities (sources, fields, potentials) is the observation that *charge is an invariant*. This is evident because atoms remain exactly neutral even when the electrons within them move at speeds approaching c.

Now if charge is invariant then charge density cannot be, since volume is subject to the length-contraction factor γ^{-1} [Eq. (16-19)]. Furthermore, charge density is associated with current density in the continuity equation. Therefore, it behooves us to look into the transformation of this equation, which reads

$$\frac{\partial J_x}{\partial x} + \frac{\partial J_y}{\partial y} + \frac{\partial J_z}{\partial z} + \frac{\partial \rho}{\partial t} = 0 \tag{16-29}$$

We substitute for the differential operators from Eqs. (16-27), obtaining

$$\gamma\left(\frac{\partial}{\partial x'} - \frac{v}{c^2}\frac{\partial}{\partial t'}\right)J_x + \frac{\partial J_y}{\partial y'} + \frac{\partial J_z}{\partial z'} + \gamma\left(\frac{\partial}{\partial t'} - v\frac{\partial}{\partial x'}\right)\rho = 0$$

On rearranging terms, we have

$$\frac{\partial}{\partial x'}(\gamma J_x - \gamma v \rho) + \frac{\partial J_y}{\partial y'} + \frac{\partial J_z}{\partial z'} + \frac{\partial}{\partial t'}\left(\gamma \rho - \frac{\gamma v J_x}{c^2}\right) = 0$$

This can be written in the *same form* as Eq. (16-29), namely

$$\frac{\partial J_x'}{\partial x'} + \frac{\partial J_y'}{\partial y'} + \frac{\partial J_z'}{\partial z'} + \frac{\partial \rho'}{\partial t'} = 0 \qquad (16\text{-}30)$$

provided we make the identifications

$$c\rho' = \gamma(c\rho - \beta J_x) \qquad (16\text{-}31\text{a})$$

$$J_x' = \gamma(J_x - \beta c\rho) \qquad (16\text{-}31\text{b})$$

$$J_y' = J_y \qquad J_z' = J_z \qquad (16\text{-}31\text{c})$$

Aside from the γ factors, these transformations are just what we would expect intuitively, since a static charge distribution in one frame becomes a charge-plus-current distribution in another frame. The γ factors are the result of the length contraction.

Since the current continuity equation takes the same form in all inertial frames, it is said to be *Lorentz covariant*. Covariance is an obvious requirement of any valid physical law in view of the principle of relativity.

It is interesting to note that the set of four quantities $(c\rho, \mathbf{J})$ transforms by exactly the same rules as the set $(ct, \mathbf{r})$. Such sets are called *four-vectors*. The mathematics of vector and tensor transformations can facilitate many of the calculations, but we will manage to survive without it.

Potentials

The Lorentz-gauge equations for the potentials, Eqs. (7-19) and (7-20), may be written (for vacuum)

$$-\frac{1}{\mu_0}\Box \mathbf{A} = \mathbf{J}$$

$$-\epsilon_0 c^2 \Box \frac{U}{c} = c\rho$$

where $\Box$ is the d'Alembertian operator defined in Example 16-9. Since this operator is an invariant and since the coefficients on the left are equal, it is clear that $(U/c, \mathbf{A})$ transforms like $(c\rho, \mathbf{J})$; in other words, it is another four-vector. Explicitly, the equations are

$$\frac{U'}{c} = \gamma\left(\frac{U}{c} - \beta A_x\right) \qquad (16\text{-}32\text{a})$$

$$A_x' = \gamma\left(-\frac{\beta}{c}U + A_x\right) \qquad (16\text{-}32\text{b})$$

$$A_y' = A_y \qquad A_z' = A_z \qquad (16\text{-}32\text{c})$$

Fields

Since the fields are expressible as derivatives of the potentials, we can proceed to calculate them directly.

Example 16-10

Derive the transformation relations for the electric field component E_x.

Solution   From Eq. (7-11),

$$E_x = -\frac{\partial U}{\partial x} - \frac{\partial A_x}{\partial t}$$

$$E'_x = -\frac{\partial U'}{\partial x'} - \frac{\partial A'_x}{\partial t'}$$

$$= -\left[\gamma\left(\frac{\partial}{\partial x} + \frac{\beta}{c}\frac{\partial}{\partial t}\right)\right]\left[\gamma(U - vA_x)\right] - \left[\gamma\left(\frac{\partial}{\partial t} + v\frac{\partial}{\partial x}\right)\right]\left[\gamma\left(A_x - \frac{\beta}{c}U\right)\right]$$

$$= -\gamma^2(1 - \beta^2)\left(\frac{\partial U}{\partial x} + \frac{\partial A_x}{\partial t}\right) = \frac{\partial U}{\partial x} + \frac{\partial A_x}{\partial t} = E_x \qquad (16\text{-}33)$$

Example 16-11

Do the same for E_y.

Solution $$E'_y = -\frac{\partial U'}{\partial y'} - \frac{\partial A'_y}{\partial t'}$$

$$= -\frac{\partial}{\partial y}[\gamma(U - vA_x)] + \gamma\left(\frac{\partial}{\partial t} + v\frac{\partial}{\partial x}\right)A_y$$

$$= \gamma\left[\left(-\frac{\partial U}{\partial y} - \frac{\partial A_y}{\partial t}\right) + v\left(\frac{\partial A_x}{\partial y} - \frac{\partial A_y}{\partial x}\right)\right]$$

$$= \gamma(E_y - vB_z) \qquad (16\text{-}34)$$

By proceeding in this way through all six field components, we find that the results can be expressed concisely as

$$E'_\parallel = E_\parallel \qquad B'_\parallel = B_\parallel \qquad (16\text{-}35\text{a})$$

$$\mathbf{E}'_\perp = \mathbf{E}_\perp + \mathbf{v} \times \mathbf{B}_\perp \qquad (16\text{-}35\text{b})$$

$$\mathbf{B}'_\perp = \mathbf{B}_\perp - \mathbf{v} \times \frac{\mathbf{E}_\perp}{c^2} \qquad (16\text{-}35\text{c})$$

where $\mathbf{v}$ is the vector velocity of frame S' relative to S, and the subscript symbols refer to components parallel and perpendicular to this direction.

It is seen that the fields are *mixed together* in this transformation. A purely electric field in one frame becomes a mixed field in other frames. In other words, the electric and magnetic fields lose their individual identities, and the two together make up a single *electromagnetic* field. For those who are familiar with the terminology, the six independent components comprise an antisymmetric second-rank four-tensor.

Example 16-12

Investigate the transformation of the quantity $(\mathbf{E}^2 - c^2\mathbf{B}^2)$.

Solution $$\mathbf{E}^2 - c^2\mathbf{B}^2 = E_\parallel^2 - c^2B_\parallel^2 + \mathbf{E}_\perp^2 - c^2\mathbf{B}_\perp^2$$

By Eqs. (16-35) the first two terms transform into themselves. The second two become

$$\mathbf{E}_{\perp}'^{2} - c^{2}\mathbf{B}_{\perp}'^{2} = \gamma^{2}[\mathbf{E}_{\perp}^{2} + (\mathbf{v} \times \mathbf{B}_{\perp})^{2} + 2\mathbf{E}_{\perp} \cdot \mathbf{v} \times \mathbf{B}_{\perp}]$$

$$-\gamma^{2}c^{2}\left[\mathbf{B}_{\perp}^{2} + \left(\frac{\mathbf{v} \times \mathbf{E}_{\perp}}{c^{2}}\right)^{2} - 2\mathbf{B}_{\perp} \cdot \frac{\mathbf{v} \times \mathbf{E}_{\perp}}{c^{2}}\right]$$

The mixed terms cancel, as can be seen by interchanging dot and cross in one of them. The squares of cross-products reduce simply to scalar products. Thus

$$\mathbf{E}_{\perp}'^{2} - c^{2}\mathbf{B}_{\perp}'^{2} = \gamma^{2}\left(1 - \frac{v^{2}}{c^{2}}\right)(\mathbf{E}_{\perp}^{2} - c^{2}\mathbf{B}_{\perp}^{2})$$

$$= (\mathbf{E}_{\perp}^{2} - c^{2}\mathbf{B}_{\perp}^{2})$$

Thus the entire result is

$$\mathbf{E}^{2} - c^{2}\mathbf{B}^{2} = \text{invariant} \qquad (16\text{-}36)$$

Because of this, the transformation of electric into magnetic fields, and vice versa, is not completely arbitrary. If $|E| < c|B|$, then B can never be reduced to zero in any frame, and conversely. In any case, only a uniform field can be reduced to zero at all points.

16-7 PLANE-WAVE FIELDS

Since the speed of light is one of the foundation pillars of relativity theory, it seems appropriate to investigate how some other aspects of light propagation are affected.

Wave Parameters

The wave function describing the variation of all fields in space and time is

$$\psi(\mathbf{r}, t) = e^{i(\mathbf{K} \cdot \mathbf{r} - \omega t)}$$

$$= e^{i[\mathbf{K} \cdot \mathbf{r} - (\omega/c)(ct)]} \qquad (16\text{-}37)$$

Now the exponent is a pure number (i.e., dimensionless), hence must be invariant to a coordinate transformation. This will be the case if the set $(\omega/c, \mathbf{K})$ transforms in the same way as $(ct, \mathbf{r})$ (i.e., as a four-vector). The explicit relations are

$$\frac{\omega'}{c} = \gamma\left(\frac{\omega}{c} - \beta K_{x}\right) \qquad (16\text{-}38a)$$

$$K_{x}' = \gamma\left(-\frac{\beta\omega}{c} + K_{x}\right) \qquad (16\text{-}38b)$$

$$K_{y}' = K_{y} \qquad K_{z}' = K_{z} \qquad (16\text{-}38c)$$

Example 16-13: Doppler Effect

Derive the frequency in frame S' of a light wave that in frame S has frequency ω and is propagating at angle θ to the x axis.

Solution   Since $K_x = (\omega/c) \cos \theta$, Eq. (16-38a) gives

$$\omega' = \gamma\omega(1 - \beta \cos \theta) \qquad (16\text{-}39)$$

Aside from the factor γ, this is much like the classical Doppler formula, which is most familiar in terms of sound waves, such as a train whistle approaching, passing, and receding from the listener. Actually, the Doppler effect is not confined to waves at all, but pertains to any periodic signal, such as machine-gun bullets (Figure 16-12). The factor γ is, of course, due to time dilation. It produces a *transverse* (i.e., $\theta = 90$ degrees) Doppler shift that is absent at low speeds.

Figure 16-12   Illustration of Doppler effect; tank moving toward target while shooting. The arrival frequency is greater than the firing frequency because successive bullets have shorter transit times.

Example 16-14: Aberration

Calculate the angle θ' of the propagation relative to S'.

Solution   For simplicity, chose the y axis such that K lies in the xy plane. Then

$$\tan \theta' = \frac{K'_y}{K'_x} = \frac{K_y}{\gamma\left[\left(\dfrac{\omega}{c}\cos\theta - \beta\dfrac{\omega}{c}\right)\right]} = \frac{\sin\theta}{\gamma(\cos\theta - \beta)} \qquad (16\text{-}40)$$

This is again the formula for stellar aberration. Comparison with Example 16-6 shows that it has nothing to do with the wave nature of the phenomenon, just with the fact that it propagates at speed c.

Field Strengths of Plane Wave Fields

Let us again choose axes as in the preceding example, and consider a wave polarized in the xy plane. Then its field-component amplitudes are

$$E_x = -E_0 \sin \theta$$

$$E_y = E_0 \cos \theta$$

$$B_z = \frac{E_0}{c}$$

The transformed electric field components are, according to Eqs. (16-35),

$$E'_x = E_x = -E_0 \sin \theta \qquad (16\text{-}41\text{a})$$

$$E'_y = \gamma(E_y - \beta c B_z) = \gamma E_0(\cos \theta - \beta) \qquad (16\text{-}41\text{b})$$

and the transformed amplitude is given by

$$\left(\frac{E_0'}{E_0}\right)^2 = \sin^2\theta + \gamma^2(\cos\theta - \beta)^2$$

$$= \gamma^2[(1 - \beta^2)\sin^2\theta + \cos^2\theta + \beta^2 - 2\beta\cos\theta]$$

$$= \gamma^2(1 - \beta\cos\theta)^2$$

$$E_0' = \gamma E_0(1 - \beta\cos\theta) \tag{16-42}$$

On comparison with Eq. (16-39), we see that

$$\frac{E_0'}{E_0} = \frac{\omega'}{\omega} \tag{16-43}$$

The magnetic field amplitude can also be transformed directly, but it is easier to note that since $cB_0 = E_0$ in a plane wave by Eq. (13-15), Eq. (16-36) tells us that

$$cB_0' = E_0' \tag{16-44}$$

Energy Transport in Plane Waves

Since energy density and flux are proportional to the square of the field amplitude, Eq. (16-43) tells us that

$$\frac{u'}{u} = \left(\frac{\omega'}{\omega}\right)^2 = \gamma^2(1 - \beta\cos\theta)^2$$

Part of this is, however, due to the change of the volume element. It is more germane to consider a specific volume element moving with the wave. Let V_0 be the *rest volume* of this element. Then, since it is contracted in the x direction in *each* frame, its volumes in S and S', respectively, are

$$V = \frac{V_0}{\gamma_u} \qquad V' = \frac{V_0}{\gamma_{u'}}$$

After a bit of algebra, the ratio can be evaluated from the velocity transformation relations, Eqs. (16-21). The result is (see Problem 16-11)

$$\frac{V'}{V} = \frac{\gamma_u}{\gamma_{u'}} = \frac{1}{\gamma_v(1 - u_x v/c^2)}$$

and since $u_x = c\cos\theta$ in the present case,

$$\frac{V'}{V} = \frac{1}{\gamma_v(1 - \beta\cos\theta)} \tag{16-45}$$

Thus the ratio of *energy contents* is

$$\frac{W'}{W} = \frac{u'V'}{uV} = \frac{\omega'}{\omega} \tag{16-46}$$

This is a most important result from the standpoint of quantum theory. There, as we discussed in Section 15-6, light is recognized to have particlelike as well as wavelike properties. The particles are called *photons*, and the energy of each is, by Eq. (15-46), *proportional to the frequency* according to the relation

$$W_{\text{photon}} = \frac{\hbar\omega}{2\pi} \tag{16-47}$$

Thus Eq. (16-46) tells us that the *number of photons* in a given volume is a Lorentz invariant, a result without which the whole photon concept would be meaningless.

16-8 POINT CHARGE FIELDS

Uniform Motion

The potentials due to a *uniformly* moving point charge are easily found. We simply place the particle at the origin of some reference frame (say S'), write the *static* potentials in that frame, and use Eqs. (16-32) to transform to the frame relative to which the particle is moving (Figure 16-13). The potentials in S' are

$$U'(x', y', z') = \frac{q/4\pi\epsilon_0}{\sqrt{x'^2 + y'^2 + z'^2}} \tag{16-48a}$$

$$A'(x', y', z') = 0 \tag{16-48b}$$

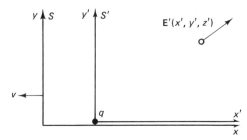

Figure 16-13 Point charge fixed at origin of S'. The field is purely electrostatic relative to (x', y', z').

Thus, after transforming and using the fact that $x' = \gamma(x - vt)$, $y' = y$, $z' = z$, we find

$$U(x, y, z, t) = \frac{\gamma q/4\pi\epsilon_0}{\sqrt{\gamma^2(x - vt)^2 + y^2 + z^2}}$$

$$= \frac{q/4\pi\epsilon_0}{\sqrt{(x - vt)^2 + (1 - \beta^2)(y^2 + z^2)}} \tag{16-49a}$$

$$A_x(x, y, z, t) = \frac{v}{c^2} U(x, y, z, t) \tag{16-49b}$$

$$A_y = A_z = 0 \tag{16-49c}$$

The field components are found by straightforward differentiation.

$$E_x = -\frac{\partial U}{\partial x} - \frac{\partial A_x}{\partial t}$$

$$= \frac{q}{4\pi\epsilon_0} \frac{(1 - \beta^2)(x - vt)}{[(x - vt)^2 + (1 - \beta_2)s^2]^{3/2}} \tag{16-50a}$$

where s is the transverse coordinate, $s^2 = y^2 + z^2$.

$$E_s = -\frac{\partial U}{\partial s} = \frac{q}{4\pi\epsilon_0} \frac{(1 - \beta^2)s}{[(x - vt)^2 + (1 - \beta^2)s^2]^{3/2}} \tag{16-50b}$$

$$B_\phi = -\frac{\partial A_x}{\partial s} = \frac{qv}{4\pi\epsilon_0} \frac{(1 - \beta^2)s}{c^2[(x - vt)^2 + (1 - \beta^2)s^2]^{3/2}} \tag{16-50c}$$

Example 16-15

Find the direction of the electric field at any time t.

Solution From Eqs. (16-50) we get

$$\frac{E_s}{E_x} = \frac{s}{x - vt} = \frac{\text{perpendicular distance}}{\text{parallel distance}} \tag{16-51}$$

Thus the field is radial with respect to the particle position at the *same time t* as the field measurement. The direction changes continuously, of course, as the particle moves (Figure 16-14).

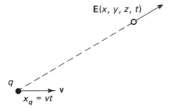

Figure 16-14 Electric field direction in relation to point charge position at same instant.

Example 16-16

For a given distance, how does the field strength vary with direction?

Solution Consider, at time t, a given distance R. If this lies in the x direction then $x - vt = R$, $s = 0$, so

$$E_x = \frac{q}{4\pi\epsilon_0} \frac{(1 - \beta^2)R}{R^3} = \frac{1}{\gamma^2} \frac{q}{4\pi\epsilon_0 R^2} \tag{16-52a}$$

If R lies in the perpendicular direction, $s = R$ and $x - vt = 0$, so

$$E_s = \frac{q}{4\pi\epsilon_0} \frac{(1 - \beta^2)R}{[(1 - \beta^2)R^2]^{3/2}} = \gamma \frac{q}{4\pi\epsilon_0 R^2} \tag{16-52b}$$

Thus the field is weakened by the factor γ^{-2} in the forward and backward directions and strengthened by the factor γ in the transverse directions. The field-line

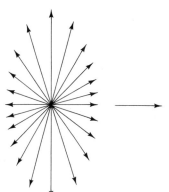

Figure 16-15 Field pattern of a uniformly moving point charge. (The length of the arrows symbolizes the strength of the field.)

pattern is changed from spherical to "pancake-shaped," but the field moves bodily along with the particle, as indicated in Figure 16-15.

Expression in Terms of Retarded Time

The foregoing expressions for the potentials and fields at any time t are given in terms of the particle position at that time. This, however, must be regarded as a mathematical artifice, since physically the fields originated at an earlier time, the retarded time. We can express the potentials explicitly in terms of the position of the particle at that time by a bit of geometry. Remember that the retarded time t_r is defined by Eq. (15-8) as

$$t - t_r = \frac{R_r}{c}$$

Thus, with reference to Figure 16-16,

$$x - vt = R_r \cos \theta_r - v(t - t_r)$$
$$= R_r(\cos \theta_r - \beta) \tag{16-53}$$

Hence the expression under the square root sign in the denominator of the potentials [Eqs. (16-49)] is

$$(x - vt)^2 + (1 - \beta^2)s^2$$
$$= R_r^2(\cos \theta_r - \beta)^2 + (1 - \beta^2)R_r^2 \sin^2 \theta_r$$
$$= R_r^2(\cos^2 \theta_r + \beta^2 - 2\beta \cos \theta_r + \sin^2 \theta_r - \beta^2 \sin^2 \theta_r)$$
$$= R_r^2(1 - \beta \cos \theta_r)^2 \tag{16-54}$$

and the potentials can be written

$$U(x, y, z, t) = \frac{q}{4\pi\epsilon_0} \frac{1}{[R(1 - \beta \cos \theta]_{t_r}} \tag{16-55a}$$

$$A_x(x, y, z, t) = \frac{q}{4\pi\epsilon_0} \frac{v/c^2}{[R(1 - \beta \cos \theta]_{t_r}} \tag{16-55b}$$

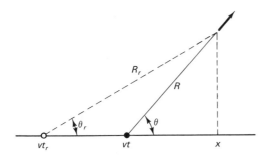

Figure 16-16   Geometry for relating field to particle position at the retarded time t_r.

where the subscript t_r means that all quantities in the brackets are to be evaluated at the retarded time.

Particle in Arbitrary Motion

We are now in a position to write the potentials due to a point charge in any kind of motion whatever. The crux of the argument is that only the state of motion at time t_r affects the field point at time t. Anything the particle did before t_r has already passed the field point, and anything it did or will do after t_r has not yet reached the field point. Thus all we have to do is specify in Eq. (16-55b) that v is the velocity at the retarded time, that is,

$$U(x, y, z, t) = \frac{q}{4\pi\epsilon_0} \left[\frac{1}{R(1 - \beta \cos \theta)} \right]_{t_r} \tag{16-56a}$$

$$\mathbf{A}(x, y, z, t) = \frac{q}{4\pi\epsilon_0 c} \left[\frac{\boldsymbol{\beta}}{R(1 - \beta \cos \theta)} \right]_{t_r} \tag{16-56b}$$

There are called the *Liènard–Wiechert potentials*.

Although the potentials appear fairly simple, the calculation of the fields is exceedingly complicated. The reason is that we need derivatives with respect to t and with respect to x, y, z at constant t, whereas the expressions are explicit functions of t_r, so that t is only an implicit argument. Rather than work through the several pages of calculations needed, we shall simply quote the results. The numerous terms can be grouped into two sets: those that contain β but not its time derivative (the *velocity-dependent* terms), and those that contain $d\beta/dt$ (the *acceleration-dependent* terms).

Velocity-dependent terms:

$$\mathbf{E}(v) = \frac{q}{4\pi\epsilon_0} \left[\frac{(1 - \beta^2)(\hat{\mathbf{n}} - \boldsymbol{\beta})}{(1 - \beta \cos \theta)^3 R^2} \right]_{t_r} \tag{16-57a}$$

$$\mathbf{B}(v) = \frac{1}{c} \hat{\mathbf{n}} \times \mathbf{E}(v) \tag{16-57b}$$

where $\hat{\mathbf{n}}$ is the unit vector from the particle toward the field point. Not surprisingly, this is exactly the same as for uniform motion, Eqs. (16-50), but expressed in terms of t_r instead of t.

Acceleration-dependent terms:

$$\mathbf{E}(a) = \frac{q}{4\pi\epsilon_0 c^2} \left[\frac{\hat{\mathbf{n}} \times [(\hat{\mathbf{n}} - \boldsymbol{\beta}) \times \mathbf{a}]}{(1 - \beta \cos\theta)^3 R} \right]_{t_r} \tag{16-58a}$$

$$B(a) = \frac{1}{c}\,\hat{\mathbf{n}} \times \mathbf{E}(a) \tag{16-58b}$$

These terms are seen to (1) fall off only as R^{-1}, not R^{-2}, (2) lie perpendicular to the line from charge point to field point, and (3) lie perpendicular to each other. They thus constitute a *radiation field*. Indeed, for low velocity ($\beta \ll 1$) sinusoidal motion, Eqs. (16-58) reduce to the dipole radiation formula of Eqs. (15-18) and (15-19). For highly relativistic particles, on the other hand, the factor $(1 - \beta \cos\theta)^3$ in the denominator leads to extreme strengthening and peaking of the radiation near the forward direction. This is called *synchrotron radiation*. It is a source of unwanted energy loss in some particle accelerators, but is also put to many valuable research uses, so much so that specially designed accelerators called *electron synchrotrons* are built to generate it.

Cerenkov Radiation

In the foregoing we have tacitly assumed that the particle was moving in vacuum, so that it could never exceed the speed of light. The result was that in uniform motion the field simply moved bodily with the particle, although distorted in shape. Thus no radiation occurred in uniform motion.

The situation is different for a particle moving through a dielectric medium. The particle speed may easily exceed the speed of light in the medium, and the field can no longer keep up with it. The result is a new kind of radiation process, called *Cerenkov radiation*, in which the field streams out behind the particle in a conical pattern much like the wake of a fast-moving motor boat (Figure 16-17). This can be easily observed as a bluish-white glow surrounding the core of a water-cooled nuclear reactor. The effect has been used in the detection of fast charged particles.

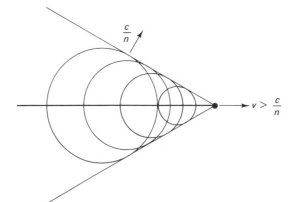

Figure 16-17 Field pattern of Cerenkov radiation.

16-9 RELATIVISTIC DYNAMICS

We close this final chapter by developing some of the relations involved in particle motion under forces. Newton's laws must, of course, be modified. But since they have proven to be exceedingly accurate for low velocities, we demand that any proposed replacements should reduce to Newton's laws in the limit $v \ll c$. In other words, at $v = 0$ Newton's second law should be exact! We exploit this by writing the equations of motion in the "instantaneous rest frame" of a particle. If this is the frame S' as in Figure 16-18, then, by definition, $v'_x = v'_y = v'_z = 0$ at $t' = 0$. Thus, if the particle is charged and is being accelerated by electric and magnetic fields, its equations of motion, in frame S' at this instant, are

$$\frac{du'_x}{dt'} = qE'_x \qquad (16\text{-}59a)$$

$$\frac{du'_y}{dt'} = qE'_y \qquad \text{etc.} \qquad (16\text{-}59b)$$

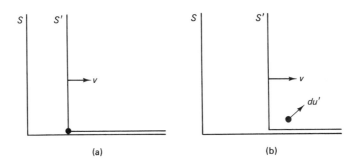

Figure 16-18   (a) Instantaneous rest frame (S') of a particle at time t'; (b) same frames at time $t' + dt'$.

These equations may now be transformed to a frame S, relative to which the particle has velocity components $u_x = v$, $u_y = 0$, $u_z = 0$ at this instant. The acceleration transformations were given in Eqs. (16-26). Their inverses, for the case $u' = 0$ which we are considering, read

$$\frac{du'_x}{dt'} = \gamma_v^3 \frac{du_x}{dt} \qquad \text{and} \qquad \frac{du'_y}{dt'} = \gamma_v^2 \frac{du_y}{dt}$$

The field transformations were given in Eqs. (16-35). On inserting all these into the foregoing equations of motion, we obtain in frame S

$$m\gamma_v^3 \frac{du_x}{dt} = qE_x \qquad (16\text{-}60a)$$

$$m\gamma_v \frac{du_y}{dt} = q[E_y - (\mathbf{v} \times \mathbf{B})_y] \qquad (16\text{-}60b)$$

Now since $\mathbf{v}$ is the instantaneous velocity of the particle relative to frame S, the terms on the right are exactly the components of the Lorentz force in S. From this, it follows that the left sides must be the *rates of change of the particle's momentum components.*

This leads us to propose a relativistic definition of particle momentum as

$$\mathbf{p} = m\gamma_u \mathbf{u} \tag{16-61}$$

Clearly, this reduces to the Newtonian expression $m\mathbf{u}$ when $u \ll c$, $\gamma_u \cong 1$. Also, the derivatives of its components are consistent with the left sides of Eqs. (16-60) since, for the instantaneous rest-frame, $v = u$, $u_x = u$, and $u_y = 0$.

Kinetic Energy

We are now in a position to calculate the kinetic energy of the particle. This is simply the total work done by the external fields (or other forces) in accelerating it from rest to its final momentum.

$$
\begin{aligned}
W_K &= \int_0^x \frac{dp}{dt}\, dx = \int_0^p dp\left(\frac{dx}{dt}\right) \\
&= \int_0^u u\, d(m\gamma_u u) = m\int_0^u u\, d\left(\frac{u}{\sqrt{1 - u^2/c^2}}\right) \\
&= m\frac{u^2}{\sqrt{1 - u^2/c^2}} - m\int_0^u \frac{u\, du}{\sqrt{1 - u^2/c^2}} \\
&= \frac{mu^2}{\sqrt{1 - u^2/c^2}} + mc^2\left[\sqrt{1 - \frac{u^2}{c^2}} - 1\right] \\
&= \frac{m}{\sqrt{1 - u^2/c^2}}\left[u^2 + c^2\left(1 - \frac{u^2}{c^2}\right)\right] - mc^2 \\
&= (\gamma - 1)mc^2 \tag{16-62}
\end{aligned}
$$

Example 16-17

Show that the kinetic energy reduces to the Newtonian value when $u \ll c$.

Solution   Insert the expression for γ and expand in a binomial series.

$$
\begin{aligned}
W_K &= mc^2\left[\left(1 - \frac{u^2}{c^2}\right)^{-1/2} - 1\right] \\
&= mc^2(1 + \tfrac{1}{2}u^2/c^2 + \cdots - 1) \\
&\cong \tfrac{1}{2}mu^2
\end{aligned}
$$

Example 16-18

Show that a particle of nonzero mass can never be accelerated up to the speed of light.

Solution   As $u \to c$, γ becomes infinite, so an infinite amount of work would have to be done.

Total Energy

Equation (16-62) displays the kinetic energy as the difference of two terms, γmc^2 and mc^2. Since the latter is the value of the former when the particle is at rest (in some reference frame), the term γmc^2 plays the role of the *total* energy, and $(\gamma - 1)mc^2$ is its *increase* due to motion relative to that frame. Thus we arrive at the famous Einstein formula, which in our notation reads

$$W = \gamma mc^2 \qquad (16\text{-}63)$$

This is more commonly written as $E = Mc^2$, where E stands for the total energy and $M = \gamma m$. M is called the *relativistic mass*. It obviously depends on the particle velocity and thus on the reference frame to which the motion is referred. On the other hand, m is an *invariant* property of the particle. It is the value of M when $\gamma = 1$ and is thus called the *rest mass*. The corresponding energy

$$W_0 = mc^2 \qquad (16\text{-}64)$$

is called the *rest* energy.

Equation (16-64) shows that whenever a system at rest undergoes a change of energy it must also undergo a proportional change of mass. This is, of course, the basis for the interconvertibility of mass and energy. For example, if an atom absorbs a photon it gets heavier. Because the factor c^2 is so huge, the mass changes are appreciable only for very energetic processes such as nuclear reactions.

Energy–Momentum Relation

Finally, we combine some of the foregoing relations to obtain a direct connection between energy and momentum (Figure 16-19). Starting with Eq. (16-63) we have

$$W^2 = \gamma^2 m^2 c^4 = \frac{m^2 c^4}{1 - u^2/c^2}$$

$$W^2\left(1 - \frac{u^2}{c^2}\right) = m^2 c^4 = W_0^2$$

$$W^2 = W_0^2 + \frac{W^2 u^2}{c^2}$$

$$= W_0^2 + \frac{(\gamma mc^2)^2 u^2}{c^2}$$

$$= W_0^2 + (\gamma mu)^2 c^2 = W_0^2 + p^2 c^2$$

$$W = \sqrt{W_0^2 + p^2 c^2} \qquad (16\text{-}65)$$

Example 16-19

Write the energy–momentum relation for a photon.

Solution A photon must be a particle with *zero rest mass*, since it travels at speed c. Hence $W_0 = 0$, and Eq. (16-65) is

$$W_{\text{photon}} = pc \qquad (16\text{-}66)$$

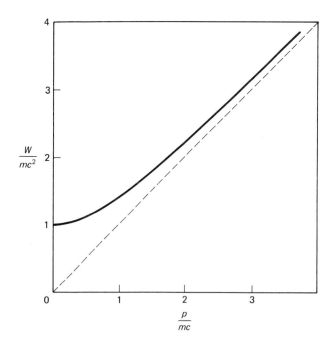

Figure 16-19 Energy–momentum relation for particles with nonzero rest mass. The asymptote (dashed line) is the curve for particles of zero rest mass (e.g., photons).

This relation has several intriguing ramifications. First, it verifies the general relation of energy and momentum for electromagnetic fields, as we discussed in Chapter 10. Second, it gives a strong hint about quantum mechanics. Since the photon energy is $h\nu$, Eq. (16-66) may be written

$$p = \frac{h\nu}{c} = \frac{h}{\lambda} \tag{16-67}$$

Simply applying this to *all* particles, not just massless ones, leads directly to the Schrödinger equation. But that is another story.

FURTHER REFERENCES

No scientific development has captured the worldwide popular imagination as has the theory of relativity. As a result there have been literally thousands of books, chapters, pamphlets, essays, papers, and every other form of prose or poetry written about it at every level from the most profound to the utterly crackpot. One of the best is a popular exposition by the master himself, A. Einstein (translated by R. W. Lawson), *Relativity, the Special and General Theory*, Henry Holt, New York, 1920. Other excellent treatments at an appropriate level are S. J. Prokhovnik, *The Logic of Special Relativity*, Cambridge University Press, Cambridge, 1967; A. P. French, *Special Relativity*, Norton, New York, 1968; and D. Bohm, *The Special Theory of Relativity*, Benjamin, New York, 1965. For deeper treatments, consult any of the advanced electricity and magnetism texts or C. Møller, *The Theory of Relativity*, Clarendon Press, Oxford, 1972.

PROBLEMS

16-1. Lightning strikes twice at the same point on a railroad track. The two strokes are 1 μs apart in the earth frame. What are the time and space intervals in a train going past at $0.746c$?

16-2. In Problem 16-1, what is the proper-time interval? The space-time interval?

16-3. Muons have a mean lifetime of about 2.2 μs when at rest. A measurement of average arrival rates of cosmic-ray muons disclosed that the rate at sea level is about 70% of that at 2000 m altitude. What was the average γ value of the muons in these measurements?

16-4. At time zero, a rocket going at $0.7c$ passes a space station. One microsecond later, the station sends out a light flash. How far away (in the station frame) is the rocket when the flash gets to it? What time does the rocket clock then read? At what station time does the reflected flash get back?

16-5. By solving Eqs. (16-12) for the unprimed coordinates, derive the inverse Lorentz transformation.

16-6. In frame S, two events are separated by 1 s in time and 2 light-seconds in distance. What is their distance separation in the frame S' wherein they are simultaneous? What is the velocity of S' relative to S?

16-7. In Problem 16-6, what are the limits on possible time intervals and space intervals between the events?

16-8. In Problem 16-6, can a proper-time interval be defined?

16-9. Much discussion has centered on Einstein's "twin paradox," in which a space traveler returns home younger than his twin brother. To illustrate this, suppose that the traveler starts on his twenty-first birthday and travels 5 years out and 5 years back by his own clock at $0.745c$, thus returning home on his thirty-first birthday. How old is his twin at that time?

16-10. What is paradoxical about the twin paradox?

16-11. Prove that for collinear velocities $\gamma_{u'} = \gamma_u \gamma_v (1 - \beta_u \beta_v)$. Generalize this to noncollinear velocities.

16-12. Two rocket ships pass a space station going in opposite directions, each at $0.8c$. What is their speed relative to each other?

16-13. One of the ships in Problem 16-12 fires a bullet at the other after they have passed. If the bullet is to approach its target at $0.1c$, what should its muzzle velocity be?

16-14. Fizeau measured the speed of light through water moving in the same direction. Analyze the experiment relativistically and show that Eq. (16-3) is correct to first order in v/c. Thus determine the "ether drag coefficient" α.

16-15. Show that for an extreme relativistic particle $2\gamma^2(1 - \beta) \cong 1$.

16-16. Show that the set $(\gamma c, \gamma \mathbf{u})$ transforms the same way as the set $(ct, \mathbf{r})$.

16-17. How does $(\rho U - \mathbf{J} \cdot \mathbf{A})$ transform?

16-18. How does $(U^2 - c^2 \mathbf{A}^2)$ transform?

16-19. In frame S there are uniform electric and magnetic fields $\mathbf{E}$ and $\mathbf{B}$ perpendicular to each other with $cB > E$. What is the vector velocity of the frame S' in which the field is purely magnetic?

16-20. An electron has a mass of 9×10^{-31} kg. What is its rest energy in electron-volts? (One electron-volt $= 1.6 \times 10^{-19}$ joule.)

16-21. What is the speed of an electron of kinetic energy 1×10^3 eV, 1×10^6 eV, 1×10^9 eV? What is the momentum of the electron in each case?

appendix

Mathematical Resumé

A-1 SCALARS AND VECTORS

Many physical quantities are fully defined by a single dimensional magnitude. Such quantities are called *scalars*. Examples include temperature, distance, time, mass density, and many others. Another class of quantities requires both magnitude and one direction for full definition. These are called *vectors*. Examples are force, displacement, velocity, acceleration, and many others. Since in n dimensions a direction is specified by $n - 1$ independent angles, a vector in n-dimensional space requires a set of n numbers for full definition. A *unit vector*, which describes only a *direction*, requires $n - 1$ numbers. There are further classes of quantities involving a magnitude and two, three, or more directions. These are called *tensors of rank two*, *three*, and so on. In these terms, scalars and vectors are tensors of rank zero and one respectively.

A-2 VECTOR ALBEGRA

It is assumed that the reader is thoroughly conversant with the rules of vector addition, multiplication, and so forth. Nevertheless, we list here a few of the possibly less familiar algebraic identities.

$$\mathbf{A} \times (\mathbf{B} \times \mathbf{C}) = (\mathbf{A} \cdot \mathbf{C})\mathbf{B} - (\mathbf{A} \cdot \mathbf{B})\mathbf{C}$$
$$\equiv \mathbf{B}(\mathbf{A} \cdot \mathbf{C}) - \mathbf{C}(\mathbf{A} \cdot \mathbf{B}) \text{ (back-cab rule)} \tag{A-1}$$
$$\mathbf{A} \cdot (\mathbf{B} \times \mathbf{C}) = (\mathbf{A} \times \mathbf{B}) \cdot \mathbf{C} \text{ (can interchange dot and cross)} \tag{A-2}$$
$$(\mathbf{A} \times \mathbf{B}) \cdot (\mathbf{C} \times \mathbf{D}) = \mathbf{A} \cdot [\mathbf{B} \times (\mathbf{C} \times \mathbf{D})] \tag{A-3}$$
$$= (\mathbf{A} \cdot \mathbf{C})(\mathbf{B} \cdot \mathbf{D}) - (\mathbf{A} \cdot \mathbf{D})(\mathbf{B} \cdot \mathbf{C})$$

Division by a vector is undefined. Nevertheless, a sort of inverse of each type of multiplication may be found by solving an algebraic equation.

Example A-1

Find the most general vector **X** such that $\mathbf{X} \times \mathbf{B} = \mathbf{A}$, where obviously $\mathbf{A} \perp \mathbf{B}$.

Solution

We can certainly express **X** in terms of the three mutually perpendicular vectors **A**, **B**, and $\mathbf{A} \times \mathbf{B}$. Hence let

$$\mathbf{X} = \alpha\mathbf{A} + \beta\mathbf{B} + \gamma\mathbf{A} \times \mathbf{B}$$

Then

$$\mathbf{A} \equiv \mathbf{X} \times \mathbf{B} = \alpha\mathbf{A} \times \mathbf{B} + \gamma(\mathbf{A} \times \mathbf{B}) \times \mathbf{B}$$
$$= \alpha\mathbf{A} \times \mathbf{B} - \gamma B^2\mathbf{A}$$

or

$$\mathbf{A}(1 + \gamma B^2) = \alpha\mathbf{A} \times \mathbf{B}$$

Since the two sides are perpendicular, this equality is possible only if both coefficients vanish. Hence

$$\alpha = 0, \quad \beta \text{ is arbitrary}, \quad \gamma = -1/B^2, \text{ i.e.,}$$
$$\mathbf{X} = (\mathbf{B} \times \mathbf{A})/B^2 + \beta\mathbf{B} \tag{A-4}$$

A-3 COORDINATE SYSTEMS

We deal in this book with three explicit coordinate systems:

1. Cartesian (Figure A-1). The coordinates are x, y, z, the projected distances along three mutually perpendicular directions.

2. Cylindrical (Figure A-2). The coordinates are

   s = radial distance from z axis

   ϕ = dihedral angle in x-y plane ($0 \leq \phi < 2\pi$)

   z = same as in Cartesian

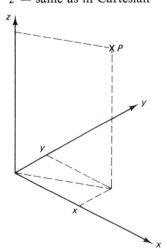

Figure A-1 Cartesian coordinates of point P.

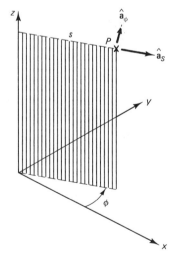

Figure A-2 Cylindrical coordinates of point P.

3. Spherical (Figure A-3). The coordinates are
   r = radial distance from origin
   θ = polar angle from z axis $(0 \leq \theta \leq \pi)$
   ϕ = same as in cylindrical

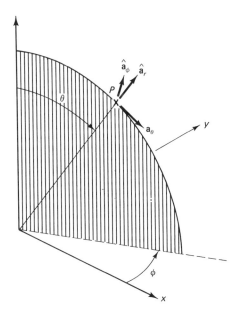

Figure A-3 Spherical (polar) coordinates of point P.

Relations among coordinates are easily obtained by inspection of the figures.

$$s = \sqrt{x^2 + y^2} \qquad\qquad x = s \cos \phi$$

$$\phi = \tan^{-1} y/x \qquad\qquad y = s \sin \phi$$

$$r = \sqrt{x^2 + y^2 + z^2} = \sqrt{s^2 + z^2} \qquad s = r \sin \theta \qquad\qquad \text{(A-5)}$$

$$\theta = \cos^{-1} \left(z/\sqrt{x^2 + y^2 + z^2} \right) \qquad z = r \cos \theta$$

In calculating partial derivatives of function of these coordinate sets, it is essential to keep clearly in mind what quantities are held constant. Unless explicitly indicated, it is always the *companion* coordinates within the same set.

Example A-2

Evaluate $\partial r/\partial z$.

Solution Since x and y are to be held constant, we start with the expression for r in Cartesian coordinates.

$$\left(\frac{\partial r}{\partial z} \right)_{x,y} = \frac{\partial}{\partial z} \sqrt{x^2 + y^2 + z^2}$$

$$= \frac{z}{\sqrt{x^2 + y^2 + z^2}} = \frac{z}{r} = \cos \theta \qquad\qquad \text{(A-6)}$$

Unit Vectors

For each coordinate system, unit vectors are defined as the directions of motion when one single coordinate is increased. For the Cartesian system they are commonly designated $\hat{\mathbf{i}}, \hat{\mathbf{j}}, \hat{\mathbf{k}}$. For cylindrical and spherical, they are indicated in Figures A-2 and A-3. All three sets are orthogonal (mutually perpendicular). Also it is important to note that except for $\hat{\mathbf{i}}, \hat{\mathbf{j}}, \hat{\mathbf{k}}$ the unit vectors are *functions of the coordinates of the point to which they pertain.*

An arbitrary differential displacement vector is the vector sum of differential displacements in the three directions.

$$\mathbf{dr} = dx\,\hat{\mathbf{i}} + dy\,\hat{\mathbf{j}} + dz\,\hat{\mathbf{k}} \tag{A-7a}$$

$$= ds\,\hat{\mathbf{a}}_s + sd\phi\,\hat{\mathbf{a}}_\phi + dz\,\hat{\mathbf{k}} \tag{A-7b}$$

$$= dr\,\hat{\mathbf{a}}_r + rd\theta\,\hat{\mathbf{a}}_\theta + r\sin\theta\,d\phi\,\hat{\mathbf{a}}_\phi \tag{A-7c}$$

Note that the increments $d\theta$ and $d\phi$ of the angular coordinates have to be multiplied by *scale factors* r and s respectively to convert them into distances. Differential area and volume elements are made up of products of the differential distances.

The vector position of a point (for example, its displacement vector from the origin) can be expressed in components in any of the basis systems.

$$\mathbf{r} = x\,\hat{\mathbf{i}} + y\,\hat{\mathbf{j}} + z\,\hat{\mathbf{k}}$$

$$= s\,\hat{\mathbf{a}}_s + z\,\hat{\mathbf{k}}$$

$$= r\,\hat{\mathbf{a}}_r \tag{A-8}$$

The basis vectors, in turn, can be expressed as derivatives of the position vector $\mathbf{r}$ with respect to distances.

Example A-3

Express $\hat{\mathbf{a}}_r$ in terms of $\hat{\mathbf{i}}, \hat{\mathbf{j}}, \hat{\mathbf{k}}$.

Solution

$$\hat{\mathbf{a}}_r = \frac{\partial}{\partial r}(x\hat{\mathbf{i}} + y\hat{\mathbf{j}} + z\hat{\mathbf{k}})$$

$$= \frac{\partial x}{\partial r}\hat{\mathbf{i}} + \frac{\partial y}{\partial r}\hat{\mathbf{j}} + \frac{\partial z}{\partial r}\hat{\mathbf{k}}$$

$$= \sin\theta\cos\phi\,\hat{\mathbf{i}} + \sin\theta\sin\phi\,\hat{\mathbf{j}} + \cos\theta\,\hat{\mathbf{k}}$$

The complete set of relations is as follows:

$$\begin{array}{ll}
\hat{\mathbf{a}}_s = \hat{\mathbf{i}}\cos\phi + \hat{\mathbf{j}}\sin\phi & \hat{\mathbf{i}} = \hat{\mathbf{a}}_s\cos\phi - \hat{\mathbf{a}}_\phi\sin\phi \\
\hat{\mathbf{a}}_\phi = -\hat{\mathbf{i}}\sin\phi + \hat{\mathbf{j}}\cos\phi & \hat{\mathbf{j}} = \hat{\mathbf{a}}_s\sin\phi + \hat{\mathbf{a}}_\phi\cos\phi
\end{array} \tag{A-10}$$

$$\begin{array}{ll}
\hat{\mathbf{a}}_r = \hat{\mathbf{a}}_s\sin\theta + \hat{\mathbf{k}}\cos\theta & \hat{\mathbf{a}}_s = \hat{\mathbf{a}}_r\sin\theta + \hat{\mathbf{a}}_\theta\cos\theta \\
\hat{\mathbf{a}}_\theta = \hat{\mathbf{a}}_s\cos\theta - \hat{\mathbf{k}}\sin\theta & \hat{\mathbf{k}} = \hat{\mathbf{a}}_r\cos\theta - \hat{\mathbf{a}}_\theta\sin\theta
\end{array} \tag{A-11}$$

We have frequent occasion to calculate the distance between two points, $r_{12} \equiv |\mathbf{r}_1 - \mathbf{r}_2|$. The safest way to do this is to write $\mathbf{r}_1$ and $\mathbf{r}_2$ in Cartesian components.

Example A-4

Find the distance between points (s_1, ϕ_1, z_1) and (s_2, ϕ_2, z_2).

Solution

$$\mathbf{r}_1 = s_1 \cos \phi_1 \,\hat{\mathbf{i}} + s_1 \sin \phi_1 \,\hat{\mathbf{j}} + z_1 \,\hat{\mathbf{k}}$$

$$\mathbf{r}_2 = s_2 \cos \phi_2 \,\hat{\mathbf{i}} + s_2 \sin \phi_2 \,\hat{\mathbf{j}} + z_2 \,\hat{\mathbf{k}}$$

$$|\mathbf{r}_1 - \mathbf{r}_2| = [(s_1 \cos \phi_1 - s_2 \cos \phi_2)^2$$

$$+ (s_1 \sin \phi_1 - s_2 \sin \phi_2)^2 + (z_1 - z_2)^2]^{1/2}$$

$$= [s_1^2 + s_2^2 - 2s_1 s_2 \cos (\phi_1 - \phi_2) + (z_1 - z_2)^2]^{1/2} \qquad (\text{A-12})$$

Note that a distance is *always positive*. The symbol $\sqrt{x}$ or $(x)^{1/2}$ conventionally means the *positive* root of the equation $x^2 - 1 = 0$. That is, $\sqrt{x^2} = |x| = x$ if $x \geq 0$, $-x$ if $x < 0$.

A-4 LINE, SURFACE, AND VOLUME INTEGRALS

We deal throughout this book with various scalar and vector *fields*. These are physical quantities that are represented mathematically by scalar or vector *point functions*, that is, scalar or vector quantities that may vary from point to point in a region of space. Examples are temperature at points in a room, water velocity at points in a river, and so on. We denote these classes of function by $f(\mathbf{r})$ and $\mathbf{F}(\mathbf{r})$, where $\mathbf{r}$ stands for any of the sets of coordinates (x, y, z), (s, ϕ, z), or (r, θ, ϕ).

Integrals are sums of values of a function over some specified continuous manifold of points. The manifold may be one-, two-, or three-dimensional, giving line, surface, or volume integrals. We designate the dimensionality (somewhat redundantly) by the number of integral signs as well as the differential element. Also, lines and surfaces may be either open or closed; we designate closed manifolds by a circle through the integral signs.

The general approach to evaluation of such integrals is to express the integrand in a coordinate system appropriate to the geometry of the domain of integration and then to reduce it to one or more ordinary integrals. The following examples illustrate some techniques.

Example A-5

Evaluate $\int_C f \, dl$ with $f = r^2$, $C = (0, 0, 0) \longrightarrow (1, 0, 0) \longrightarrow (1, 1, 0)$.

Solution The contour C is best described in Cartesian coordinates, so we write $f(\mathbf{r}) = x^2 + y^2 + z^2$. The contour falls into two segments with (1) $y = 0$, $z = 0$, $dl = dx$, (2) $z = 0$, $x = 1$, $dl = dy$. Hence

$$\int_C f(\mathbf{r}) \, dl = \int_0^1 x^2 \, dx + \int_0^1 (1 + y^2) \, dy = \tfrac{5}{3}$$

Example A-6

Take the same $f(\mathbf{r})$ but let C be the straight line $(0, 0, 0) \rightarrow (1, 1, 0)$.

Solution   Now $y = x$, $dy = dx$, $dl = \sqrt{(dx)^2 + (dy)^2} = \sqrt{2}\ dx$

$$\int_C f(\mathbf{r})\ dl = \int_0^1 (2x^2)(\sqrt{2}\ dx) = 2\sqrt{2}/3$$

Note that although the function and end-points are the same as in the previous example, the integral has a different value.

Example A-7

Evaluate $\displaystyle\int_C \mathbf{F}(\mathbf{r}) \cdot \mathbf{dl}$ with $\mathbf{F}(\mathbf{r}) = 3x^2\ \hat{\mathbf{i}} + 5xy\ \hat{\mathbf{j}}$ and $C - (0, 0, 0) \rightarrow (1, 0, 0)$ $\rightarrow (1, 1, 0)$.

Solution   On the two segments of C we have (1) $y = 0$, $\mathbf{dl} = dx\ \hat{\mathbf{i}}$, $\mathbf{F} \cdot \mathbf{dl} = 3x^2\ dx$; (2) $x = 1$, $\mathbf{dl} = dy\ \hat{\mathbf{j}}$, $\mathbf{F} \cdot \mathbf{dl} = 5y\ dy$

$$\int_C \mathbf{F} \cdot \mathbf{dl} = \int_0^1 3x^2\ dx + \int_0^1 5y\ dy = \tfrac{7}{2}$$

Example A-8

Take the same $\mathbf{F}(\mathbf{r})$ but let C now be the closed contour made by adding two more sides to form a square.

Solution   Added to the foregoing result will be the segments (3) $y = 1$, $\mathbf{dl} = dx\ \hat{\mathbf{i}}$; (4) $x = 0$, $\mathbf{dl} = dy\ \hat{\mathbf{j}}$. Note that we *do not* write $\mathbf{dl} = -dx\ \hat{\mathbf{i}}$ and $\mathbf{dl} = -dy\ \hat{\mathbf{j}}$; the limits of integration will take care of the directions of progression along these segments. Thus

$$\oint \mathbf{F} \cdot \mathbf{dl} = \tfrac{7}{2} + \int_1^0 3x^2\ dx + \int_1^0 0\ dy$$

$$= \tfrac{7}{2} + (-1) + 0 = \tfrac{5}{2}$$

Example A-9

Evaluate $\displaystyle\iint_S \mathbf{F} \cdot \hat{\mathbf{n}}\ dS$ with $\mathbf{F} = x\hat{\mathbf{i}}$, $S = 90$-degree cylinder sector of radius a, height b.

Solution   The domain of integration obviously demands cylindrical coordinates so we express everything else in this system.

$$\mathbf{F}(\mathbf{r}) = (a \cos \phi)\ \hat{\mathbf{i}}$$

$$\hat{\mathbf{n}}(\mathbf{r}) = \cos \phi\ \hat{\mathbf{i}} + \sin \phi\ \hat{\mathbf{j}}$$

$$dS = (a\ d\phi)(dz)$$

$$\mathbf{F} \cdot \hat{\mathbf{n}}\ dS = (a \cos^2 \phi)(a\ d\phi)(dz)$$

$$\iint \mathbf{F} \cdot \hat{\mathbf{n}}\ dS = a^2 \int_0^b dz \int_0^{\pi/2} \cos^2 \phi\ d\phi$$

$$= a^2 b(\pi/4)$$

Note that we chose $\hat{\mathbf{n}}$ to be in the outward radial direction. This is arbitrary for an open surface. For closed surfaces, $\hat{\mathbf{n}}$ is conventionally always taken to be in the outward sense.

A-5 GRADIENT OF A SCALAR POINT FUNCTION

The rate of change, with respect to distance, of a function f at a point $\mathbf{r}$ will vary with the *direction* of the incremental distance from $\mathbf{r}$. It will have a *maximum* value in some particular direction, namely that direction in which the distance to the neighboring contour surface, on which $f = [f(\mathbf{r}) + df]$, is the *smallest*. This maximum rate of change, multiplied by the unit vector in the corresponding direction, is called the *gradient* of f. Thus

$$\operatorname{grad} f(\mathbf{r}) = \frac{df}{dr_0}\hat{\mathbf{a}}_0 \tag{A-13}$$

where dr_0 is the smallest incremental distance to the surface on which $f = [f(\mathbf{r}) + df]$, and $\hat{\mathbf{a}}_0$ is the direction perpendicular to this surface. If we dot both sides with an arbitrary incremental vector $d\mathbf{r}$, we obtain

$$(\operatorname{grad} f) \cdot d\mathbf{r} = \frac{df}{dr_0}(\hat{\mathbf{a}}_0 \cdot d\mathbf{r}) = df \tag{A-14}$$

From this we easily obtain the explicit expression for $\operatorname{grad} f$ in any coordinate system by writing $d\mathbf{r}$ and df in that system.

Example A-10

Write $\operatorname{grad} f$ in cylindrical coordinates.

Solution   From Eq. (A-7b)

$$(\operatorname{grad} f)\cdot(ds\,\hat{\mathbf{a}}_s + s\,d\phi\,\hat{\mathbf{a}}_\phi + dz\,\hat{\mathbf{k}}) = df = \frac{\partial f}{\partial s}\,ds + \frac{\partial f}{\partial \phi}\,d\phi + \frac{\partial f}{\partial z}\,dz$$

Thus, by inspection

$$\operatorname{grad} f = \frac{\partial f}{\partial s}\hat{\mathbf{a}}_s + \frac{1}{s}\frac{\partial f}{\partial \phi}\hat{\mathbf{a}}_\phi + \frac{\partial f}{\partial z}\hat{\mathbf{k}} \tag{A-15}$$

Note that $[\operatorname{grad} f(\mathbf{r})]$ is a vector function generated from the scalar function $f(\mathbf{r})$.

A-6 DIVERGENCE OF A VECTOR FUNCTION

As a preliminary, we define the "outflow" of a vector function $\mathbf{F}(\mathbf{r})$ through a closed surface S as

$$O_F(S) \equiv \oiint_S \mathbf{F} \cdot \hat{\mathbf{n}}\, dS \tag{A-16}$$

where $\hat{\mathbf{n}}$ is the outward unit normal at each point of S. If $\mathbf{F}$ represents fluid velocity, $O_F(S)$ is the net volume rate of flow out through S.

The *divergence* of **F** at point **r** is defined as the *outflow density* (outflow per unit volume) around that point.

$$\text{div } \mathbf{F}(\mathbf{r}) \equiv \frac{1}{\Delta V} \oiint_{\Delta S} \mathbf{F} \cdot \hat{\mathbf{n}} \, dS \tag{A-17}$$

where ΔV is the volume enclosed by ΔS, both taken to be infinitesimal in size. To evaluate this expression in any coordinate system, we take a differential volume enclosing the point **r** and calculate the contribution to the outflow from the three pairs of faces.

Example A-11

Calculate div **F**(**r**) in cylindrical coordinates.

Solution   The differential volume around point (s, ϕ, z) is enclosed by pairs of surface elements:

1. $\perp \hat{\mathbf{a}}_s$ at $(s - ds/2)$ and $(s + ds/2)$
2. $\perp \hat{\mathbf{a}}_\phi$ at $(\phi - d\phi/2)$ and $(\phi + d\phi/2)$
3. $\perp \hat{\mathbf{k}}$ at $(z - dz/2)$ and $(z + dz/2)$

For pair 1, we have, respectively

$$\hat{\mathbf{n}} = -\hat{\mathbf{a}}_s \text{ and } \hat{\mathbf{a}}_s$$

$$\mathbf{F} \cdot \hat{\mathbf{n}} = -\left[F_s(r) - \frac{\partial F_s}{\partial s} \frac{ds}{2}\right] \quad \text{and} \quad F_s(r) + \frac{\partial F_s}{\partial s} \frac{ds}{2}$$

$$dS = \left(s - \frac{ds}{2}\right) d\phi \, dz \quad \text{and} \quad \left(s + \frac{ds}{2}\right) d\phi \, dz$$

Multiplying the last two lines together, adding, and dropping higher-order differential terms gives, for this pair of faces,

$$\left[\left(s \frac{\partial F_s}{\partial s} + F_s\right) ds\right] d\phi \, dz = \frac{\partial}{\partial s}(sF_s) \, ds \, d\phi \, dz$$

Similarly the other two pairs of faces give, respectively,

$$\frac{\partial F_\phi}{\partial \phi} d\phi \, ds \, dz \quad \text{and} \quad \frac{\partial F_z}{\partial z} dz \, (sd\phi)(ds)$$

Eq. (A-17) thus gives, since $\Delta V = ds\,(s\,d\phi)\,dz$,

$$\text{div } \mathbf{F}(\mathbf{r}) = \frac{1}{s} \frac{\partial}{\partial s}(sF_s) + \frac{1}{s} \frac{\partial F_\phi}{\partial \phi} + \frac{\partial F_z}{\partial z} \tag{A-18}$$

Note that the scale factor for $d\phi$ enters both the area and volume elements, and does not cancel out!

We see that [div **F**(**r**)] is a scalar function generated from the vector function **F**(**r**).

Divergence Theorem

If in Eq. (A-17) we multiply through by ΔV then add up many volume elements to make a macroscopic volume V, all the interior surface segments will cancel out, since they have opposite $\hat{\mathbf{n}}$'s for adjacent cells. This leaves

$$\iiint_V \text{div } \mathbf{F}(\mathbf{r}) \, dV = \oiint_{S(V)} \mathbf{F} \cdot \hat{\mathbf{n}} \, dS \qquad \text{(A-20)}$$

A-7 LAPLACIAN OF A SCALAR FUNCTION

The Laplacian is defined as

$$\nabla^2 f \equiv \text{div (grad } f) \qquad \text{(A-21)}$$

Explicit expressions are obtained from those for the div and grad operations.

Example A-12

Derive the expression for $\nabla^2 f$ in cylindrical coordinates.

Solution By putting the components of Eq. (A-15) into (A-18) we get

$$\nabla^2 f = \frac{1}{s} \frac{\partial}{\partial s} \left(s \frac{\partial f}{\partial s} \right) + \frac{1}{s^2} \frac{\partial^2 f}{\partial \phi^2} + \frac{\partial^2 f}{\partial z^2} \qquad \text{(A-22)}$$

Note that $[\nabla^2 f(\mathbf{r})]$ is a new scalar function generated from $f(\mathbf{r})$.

A-8 LAPLACIAN OF A VECTOR FUNCTION

Since the operator ∇^2 is a scalar differential operator, it may also be applied to a vector function $\mathbf{F}(\mathbf{r})$ to generate a new vector function $\nabla^2 \mathbf{F}$. In doing so, it must be borne in mind that the unit vectors may be functions of the coordinates, hence subject to differentiation. Only in Cartesian coordinates are they constant. Hence, it is safe to write

$$\nabla^2 \mathbf{F}(\mathbf{r}) = \nabla^2 [F_x(\mathbf{r}) \, \hat{\mathbf{i}} + F_y(\mathbf{r}) \, \hat{\mathbf{j}} + F_z(\mathbf{r}) \, \hat{\mathbf{k}}]$$
$$= \hat{\mathbf{i}} \, \nabla^2 F_x(\mathbf{r}) + \hat{\mathbf{j}} \, \nabla^2 F_y(\mathbf{r}) + \hat{\mathbf{k}} \, \nabla^2 F_z(\mathbf{r}) \qquad \text{(A-23)}$$

Of course the Laplacians of the components (scalar functions) may be calculated in any convenient coordinate system.

Example A-13

Calculate the Laplacian of the function $r^3 \hat{\mathbf{a}}_r$.

Solution

$$\mathbf{F}(\mathbf{r}) = r^3 \hat{\mathbf{a}}_r = r^2 \mathbf{r} = (x^2 + y^2 + z^2)(x\hat{\mathbf{i}} + y\hat{\mathbf{j}} + z\hat{\mathbf{k}})$$
$$\nabla^2 \mathbf{F} = \hat{\mathbf{i}} \, \nabla^2 (x^3 + xy^2 + xz^2) + \cdots$$
$$= \hat{\mathbf{i}} \, (6x + 2x + 2x) + \cdots$$
$$= 10(x\hat{\mathbf{i}} + y\hat{\mathbf{j}} + z\hat{\mathbf{k}}) = 10 \, \mathbf{r}$$

A-9 CURL OF A VECTOR FUNCTION

As a preliminary, we define the "circulation" of a vector function $\mathbf{F}(\mathbf{r})$ around a closed surface S as

$$\mathbf{C_F}(S) = \oiint_S \hat{\mathbf{n}} \times \mathbf{F}\, dS \tag{A-24}$$

This is an obvious companion of the outflow, Eq. (A-16), but since the *tangential* component at each point of S is summed, it is a measure of the circulatory flow.

The curl of $\mathbf{F}$ at point $\mathbf{r}$ is defined as the *circulation density* (circulation per unit volume) around that point

$$\text{curl } \mathbf{F}(\mathbf{r}) \equiv \frac{1}{\Delta V} \oiint_{\Delta S} \hat{\mathbf{n}} \times \mathbf{F}\, dS \tag{A-25}$$

Since curl $\mathbf{F}$ is itself a vector point function, it is useful to write an expression for any component of it. Let $\hat{\mathbf{N}}$ be an arbitrary *constant* unit vector. Taking its dot product with curl $\mathbf{F}$ gives the component of the latter in the direction of $\hat{\mathbf{N}}$.

$$\mathbf{N} \cdot \text{curl } \mathbf{F}(\mathbf{r}) \equiv \frac{1}{\Delta V} \oiint_{\Delta S} \hat{\mathbf{N}} \cdot \hat{\mathbf{n}} \times \mathbf{F}\, dS$$

$$= \frac{1}{\Delta V} \oiint_{\Delta S} (\hat{\mathbf{N}} \times \hat{\mathbf{n}}) \cdot \mathbf{F}\, dS \qquad \text{by Eq. (A-2)}$$

We apply this to a volume of the type shown in Figure A-4, namely a thin slice with faces normal to $\hat{\mathbf{N}}$. Then the integrand vanishes on the two faces and only the narrow strip contributes. On this strip, $\hat{\mathbf{N}} \times \hat{\mathbf{n}} = \hat{\mathbf{t}}$ and $dS = \Delta\xi\, dl$, so $(\hat{\mathbf{N}} \times \hat{\mathbf{n}})\, dS = \Delta\xi\, \mathbf{dl}$ and

$$\hat{\mathbf{N}} \cdot \text{curl } \mathbf{F} = \frac{\Delta\xi}{\Delta V} \oint_C \mathbf{F} \cdot \mathbf{dl}$$

$$= \frac{1}{\Delta S} \oiint_{C(\Delta S)} \mathbf{F} \cdot \mathbf{dl} \tag{A-26}$$

This may be interpreted as the *surface density* of circulation in a plane normal to $\hat{\mathbf{N}}$.

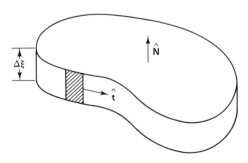

Figure A-4   Volume for evaluation of a component of curl $\mathbf{F}$.

Either Eq. (A-25) or (A-26) provides a recipe for calculation, but the latter is easier.

Example A-14

Calculate the z-component of curl **F** in cylindrical coordinates.

Solution Let the open surface ΔS be bounded by elementary arcs and radial lines as in Figure A-5. Thus the integral around the bounding contour is made up of four pieces:

1. $\mathbf{dl} = ds\,\hat{\mathbf{a}}_s$ $\qquad \mathbf{F}\cdot\mathbf{dl} = \left[F_s(\mathbf{r}) - \dfrac{\partial F_s}{\partial\phi}\dfrac{d\phi}{2}\right]ds$

2. $\mathbf{dl} = \left(s + \dfrac{ds}{2}\right)d\phi\,\hat{\mathbf{a}}_\phi$ $\qquad \mathbf{F}\cdot\mathbf{dl} = \left[F_\phi(\mathbf{r}) + \dfrac{\partial F_\phi}{\partial s}\dfrac{ds}{2}\right]\left(s + \dfrac{ds}{2}\right)d\phi$

3. $\mathbf{dl} = -ds\,\hat{\mathbf{a}}_s$ $\qquad \mathbf{F}\cdot\mathbf{dl} = -\left[F_s(\mathbf{r}) + \dfrac{\partial F_s}{\partial\phi}\dfrac{d\phi}{2}\right]ds$

4. $\mathbf{dl} = \left(s - \dfrac{ds}{2}\right)d\phi\,(-\hat{\mathbf{a}}_\phi)$ $\qquad \mathbf{F}\cdot\mathbf{dl} = -\left[F_\phi(\mathbf{r}) - \dfrac{\partial F_\phi}{\partial s}\dfrac{ds}{2}\right]\left(s - \dfrac{ds}{2}\right)d\phi$

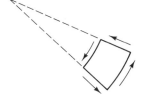

Figure A-5 Area for calculation of z component of curl **F** in cylindrical coordinates.

The sum, out to lowest-order differentials, is

$$\oint_c \mathbf{F}\cdot\mathbf{dl} = \left[-\frac{\partial F_s}{\partial\phi} + F_\phi + s\frac{\partial F_\phi}{\partial s}\right]ds\,d\phi$$

$$= \left[-\frac{\partial F_s}{\partial\phi} + \frac{\partial}{\partial s}(sF_\phi)\right]ds\,d\phi$$

Since the area is $s\,ds\,d\phi$, we obtain

$$(\text{curl } F)_z = \frac{1}{s}\left[\frac{\partial}{\partial s}(sF_\phi) - \frac{\partial F_s}{\partial\phi}\right] \tag{A-27}$$

Some Identities

The defining equations lead directly to the following identities:

$$\iiint_V \text{curl } \mathbf{F}\,d^3\mathbf{r} = \oiint_{S(V)} \hat{\mathbf{n}}\times\mathbf{F}\,dS \tag{A-28}$$

$$\iint_S \text{curl } \mathbf{F}\cdot\hat{\mathbf{n}}\,dS = \oint_{C(S)} \mathbf{F}\cdot\mathbf{dl} \qquad \text{(Stokes' theorem)} \tag{A-29}$$

Also, by direct calculation,

$$\text{curl curl } \mathbf{F} = \text{grad div } \mathbf{F} - \nabla^2 \mathbf{F} \tag{A-30}$$

$$\text{curl grad } f = 0 \tag{A-31}$$

$$\text{div curl } \mathbf{F} = 0 \tag{A-32}$$

A-10 THE OPERATOR ∇

The differential operations grad, div, and curl can all be written in terms of a single *vector operator* symbolized by an inverted delta and usually called *del*. Its expression in Cartesian components is

$$\nabla = \hat{\mathbf{i}} \frac{\partial}{\partial x} + \hat{\mathbf{j}} \frac{\partial}{\partial y} + \hat{\mathbf{k}} \frac{\partial}{\partial z} \tag{A-33}$$

In the curvilinear coordinate systems, the operator is rather more complicated and we shall not use it.* In terms of the del operator,

$$\text{grad } f = \nabla f$$
$$\text{div } \mathbf{F} = \nabla \cdot \mathbf{F} \tag{A-34}$$
$$\text{curl } F = \nabla \times \mathbf{F}$$

A-11 DERIVATIVES OF PRODUCTS

Derivatives of products of two functions are readily obtained by (1) summing the effects of operation separately on each of the factors, while (2) treating the del operator as an algebraic vector.

Example A-15

Evaluate grad(fg).

Solution

$$\begin{aligned}
\text{grad}(fg) &= \nabla(fg) \\
&= \nabla_f(fg) + \nabla_g(fg) \\
&= (\nabla f)g + f(\nabla g) \\
&= g \text{ grad } f + f \text{ grad } g
\end{aligned} \tag{A-35}$$

Example A-16

Evaluate div ($f\mathbf{F}$).

*Further discussion may be found in H. B. Phillips, *Vector Analysis*, John Wiley & Sons, Inc., New York, 1933, and in H. T. Yang, *Am. J. Phys.* **40** (1978), 109.

Solution

$$\text{div}(f\mathbf{F}) = \nabla \cdot (f\mathbf{F})$$
$$= \nabla_f \cdot (f\mathbf{F}) + \nabla_\mathbf{F} \cdot (f\mathbf{F})$$
$$= (\nabla f) \cdot \mathbf{F} + f(\nabla \cdot \mathbf{F})$$
$$= \mathbf{F} \cdot \text{grad} f + f \, \text{div} \, \mathbf{F} \tag{A-36}$$

Example A-17

Evaluate div ($\mathbf{F} \times \mathbf{G}$).

Solution

$$\text{div} (\mathbf{F} \times \mathbf{G}) = \nabla \cdot (\mathbf{F} \times \mathbf{G})$$
$$= \nabla_\mathbf{F} \cdot (\mathbf{F} \times \mathbf{G}) + \nabla_\mathbf{G} \cdot (\mathbf{F} \times \mathbf{G})$$
$$= \nabla_\mathbf{F} \times \mathbf{F} \cdot \mathbf{G} - \nabla_\mathbf{G} \times \mathbf{G} \cdot \mathbf{F}$$
$$= \mathbf{G} \cdot \text{curl} \, \mathbf{F} - \mathbf{F} \cdot \text{curl} \, \mathbf{G} \tag{A-37}$$

PROBLEMS

A-1. Given $\mathbf{X} \cdot \mathbf{B} = C$, solve for $\mathbf{X}$.

A-2. Verify the back-cab rule by writing out both sides in Cartesian components.

A-3. Derive the coordinate relations, Eqs. (A-5).

A-4. Show that $\partial s / \partial x = \cos \phi$.

A-5. Calculate $\partial r / \partial y$.

A-6. Calculate $\partial z / \partial r$. Compare with Eq. (A-6) and explain.

A-7. In spherical coordinates, write the volume element and area elements perpendicular to the three unit vectors.

A-8. Derive Eqs. (A-10) and (A-11).

A-9. Calculate the partial derivatives of $\hat{\mathbf{a}}_s$ with respect to the three cylindrical coordinates.

A-10. Find the distance between points (r_1, θ_1, ϕ_1) and (r_2, θ_2, ϕ_2).

A-11. Evaluate $\int_0^\pi \hat{\mathbf{a}}_s \, d\phi$.

A-12. Evaluate $\int_0^{2\pi} \hat{\mathbf{a}}_r \, d\phi$.

A-13. If $f(\mathbf{r}) = x^2 + y^2$, evaluate $\int_C f \, dl$ on contours:

    (a) $(0, 0, 0) \rightarrow (1, 0, 0) \rightarrow (1, 1, 0)$
    (b) straight line $(0, 0, 0) \rightarrow (1, 1, 0)$

A-14. If $\mathbf{F}(\mathbf{r}) = 3x^2 \hat{\mathbf{i}} + 5xy \hat{\mathbf{j}}$, evaluate $\int_C \mathbf{F} \cdot d\mathbf{l}$ on the same contours as in Problem A-13

and also on (c) the closed contour made by completing the square started in (a).

A-15. If $\mathbf{F}_1(\mathbf{r}) = x\hat{\mathbf{i}} + y\hat{\mathbf{j}}$ and $\mathbf{F}_2(\mathbf{r}) = y\hat{\mathbf{i}} - x\hat{\mathbf{j}}$, evaluate $\int \mathbf{F} \cdot \mathbf{dl}$ for both functions along the same two contours as in Problem A-13, and also on the parabola $y = x^2$ from $(0, 0, 0)$ to $(1, 1, 0)$.

A-16. In Problem A-15, all three integrals of $\mathbf{F}_1$ are equal. Is this a coincidence?

A-17. In Problem A-15, the integrals of $\mathbf{F}_2$ along the different contours are not equal. How are they related?

A-18. For the same two vector functions as in Problem A-15, evaluate $\iint\limits_S \mathbf{F} \cdot \hat{\mathbf{n}} \, dS$ on open surfaces:
  (a) square with corners at $(0, 0, 0)$, $(0, 1, 0)$, $(0, 1, 1)$, and $(0, 0, 1)$
  (b) rectangle with corners at $(1, 0, 0)$, $(0, 1, 0)$, $(0, 1, 1)$, $(1, 0, 1)$

A-19. Again for the same two vector functions as in Problem A-15, evaluate $\oiint \mathbf{F} \cdot \hat{\mathbf{n}} \, dS$ on closed surfaces:
  (a) cube, center at origin, edge length $2a$
  (b) sphere, center at origin, radius a

A-20. Repeat Problem A-19 by use of the divergence theorem.

A-21. Show that the function $\mathbf{F}_2 = y\hat{\mathbf{i}} - x\hat{\mathbf{j}}$ is a curl, for example, that $\mathbf{F}_2 = \operatorname{curl} \mathbf{G}$. Show that $\mathbf{G}_1 = \frac{1}{2}(x^2 + y^2)\,\hat{\mathbf{k}}$ and $\mathbf{G}_2 = -z(x\hat{\mathbf{i}} + y\hat{\mathbf{j}})$ both satisfy this equation. Show that they differ by an exact gradient; write it explicitly. Write a few more $\mathbf{G}$'s that also satisfy the equation.

A-22. Use Stokes' theorem along with $\mathbf{G}_1$ and $\mathbf{G}_2$ of Problem A-21 to evaluate the integral of $\mathbf{F}_2$ over surface (b) in Problem A-18.

Answers
to Odd-Numbered Problems

Chapter 1

1-1. 9×10^5 newtons

1-3. $-K_f\left[\dfrac{1}{z^2} + \dfrac{1}{(z-1)^2} - \dfrac{2}{(z-2)^2}\right]\hat{\mathbf{k}}$

1-5. $\mathbf{E}_S(s, \phi, z)$
$$= K_f\lambda\left\{\left[\frac{z+L/2}{[s^2+(z+L/2)^2]^{1/2}}\right.\right.$$
$$\left.- \frac{z-L/2}{[s^2+(z-L/2)^2]^{1/2}}\right]\frac{\hat{\mathbf{a}}_s}{s}$$
$$- \left[s^2+(z+L/2)^2\right]^{-1/2}$$
$$\left.- [s^2+(z-L/2)^2]^{-1/2}\right]\hat{\mathbf{k}}\right\}$$

1-7. $\mathbf{E}_S(0, 0, z)$
$$= \left[\frac{K_f\lambda}{(a^2+z^2)^{3/2}}\right](\pi a z\hat{\mathbf{k}} - 2a^2\hat{\mathbf{j}})$$

1-9. $\mathbf{E}_S(0, 0, z)$
$$= 2\pi K_f\sigma\frac{z}{|z|}\left(1 - \sqrt{\frac{z^2}{a^2+z^2}}\right)\hat{\mathbf{k}}$$

1-11. for $z > L/2$, $\mathbf{E}_S(0, 0, z)$
$$= 2\pi K_f\rho\left[L + \sqrt{\left(z-\frac{L}{2}\right)^2 + a^2}\right.$$
$$\left.- \sqrt{\left(z+\frac{L}{2}\right)^2 + a^2}\right]\hat{\mathbf{k}}$$
for $0 < z < L/2$, $\mathbf{E}_S(0, 0, z)$
$$= 2\pi K_f\rho\left[2z + \sqrt{\left(z-\frac{L}{2}\right)^2 + a^2}\right.$$
$$\left.- \sqrt{\left(z+\frac{L}{2}\right)^2 + a^2}\right]\hat{\mathbf{k}}.$$
Similar expressions with $-\hat{\mathbf{k}}$
replacing $\hat{\mathbf{k}}$ hold for negative values
of z.

1-13. $\mathbf{E}_S(0, 0, z)$
$$= K_fC\left(\ln\frac{z-L}{z} + \frac{L}{z-L}\right)\hat{\mathbf{k}}$$

1-15. Point charge, $q = A/K_f$, located at
the origin.

1-17. $\mathbf{E}_S(s) = 2\pi K_f\rho_0\hat{\mathbf{a}}_s\left[\dfrac{a^2}{s} - \dfrac{(a/2)^2}{s-a/2}\right]$

1-19. $\mathbf{E}_S(r) = (K_fq/r^2)\hat{\mathbf{a}}_r \qquad$ for $r > a$
$\qquad\quad = (K_fq/a^3)\mathbf{r} \qquad$ for $r < a$

1-21. $\mathbf{E}_S = 2\pi\lambda K_f\hat{\mathbf{a}}_s/s$

1-23. $U(0, 0, z) = K_fq/\sqrt{a^2+z^2}$

1-25. $U(r_2) = 4\pi K_f \displaystyle\int_{r_2}^{\infty} \frac{dr_1}{r_1^2} \int_0^{r_1} \rho(r) r^2 \, dr$

1-27. $K_f \lambda a \displaystyle\int_0^{\pi} \frac{d\phi'}{\sqrt{(x - a\cos\phi')^2 + (y - a\sin\phi')^2 + z^2}}$

1-31. $\mathbf{p} = -3\hat{\mathbf{k}}$ coulomb-meters

1-35. About 2×10^{-13}

Chapter 2

2-1. $\dfrac{I}{\pi(b^2 - a^2)}$

2-3. $\frac{1}{2} C a^3 \cos^2\theta$

2-5. $(mv/qB) \sin\theta$

2-7. $\dfrac{\mu_0 I a}{\pi} \dfrac{1}{(z^2 + a^2/4)\sqrt{z^2 + a^2/2}} \hat{\mathbf{k}}$

2-9. $\dfrac{\mu_0 NI}{2L}\left(\dfrac{a}{\sqrt{a^2 + 2L^2}} - \dfrac{a}{\sqrt{a^2 + L^2}}\right)\hat{\mathbf{k}}$

2-11. $\dfrac{\mu_0 I}{2\pi} \ln \dfrac{r_1}{r_2} \hat{\mathbf{k}}$

2-13. $(Ca^3/s)\hat{\mathbf{a}}_\phi$

2-15. $\cos^{-1}\left(\frac{1}{3}\right)$

2-17. $\mu_0 N^2(b - \sqrt{b^2 - a^2})$

Chapter 3

3-1. Hm/q volts

3-3. $\frac{1}{2}\omega BL^2$

3-7. Counterclockwise as seen from above

3-9. $\dfrac{\mu_0 NC}{2L} s\hat{\mathbf{a}}_\phi$

3-11. $\dfrac{C}{4\pi r^2}\hat{\mathbf{a}}_r$ between the spheres, zero elsewhere

Chapter 4

4-1. $\dfrac{P}{\epsilon_0}\left(1 - \dfrac{L}{2\sqrt{a^2 + L^2/4}}\right)$

4-3. $\dfrac{1}{2}P\left(\dfrac{z + L/2}{\sqrt{a^2 + (z + L/2)^2}} - \dfrac{z - L/2}{\sqrt{a^2 + (z - L/2)^2}}\right)$

4-5. $\mathbf{D} = (q/4\pi r^2)\hat{\mathbf{a}}_r$

$\mathbf{E} = \begin{cases} (q/4\pi\kappa\epsilon_0 r^2)\hat{\mathbf{a}}_r, & \text{for } r < a \\ (q/4\pi\epsilon_0 r^2)\hat{\mathbf{a}}_r, & \text{for } r > a \end{cases}$

$\mathbf{P} = \begin{cases} (q/4\pi r^2)(1 - \kappa^{-1})\hat{\mathbf{a}}_r, & \text{for } r < a \\ 0 & \text{for } r > a \end{cases}$

4-7. 2×10^{-40} farad-meter^2

4-9. $\tan^{-1}\delta$

4-11. $\epsilon_0 E_0(\sqrt{3} + i) e^{-i\omega t}$ ("Re" understood)

4-13. $\kappa_{\mathrm{re}} = 1 - \dfrac{nq^2}{m\epsilon_0} \dfrac{1}{\omega^2 + \gamma^2}$

$\kappa_{\mathrm{im}} = \dfrac{nq^2}{m\epsilon_0} \dfrac{\gamma/\omega}{\omega^2 + \gamma^2}$

4-15. 2

4-17. $\dfrac{4\pi\epsilon_0}{a^{-1} - b^{-1}}$

Chapter 5

5-1. 31 ohm-cm $= 0.31$ ohm-m

5-3. 2×10^5 and 9×10^4 cm/s

5-5. $4.6 k_B T (0.12$ electron-volt)

5-7. $E = \left(\dfrac{U_0}{L_{TF}}\right) \exp\left(\dfrac{-z}{L_{TF}}\right)$

5-9. 9×10^{-5} C/m^2; 9×10^6 C/m^3

5-11. 6×10^{-14} s; 8×10^{-8} m $= 800$ Å

5-13. -6.3×10^4 cm^3/C

5-15. $\omega = 1/\tau = \frac{1}{6} \times 10^{14}$ s^{-1}

5-17. $(g_0/\epsilon_0\tau)^{1/2} \cong 10^{16}$ s^{-1}

Chapter 6

6-1. $(4/3)\pi a^3 \mathbf{M}_0$

6-3. $\mathbf{J}_M = 2Cx\hat{\mathbf{j}}$
$\mathbf{j}_M = Ca^2 \sin^3\theta \cos^2\phi \,\hat{\mathbf{a}}_\phi$

6-5. $\mathbf{B} = \frac{1}{2}\mu_0 M_0 \hat{\mathbf{i}}$ $\mathbf{H} = -\frac{1}{2}M_0\hat{\mathbf{i}}$

6-7. $\mathbf{j}_M = \dfrac{NI}{L}\left(\dfrac{\mu}{\mu_0} - 1\right)\hat{\mathbf{a}}_\phi$ on cylinder surface

6-9. $M = \dfrac{k_B T}{\mu_0 m_0 \gamma} X$
$M = nm_0 \mathfrak{B}(X)$

6-11. 2×10^{-3} A/m upward

Chapter 7

7-3. Transformations in which f satisfies the Laplace equation; then div $\mathbf{A}$ is unchanged.

7-5. $\mathbf{A} = \text{curl } \mathbf{Z}'$, $U = 0$

7-7. $\{\text{acc}\} = T^{-1}$, $\{\text{mass}\} = T$

7-9. $\mathbf{E} = \left(\dfrac{q}{r^2}\right)\hat{\mathbf{a}}_r$, $U = \dfrac{q}{r}$

7-11. Centimeters

Chapter 8

8-1. 90 J

8-3. 4.5×10^4 J

8-5. $\dfrac{q^2}{32\pi^2\epsilon_0 r^4}$ for $r > a$

      0   for $r < a$

8-9. 3.52 N/m²

8-11. $-\left(\dfrac{p_1 p_2}{a^4}\right)\hat{\mathbf{j}}$

8-13. 3.8×10^5 m/s²

8-15. $\pm\dfrac{p_1 p_2}{4\pi\epsilon_0 R^3}$

8-17. $\pm\frac{3}{2}\dfrac{p_1 p_2}{4\pi\epsilon_0 R^3}\sin 2\theta\,\hat{\mathbf{k}}$

8-19. $\mathbf{F} = \alpha(\mathbf{E}' \cdot \text{grad})\mathbf{E}' = \frac{1}{2}\alpha\,\text{grad}\,(E'^2)$
Note that the force is always in the direction of increasing field strength because the induced dipole is always aligned with the field.

8-21. $\frac{1}{2}R_1 R_2 \cos(\phi_1 - \phi_2)$

Chapter 9

9-1. 162×10^{-9} J

9-3. Increase

9-5. $2\mathfrak{U}_{max}$ per half-revolution or $4LI_1I_2$ per revolution

9-7. $\left(\dfrac{Km}{r^3}\right)(-2\cos\theta_0\,\hat{\mathbf{a}}_r + \sin\theta_0\,\hat{\mathbf{a}}_\theta)$

9-9. This measures the energy-storage capability.

Chapter 10

10-1. $\left(\dfrac{I^2}{2\pi^2 a^4 g}\right)(-\hat{\mathbf{a}}_s)$

10-3. Across the solenoid parallel to the plates

10-5. $\left(\dfrac{\kappa E_0 H}{c^2}\right) \times$ volume of capacitor, in direction of $\mathbf{E}_0 \times \mathbf{H}$

10-7. $(1/4\pi c)(\mathbf{E} \times \mathbf{H})$

Chapter 11

11-1. $K\hat{\mathbf{n}} + E_2\hat{\mathbf{t}}$, where K is arbitrary

11-3. $(q/4\pi\epsilon_0)\{[(x - x_1)^2 + (y - y_1)^2 + (z - z_1)^2]^{-1/2}$
$+ [(x + x_1)^2 + (y + y_1)^2 + (z - z_1)^2]^{-1/2}$
$- [(x - x_1)^2 + (y + y_1)^2 + (z - z_1)^2]^{-1/2}$
$- [(x + x_1)^2 + (y - y_1)^2 + (z - z_1)^2]^{-1/2}\}$

11-5. $U(x, 0, 0) = (q/4\pi\epsilon_0)\left(\dfrac{1}{x} - \dfrac{1}{2a - x}\right.$
$\left.- \dfrac{1}{2a + x} + \dfrac{1}{4a - x} + \dfrac{1}{4a + x}\right.$
$\left.+ \cdots\right)$ for $-a \le x \le a$
$U = 0$ for $|x| > a$

11-7. $\dfrac{-q^2 a z_1 \hat{\mathbf{a}}_r}{4\pi\epsilon_0(z_1^2 - a^2)}$

11-9. $\dfrac{\epsilon_2 - \epsilon_1}{\epsilon_2 + \epsilon_1}\dfrac{q^2}{4\pi\epsilon_0(2z_1)^2}(-\hat{\mathbf{k}})$

11-11. Three

11-15. $\dfrac{\partial}{\partial z}(x^2 + y^2 + z^2)^{-1/2} = z/r^3$
$= r^{-2}\cos\theta$
$\dfrac{\partial^2}{\partial z^2}(x^2 + y^2 + z^2)^{-1/2}$
$= -r^{-3}(3\cos^2\theta - 1)$
$\dfrac{\partial^n}{\partial z^n}(x^2 + y^2 + z^2)^{-1/2}$
$= \text{const.} \times r^{-(n+1)}P_n(\theta)$

11-17. $U(s, \phi) = E_0\left(\dfrac{a^2}{s} - s\right)\cos\phi$

11-19. $U(r, \theta) =$
$$\frac{q}{4\pi\epsilon_0} \sum_{m=0}^{\infty} \frac{(-1)^m(2m-1)!!}{2^m m!} \frac{a^{2m}}{r^{2m+1}} P_{2m}(\cos\theta)$$
where
$$n!! \equiv n(n-2)(n-4)\cdots 1$$

11-21. They diverge.

Chapter 12

12-1. $B_1\hat{n} + K_1\hat{t} + K_2(\hat{n} \times \hat{t})$, where K_1 and K_2 are arbitrary

12-3. $\displaystyle\int_{-a/2}^{a/2} \int_{-a/2}^{a/2} \frac{z\,dx'\,dy'}{[(x-x')^2 + (y-y')^2 + z^2]^{3/2}}$

12-5. (a) Image loop of same sense;
(b) image loop of opposite sense

12-7. 0.47 T

Chapter 13

13-3. Elliptically polarized
$$\mathbf{E} = (E_1\hat{i} - iE_2\hat{j})e^{i(Kz-\omega t)}$$
$$\mathbf{B} = \frac{1}{c}(E_1\hat{j} + iE_2\hat{i})e^{i(Kz-\omega t)}$$
$$\mathbf{S} = \tfrac{1}{2}\epsilon(E_1^2 + E_2^2)\hat{k}$$

13-5. In a circularly polarized wave, the variation is orderly and periodic; in an unpolarized wave, it is random.

13-9. $k(\omega_2) = n(\omega_1)$, and vice versa

13-11. $S \cong 1.326 \times 10^{-3} E_0^2$ watts/meter^2 for E_0 in volts/meter

13-13. 10^5 V/m

13-15. $v_g = c\sqrt{1 - \dfrac{\omega_p^2}{\omega^2}}$

Chapter 14

14-1. $\dfrac{2}{n+1}E_0$ and $\dfrac{2n}{n+1}\dfrac{E_0}{c}$

14-3. Yes. Since $\theta_B = \tan^{-1}(n_2/n_1)$ while $\theta_C = \sin^{-1}(n_2/n_1)$, θ_B is always less than θ_C.

14-5. $E_{0t} = \dfrac{2}{1 + c\sqrt{\dfrac{i\mu g}{\omega}}} E_0 \ (\ll E_0)$

$B_{0t} = \dfrac{2}{c + \sqrt{\dfrac{\omega}{i\mu g}}} E_0 \cong 2B_0$

14-7. $\mathbf{E} = 2E_0(-S\cos KCz\,\hat{k}$
$\qquad + iC\sin KCz\,\hat{i})e^{iKSx}$
$\mathbf{B} = (2E_0/c)\cos KCz\,e^{iKSx}\,\hat{j}$
$\mathbf{E} \times \mathbf{H} = (2E_0^2/\mu_0 c)\,S\cos^2 KCz\,\hat{i}$
$\sigma = (-2E_0 S\epsilon_0)e^{iKSx}$
$\mathbf{j} = (2E_0/\mu_0 c)e^{iKSx}\,\hat{i}$
where $S = \sin\theta_i$, $C = \cos\theta_i$

14-9. $R = \dfrac{2r(1 - \cos 2\omega nd/c)}{1 + r^2 - 2r\cos 2\omega nd/c}$
where $r = \left(\dfrac{n-1}{n+1}\right)^2$

14-11. $TE_{01}, TE_{02}, TE_{10}, TE_{11}, TM_{11}, TE_{20}, TE_{21}, TM_{21}$

Chapter 15

15-1. $t_r = t/(1 - \beta)$ for $t < 0$, $t/(1 + \beta)$ for $t > 0$, where $\beta = v/c$

15-3. $E = \dfrac{\pi p_0}{\epsilon\lambda^2}\left[\dfrac{e^{iK(R+l)}}{R+l} - \dfrac{e^{iK(R-l)}}{R-l}\right]$

15-5. No. The current distribution is different.

15-7. 5.03 A, 1.19×10^{-7} W/m^2

15-9. $\dfrac{2}{3}\sqrt{\dfrac{\mu}{\epsilon}}\dfrac{\omega q^2}{mc^2}$

15-11. No. No dipole or higher moments due to spherical symmetry.

15-13. $1/n\sigma$

15-15. Parallel and antiparallel to the incident beam

Chapter 16

16-1. 1.5 μs, 1.12 light-μs $= 336$ m

16-3. 8.5

16-5. $ct = \gamma(ct' + \beta x')$
$x = \gamma(\beta ct' + x')$
$y = y'; z = z'$

16-7. $-\infty < \Delta t < \infty$
$\sqrt{3}$ light-sec $\leq \Delta x < \infty$

16-9. 36 years old

16-11. $\gamma_{u'} = \gamma_u \gamma_v (1 - \boldsymbol{\beta}_u \cdot \boldsymbol{\beta}_v)$

16-13. $0.980c$

16-17. Invariant

16-19. $c^2(\mathbf{E} \times \mathbf{B})/B^2$

16-21. $v = 0.06, 0.942, (1 - 1.25 \times 10^{-7})c$
$p = 0.06, 2.81, 2000mc$

Appendix

A-1. $(C/B^2)\mathbf{B} +$ any vector
perpendicular to B

A-5. $\sin \theta \sin \phi$

A-7. $r^2 \sin \theta \, dr \, d\theta \, d\phi$; $r^2 \sin \theta \, d\theta \, d\phi$;
$r \sin \theta \, dr \, d\phi$; r $d\theta \, dr$

A-9. $\dfrac{\partial \hat{\mathbf{a}}_s}{\partial s} = 0, \dfrac{\partial \hat{\mathbf{a}}_s}{\partial \phi} = \hat{\mathbf{a}}_s, \dfrac{\partial \hat{\mathbf{a}}_s}{\partial z} = 0$

A-11. $2\hat{\mathbf{j}}$

A-15. For $\mathbf{F}_1$: 1 for all contours.
For $\mathbf{F}_2$: (a) 1; (b) 0; (c) $-1/3$

A-17. curl $\mathbf{F}_2 = -2\hat{\mathbf{k}}$;
$\displaystyle \int_{c_a} - \int_{c_b} = (-2) \times$ area encircled

A-19. For $\mathbf{F}_1$: (a) $16a^3$; (b) $I\pi a^3/3$. For
$\mathbf{F}_2$: 0 for both.

Index

A

Abampere, 149–150, 154
Aberration, stellar, 350
Acceleration transformation, 345–46
Action-at-a-distance point of view, 5
Alfven waves, 280
Alternating current (ac), 61
Ampere, 33, 148, 153, 154
Ampere-meters, 34
Ampere's circuital law, 43–47, 67, 243
Amperes per meter, 34, 123
Angular frequency, 262
Angular momentum, 332
Antennas, radiation from, 324–28
Antiferromagnetism, 132
Associated Legendre functions, 245
Attenuation in waveguides, 307

B

Barkla, 330
Bohr, Niels, 331, 332
Bohr magneton, 119
Boltzmann constant, 83
Boundaries, wave fields with (*see* Wave
    fields with boundaries)
Boundary conditions on the fields:
    electrostatic, 209–12
    magnetostatic, 237–42
Boundary point, 213

Boundary-value problems, 19 (*See also*
    Electrostatic boundary-value problems;
    Magnetostatic boundary-value problems;
    Wave fields with boundaries)
Bound charge, 71–73
Bounding contour, 50
Brewster angle, 288–89
Brewster effect, 288–89
Brillouin functions, 129

C

Cartesian coordinates, 364
    solution of Laplace equations in, 223–26
Capacitance, 91–92
Capacitors, 91–92
Cathodes, 107–9
Cerenkov radiation, 356
Chain rule, 169
Charged conductors, 164–73
Charge, 1–29 (*See also* Electrostatic
    boundary-value problems; Electrostatic
    energy)
    basic concepts, 1–2
    Coulomb's law, 4–5, 28–29
    effective, 71–73
    electric field, 5–11
    electrostatic scalar potential, 14–19
    equipotential surfaces, 23–24
    field lines, 23–24
    Gauss's law, 11–14, 66, 67
    ideal conductors, 24–29